Land Use and Water Quality: The Impacts of Diffuse Pollution

Aerial photo of a flood plume from the Burdekin River, Queensland Australia, extending out to the mid-shelf of the central Great Barrier Reef, approximately 32 nautical miles (or 60 km) offshore, on 13th February 2019. Photo courtesy of Matt Curnock and TropWATER.

Scientific and Technical Report Series No. 28

Land Use and Water Quality: The Impacts of Diffuse Pollution

Edited by

Puangrat Kajitvichyanukul and Brian D'Arcy

IWA PUBLISHING

Published by **IWA Publishing**
Unit 104–105, Export Building
1 Clove Crescent
London E14 2BA, UK
Telephone: +44 (0)20 7654 5500
Fax: +44 (0)20 7654 5555
Email: publications@iwap.co.uk
Web: www.iwapublishing.com

First published 2022

British Library Cataloguing in Publication Data
A CIP catalogue record for this book is available from the British Library

ISBN: 9781789061116 (Paperback)
ISBN: 9781789061123 (eBook)
ISBN: 9781789061130 (ePub)

Contents

Foreword . . . . **xiii**
Preface . . . . **xv**

Chapter 1
Land use and diffuse pollution: are perceptions part of the problem? **1**
B. J. D'Arcy, D. Kay, F. Napier, P. Haygarth, Yan Yan Sun and Keiko Wada

1.1 Introduction 1
1.1.1 Land use and water quality – historical perspectives 1
1.2 Diffuse Pollution 2
1.2.1 Diffuse pollution – back to basics 2
1.2.2 Evidence of confusion 3
1.2.3 A new definition 6
1.2.4 Diffuse pollution hotspots 7
1.3 Quantifying Diffuse Pollution – UK Examples 8
1.3.1 High flows and water quality 8
1.3.2 Environmental quality standards (EQS) and river concentrations 11
1.3.3 Diffuse pollutants which are not easily managed by EQS 13
1.4 International Overview 14
1.4.1 Inter-connected environmental problems 14
1.4.2 Climate change, diffuse pollution and the united nations sustainable development goals 15
1.5 Conclusions and Recommendations 16
References 17

Chapter 2
Pesticide, diffuse pollution and food safety **25**
Puangrat Kajitvichyanukul and Thirasant Boonupara

2.1 History and Emerging of Pesticide Diffuse Pollution 25

2.2 Growth of Pesticide Consumption . . . 26
2.3 Environmental Contamination and Food Safety . . . 30
2.4 Challenges and Recommendations . . . 35
Acknowledgement . . . 39
References . . . 40

Chapter 3
***Long term studies in Lake Pátzcuaro, eutrophication and recovery efforts during one decade* . . . 45**
L. Bravo-Inclán, J. J. Sánchez-Chávez, A. C. Tomasini-Ortiz, R. González-Villela, M. Mijangos-Carro and F. Bernal-Brooks

3.1 Introduction . . . 45
 3.1.1 Study area . . . 46
3.2 Materials and Methods . . . 47
3.3 Results and Discussion . . . 47
3.4 Conclusions . . . 52
Acknowledgements . . . 52
References . . . 52

Chapter 4
***Impacts of oil spill incidences on levels of polycyclic aromatic hydrocarbons (PAHs) in coastal areas of Asian Countries* . . . 55**
S. Pongpiachan

4.1 Introduction . . . 55
4.2 Materials and Methods . . . 56
 4.2.1 Sampling equipment . . . 56
 4.2.2 Chemical extraction of PAHs . . . 57
4.3 Results and Discussion . . . 58
 4.3.1 Chemical characterization of PAHs . . . 58
 4.3.2 Diagnostic binary ratios of PAHs . . . 60
 4.3.3 Risk assessment of PAHs . . . 64
4.4 Conclusion . . . 65
Acknowledgement . . . 66
References . . . 66

Chapter 5
***Microbiological impact of diffuse pollution sources on water quality* . . . 73**
Mats Leifels, Kwanrawee Sirikanchana and James Ebdon

5.1 Introduction . . . 73
5.2 Waterborne Diseases . . . 73
5.3 Traditional Faecal Indicators and Their Limitations . . . 75
5.4 Agricultural Sources of Microbiological Pollution . . . 76
5.5 Catchment-Scale Models . . . 77
5.6 Mitigating Diffuse Sources of Microorganisms . . . 77
5.7 Diffuse Pollution Case Studies . . . 77

5.7.1 Case study 1: assessment of stream crossing modification on cattle-derived microbial input to source water catchments (Australia) 78
5.7.2 Case study 2: prediction of spatial and temporal E. coli burden on pasture from grazing livestock in Scotland (UK) 78
5.7.3 Case study 3: tracking sources of faecal pollution in a mixed-agricultural catchment (Thailand) 78
References 79

Chapter 6
Emerging pollutants from various land use types and their impacts on water resources **83**
N. J. Reyes and L. H. Kim

6.1 Emerging Pollutants (EPS) in the Environment 83
6.1.1 History and progress in EP research 83
6.1.2 Classifications of EPs 85
6.2 Sources, Pathways and Presence of EPS in Different Land-Use Types 86
6.2.1 Point sources 86
6.2.2 Non-point sources 88
6.3 Impacts of EPS on Water Resources 89
6.4 Challenges, Opportunities and Future Concerns 91
References 92

Chapter 7
Marine debris: the Australian perspective **99**
Daniel Burns, Wally Smith and Heidi Tait

7.1 Introduction 99
7.2 Australian Marine Debris Initiative 99
7.3 Land Sea Source Index and the Marine Debris Source Index 104
Acknowledgements 109
References 109

Chapter 8
Diffuse water pollution in the United States: trends, regulations and outlooks **111**
Mi-Hyun Park, Eakalak Khan, Xavier Swamikannu, Michael K. Stenstrom and Joo-Hyon Kang

8.1 Introduction 111
8.2 Nonpoint Source Types 112
8.3 Nonpoint Source Pollutants 117
8.4 Impacts of Stormwater Pollution 123
8.4.1 Impact on surface water 123
8.4.2 Public health impacts 127
8.4.3 Groundwater impact 127
8.5 Regulations for Stormwater Pollution 128
8.6 Emerging Issues and Future Outlook 129
8.7 Conclusions 130
References 130

Chapter 9
Rural diffuse pollution in Scotland: impacts and management **139**
F. Napier, I. Sime, M. Aitken and K. Kortland

9.1 Introduction 139
9.2 Water Regulation and Diffuse Pollution 140
9.3 Diffuse Pollution Impacts 140
 9.3.1 Current status 141
9.4 Case Studies 142
 9.4.1 Diffuse pollution priority catchments 142
 9.4.2 Loch of strathbeg 142
 9.4.3 Ythan catchment 145
 9.4.4 Freshwater pearl mussels 146
9.5 Conclusions 149
References 149

Chapter 10
Implications of land use and land-use changes to water quality in Korea: status, management and policy development **151**
F. K. Geronimo and Lehyung Kim

10.1 Introduction 151
10.2 Urban, Agricultural and Livestock Landuses 154
 10.2.1 History, status and management strategies for urban land uses in Korea 154
 10.2.2 Agricultural and livestock industries in Korea 156
10.3 Impact of Luluc on Water Quality in Large Catchments: Case of Jeonju Stream and Seongju Icheon Stream 159
10.4 Other Pollutants of Emerging Concerns in Different Land Uses 164
10.5 Policies and Governance 166
References 167

Chapter 11
Protected areas and the diffuse pollution problem: rivers of the Kruger National Park, South Africa **169**
E. S. Riddell, L. Boyd, R. M. Petersen and R. G. M. Heath

11.1 Introduction 169
 11.1.1 The Kruger National Park landscape 170
 11.1.2 Crocodile River 172
 11.1.3 Sabie River 173
 11.1.4 Olifants River 174
 11.1.5 Letaba River 176
 11.1.6 Luvuvhu River 177
11.2 Regional Diffuse Sources 178
11.3 Synthesis 179
 11.3.1 Policy Status Quo 180
 11.3.2 Institutional constraints 180
 11.3.3 Partnerships and constituency 180
11.4 Conclusion 182
References 182

Chapter 12
Determination of changes in neonicotinoid pesticide concentrations and assessment of the risk posed to aquatic organisms in rivers based on high-frequency sampling...... 187
Yuki Kotani, Yoshinori Yabuki, Arisa Banno, Takashi Daifuku and Yukio Komai

12.1 Introduction 187
12.2 Method 188
12.2.1 Sampling and analysis procedure 188
12.2.2 Species sensitivity distribution (SSD) 189
12.3 Results and Discussion 190
12.3.1 Characteristics of concentrations and frequencies of neonicotinoid pesticides 190
12.3.2 Relation between neonicotinoid pesticides detected and land use near the survey sites 191
12.3.3 Risk assessment by SSD 193
12.4 Conclusions 194
References 195

Chapter 13
Addressing adverse water quality in Strathclyde Loch, Scotland 197
Mark D. Wyer, David Kay, Calum McPhail, Chris Kay, Carol Francis, Cheryl Davies, John Watkins and Julie Bolland

13.1 Introduction 197
13.2 Materials and Methods 199
13.2.1 Sampling programmes 199
13.2.2 Hydrometric monitoring and catchment analysis 200
13.2.3 Laboratory preparation and analysis 200
13.2.4 Statistical analysis 201
13.2.5 Discharge component and sample categorization 202
13.2.6 Faecal indicator organism load and budget calculations 202
13.3 Results and Discussion 202
13.3.1 Hydrometric monitoring and discharge estimates 202
13.3.2 Faecal indicator organism concentrations 203
13.3.3 Discharge and faecal indicator organism budgets 206
13.3.4 Swim zone curtain barrier efficacy study using a bacteriophage tracer 206
13.4 Summary and Conclusions 209
Acknowledgements 211
References 212

Chapter 14
Socio-economic impacts of pollution and restoration: Loch Leven case study, UK 213
B. J. D'Arcy, L. Kuhfuss, L. May and R. Audsley

14.1 Introduction 213
14.1.1 Economic costs of pollution 213
14.1.2 Loch Leven 213
14.1.3 Research opportunity 214
14.2 Methods 215
14.2.1 Identification of businesses in relation to Loch Leven improvements 215
14.2.2 Semi-structured interviews with local businesses 215

14.2.3 Focus groups. . . . 216
14.2.4 Review of planning applications 217
14.2.5 Assessment of the impact of environmental regulations. . . . 217
14.3 Results. . . . 218
14.3.1 Growth in numbers of tourist businesses around Loch Leven. . . . 218
14.3.2 Semi-structured interviews with local businesses (2019–20) 218
14.3.3 Focus groups. . . . 220
14.3.4 The details submitted with planning applications. . . . 221
14.3.5 The impact of regulatory changes 222
14.4 Discussion. . . . 222
14.4.1 Economic impacts of pollution. . . . 222
14.4.2 The economic importance of tourism and informal recreation 222
14.4.3 Diffuse pollution and motivation to address it. . . . 223
14.4.4 Good news is not news 224
14.5 Conclusions 224
Acknowledgements 225
References. . . . 225

Chapter 15
Diffuse pollution and contamination of pesticides in natural water and water supply: case study of Northern Thailand ***227***
Puangrat Kajitvichyanukul, Suthatip Sinyoung and Ruediger Anlauf

15.1 Introduction 227
15.2 Study Sites 228
15.2.1 General framework 229
15.2.2 Water treatment. . . . 230
15.2.3 Water sampling and analysis 230
15.3 Pesticide Contamination. . . . 231
15.3.1 Pesticide loadings and environmental contamination in soil, sediment, surface water and groundwater 231
15.3.2 Pesticide residues in treated waters. . . . 236
15.3.3 Management practice and mitigation approach. . . . 237
15.4 Summary. . . . 238
Acknowledgement 238
References. . . . 238

Chapter 16
Recognition of diffuse pollution impacts on water utilities ***241***
T. Kawata, S. Fujii and H. Li

16.1 Introduction 241
16.2 Material and Methods. . . . 241
16.3 Results and Discussion 243
16.3.1 Respondents (available answers (***AA***)/total responses (***TR***) = 97/97, Q1, Q2) 243
16.3.2 Characteristics of water utilities 243
16.3.3 Effects of land use on diffuse pollution. . . . 246
16.3.4 Influences of each pollutant on water utilities 247
16.4 Conclusion 249

Acknowledgement 250
References 250

Chapter 17
Diffuse pollution and drinking water quality – cost–benefit of reducing the source to improve the environment ***251***
J. F. S. Knightbridge, I. V. Ribeiro and A. L. Cooke

17.1 Introduction 251
17.2 Methodology 252
17.2.1 Eutrophication assessment 257
17.3 Results and Discussion 257
17.4 Conclusions 263
Acknowledgements 263
References 263

Chapter 18
Landscapes for pollution risk management ***265***
B. J. D'Arcy, P. Morison, B. C. Braskerud and P. Kajitvichyanukul

18.1 Introduction 265
18.2 Land Use and Pollution Control 266
18.3 Pollution Control and Landscapes 267
18.4 Conclusions 275
18.4.1 Pollution prevention 275
18.4.2 Landscapes and inter-related problems 275
Acknowledgement 276
References 276

Foreword

At a time when the world faces the twin existential threats of climate change and biodiversity loss, greater attention is rightly being placed by scientists, politicians, businesses and civic society on the importance of taking an integrated approach to what are complex 'wicked problems'. As we move forward, both these challenges will undoubtedly impact the causes and potential solutions of diffuse pollution, and in a manner that will be increasingly difficult to predict and control. But already, there is little doubt that the drivers, pressures and impacts of diffuse pollution in respect of its origins in land use and land management practices, its impacts on water quality and in the development of potential responses by diverse stakeholders fall firmly in this category of a 'wicked problem'.

Diffuse pollution is huge in scope and extent. This is not simply an environmental problem, rather it encompasses social, cultural and economic aspects, involving diverse groups of stakeholders. Both the drivers and the impacts operate over different scales (of time and space), and potential solutions require not only a range of disciplines, but also novel forms of governance to deliver an integrated policy framework and partnerships on the ground to tackle the many issues involved. Not least of these will be how to transparently and 'fairly' resolve trade-offs between producers and consumers; between 'upstream' land-based providers of ecosystem services and those 'downstream' who benefit or are adversely impacted by land-use choices. The quality of our water, after all, is entirely dependent on what enters it upstream, largely from land-based activities.

Recognition of the importance of Integrated Water Resource Management (IWRM) is not new, but the growth of interest in nature-based solutions and other catchment intervention measures taken at the landscape scale is not just to help stop physical losses of soil (and attached nutrients and pesticides) and in doing so protect water quality, but also to reduce flood risk, to manage carbon, to improve biodiversity, and build socio-economic resilience for people. The introduction of ever wider and more widespread buffer strips along watercourses, of improved tillage practices, better targeted pesticides and nutrients, and compatible measures such as natural flood risk management, are all examples of such results. Along with this has come increased interest in not only protection of these upstream water gathering grounds, but in who should pay for this, with the focus on the 'polluter pays' principle being complemented by the notion of Payments for Ecosystem Services (PES) and the development of novel PES schemes.

PES Schemes are being developed across the world, with perhaps one of the earliest and best-known (though not originally described as such!) being the payment by the residents of New York City to the land managers of the Catskills some 100 miles upstream (and out of state) from whence comes their water supply. Even the most simple of economics showed that payments of $ millions/year to the farmers in the watershed to manage their land in a manner which protected the water supply was a cheaper and more sustainable option than building $ billion water treatment works downstream. Since then, many other schemes have developed worldwide with several water utilities in the UK for example providing grants and holding 'reverse auctions' to attract farmers to change their land management practices and improve water quality. Of course, the same processes of weather-mobilised movement of contaminants from landscapes into the water environment are equally challenging in the built environment; often involving huge amounts of sediment from disturbed ground in the construction of new developments, as well as chronic mobilisation of everything from litter and microplastics, to oil, PAHs and toxic metals from roads and traffic.

In all of this, there are many lessons to be learned, failures and successes to be analysed and shared. In this context, the International Water Association (IWA) is to be congratulated in bringing together this important production, deriving as it does experiences and insights not just from across the world, but from the different perspectives of the many diverse players and stakeholders involved.

Chris Spray

Professor Chris J Spray MBE, FRSA, FCIEEM

Chris Spray is a Senior Research Fellow and Emeritus Professor of Water Science & Policy at the UNESCO Centre for Water Law, Policy & Science at Dundee University, UK. Chris was Director of Science for the Scottish Environment Protection Agency (2004–2009) and earlier Director of Science for Northumbrian Water Group. With a 1st class degree in Geography and General Natural Sciences from Cambridge University, Chris got his PhD in Zoology from Aberdeen University, working as a research ecologist with the university for seven years, before moving into a career in the water industry from a variety of perspectives. As past chair of the Freshwater Biological Association, president of the Chartered Institute of Ecology & Environmental Management and chair of Tweed Forum, Chris has extensive research and practical experience of the importance of partnership working in support of integrated land and water management at the catchment scale.

doi: 10.2166/9781789061123_xv

Preface

The definition of diffuse pollution adopted from three IWA international conferences (from the outset of the initiative: 2017, 2018 and 2019) is 'weather-driven movement of pollutants from land and air into the aquatic environment'. Thus, diffuse pollution is huge in scope and extent, encompassing everything from nutrients and sediment to micro- and nanoparticles, toxic metals, pesticides, faecal pathogens and even water-borne litter. Taking its most important defining characteristic as being the weather-driven mobilisation and transport of the contaminants underlines the links with climate change and flood risk management, as well as with topography, soil type, land-use and drainage. That in turn explains the vital importance of landscape interventions to prevent/trap/stabilise or capture mobilised contaminants, as close to source as possible (as well as the various other good practice measures which can be taken).

This book is the output of several years aiming to clearly present evidence on the nature and/or extent of the impact of diffuse pollutants on the water environment. The substantial evidence of the impacts of diffuse pollution are displayed as a basis for action at local, national and international levels. This book also adopts an interdisciplinary approach ranging over environmental, socio-economic, and policy reports for the evidence of environmental problems from diffuse pollution.

The book is combined of review articles and original papers presented as nineteen chapters that are grouped into five categories:

- High-level state of environment reports – estimating the national or international extent and severity of diffuse pollution impacts on the water environment.
- Ecological impacts – case studies at any level, from river basin and coastal waters to local creeks and lakes, or particular species; ecosystems, from coral reefs to farmland; bioaccumulation.
- Resource impairment and socio-economic impacts – for example surface water and groundwater unfit for use by industries needing to abstract clean water (such as food and drinks, textiles, and high -quality paper businesses); recreational impacts (e.g., angling, closure of bathing beaches, algal blooms/red tides and reputational damage); risk of national/international regulatory actions.
- Water utility impacts – challenges to meet statutory water quality requirements for raw potable supplies, for surface water discharges from public networks, and impacts on performance of water treatment plants in relation to permits too, including: filtration plant running costs associated with sediment, algal blooms and blocked filters; water and health, blue- green/

Editor(s): Puangrat Kajitvichyanukul and Brian D'Arcy

cyanobacteria toxins in potable supplies, or *Cryptosporidium* outbreaks; colour and flavours in potable supplies, including turbidity and hydrocarbons respectively etc.

- Impacts of specific contaminants – evidence/proofs of specific contaminants (pesticides, polycyclic aromatic hydrocarbon, heavy metals, etc.) on the environmental problems.

This book is intended to raise awareness of the links between land use and water quality, as a basis for stimulating more effective action to address diffuse sources of pollution in international level. The authors of the papers in the book are from five continents, including USA, Mexico, Britain, Korea, Thailand, Japan, South Africa, and Australia. They cite research and case studies from all the developed continents, including studies, trends, regulations, and outlooks from USA to Australia, and the best practice for water quality protection from South Africa to Asia.

This publication aims to bring the attention of the next generations of industrial researchers and practitioners, whether organic chemists, engineers, trainee managers and economists, or current plant managers and site operations. The book would not have been possible without the support of the International Water Association (IWA), especially the Diffuse Pollution Specialist Group. The editors are grateful to all the contributing authors and reviewers. We are also grateful to the sponsor of the book. Finally, we are grateful to Mark Hammond and colleagues at IWA Publishing for their patience and help getting this book completed.

Puangrat Kajitvichyanukul and Brian J. D'Arcy
Editors
November 2021

doi: 10.2166/9781789061123_001

Chapter 1

Land use and diffuse pollution: are perceptions part of the problem?

B. J. D'Arcy[1], D. Kay[2], F. Napier[3], P. Haygarth[4], Yan Yan Sun[5] and Keiko Wada[6]

[1]Independent Environmental Consultant, Kinross KY13 9JR, UK
[2]CREH, University of Wales, Lampeter, Ceredigion, SA48 8HU Wales, UK
[3]SEPA, Scotland, UK
[4]Lancaster University
[5]IWA Netherlands
[6]Institute of Lake Biwa, Yodo River Water Environmental Research, Japan

Keywords: Diffuse pollution; nonpoint source; storm events; faecal indicators; nutrients; catchment management; environmental regulation; definitions

1.1 INTRODUCTION

1.1.1 Land use and water quality – historical perspectives

The aims of this paper are threefold:

- highlight the inter-relationships between land use and water quality;
- set out a rationale for an inclusive simple definition of diffuse pollution;
- share example insights and well established evidence from different specialists with the wider environmental science research community, policy makers and – hopefully – environmental regulators too.

Water quality is fundamentally linked to land use. A watercourse often originates as a mountain seepage, perhaps from snow melt or a glacier, or as a spring where groundwater issues at the surface where porous strata meet impervious rock. The quality of the groundwater and of surface runoff determines the quality of the resulting watercourses and downstream waterbodies. In a natural system, the local geology determines such characteristics as alkalinity, hardness, pH and trace minerals and nutrients. Nitrogen-fixing plants in the vegetation import nitrogen (N) from the atmosphere. The land cover and topography determine the extent to which natural contaminants (sediment, nutrients, faecal pathogens) are exposed to the weather and associated erosion and leaching. Vegetation protects the underlying soil and subsoil from exposure and erosion by wind or rain. The extent of natural land cover has been steadily reducing for millennia as human activities have transformed landscapes and catchments, exposing soils to erosion (by cultivation or overgrazing), and adding higher levels of nutrients which in turn can be leached or mobilised with soil erosion. The adverse environmental impacts of agriculture have been well documented, varying in scale from sedimentation of the spawning

Editor(s): Puangrat Kajitvichyanukul and Brian D'Arcy

tributaries of rivers for salmonid fish and pearl mussels (Cosgrove *et al.*, 2017; Degerman *et al.*, 2013; Ferrier & Ellis, 2000), to problems such as the development of anoxic coastal waters, for example in the Gulf of Mexico (associated with nitrates from farmland in the Mississippi catchment; see Hypoxia Task Force, 2018; Rabalais *et al.*, 2002; Rabotyagov *et al.*, 2014), and sediment, nutrient and pesticide impacts on the Great Barrier Reef in Australia (Great Barrier Reef Water Science Taskforce, 2016).

As human settlements increased in size and number, urban pollution sources became important. In modern megacities unsewered settlements provide a severe example of high-impact urban development (Chikozho *et al.*, 2019; Edokpayi *et al.*, 2017; Iwugo *et al.*, 2003; Wilson, 2018). The problems are worse in wet weather when pollutants are mobilised and washed into watercourses (Wilson, 2018). Historically, the advent of sewers was effective at taking dry-weather flow foul drainage away from centres of population, but without adequate treatment prior to discharge, sewage effluents are significant, dominating the water quality of local rivers, and often including industrial effluents (Hynes, 1971; Jones, 2000; Porter, 1973).

Manufacturing and other industrial developments are important land-use changes from pristine conditions in a catchment. As well as the impacts of process effluents, industrial pollution is often caused by spills and accidents, exacerbating the background contamination washed off surfaces in rainfall runoff (D'Arcy *et al.*, 2018; Kim *et al.*, 2018). Industrial and commercial process effluents can usefully be categorised as aqueous waste streams. Sources range from brewery and papermill effluents to silage liquor and livestock slurry on farms, and include wash-waters such as steam cleaning, rinse waters, vegetable washeries and other processes which generate an effluent discharge. Sewage is obviously associated with settlements, as are many industrial estates. But industries are also found in rural areas and can contribute to water pollution alongside agricultural or forestry land uses (e.g., sugar refineries, paper mills, distilleries) where locations are close to sources of raw materials, energy or distribution centres for the products. Land-use planning increasingly seeks to locate compatible industries together, so a waste from one is a resource for another in a complex web of inter-linked processes and production: industrial ecology (Kim & Maniquiz-Redillas, 2018; OECD, 2010; O'Keeffe, 2018; World Economic Forum, 2021). Figure 1.1 indicates the importance of land-use change, from a natural condition to the diverse human uses of the land. Note that pollution loads are generated under natural conditions and especially in the transition processes (indicated in italics), as well as in the final land uses.

Land-use change, especially diffuse pollution generated by those changes, is a global problem (see Allen *et al.*, 2012; Novotny, 2003; OECD, 2017; Wada *et al.*, 2020). Extensive changes from natural conditions can have severe impacts, for example urban streams (Wang *et al.*, 2014; Wilson *et al.*, 2005). The economic costs of water pollution are significant, as well as the impacts on public health, ecosystems and impairment of water resources. Example annual costs cited in OECD (2017) ranged from $M167–264 for drinking-water treatment costs to $M2100–1380 for the costs of gastrointestinal illnesses attributed to drinking water. Water quality is a major component of the damage and costs associated with flooding, too – thousands of tonnes of mobilised sediment; faecal pathogens and disease risks; oil and chemicals too, depending on the catchment.

1.2 DIFFUSE POLLUTION

1.2.1 Diffuse pollution – back to basics

Whilst land use determines the types of pollutants which could be present, the landscape determines the extent to which pollutants associated with a particular land use are likely to be exposed and mobilised by weather conditions, entering the water environment as diffuse pollution. For industry, agriculture, forestry and human settlements, the land-use/landscape interface determines water quality once point sources (i.e., major process effluents) are properly controlled. The unresolved challenge is diffuse pollution (D'Arcy *et al.*, 2000; Novotny, 2003; SEPA, 2006; USEPA, 2020). Figure 1.2 highlights relationships of climate and weather, land use, landscape and water quality, for pollutant delivery to water environment.

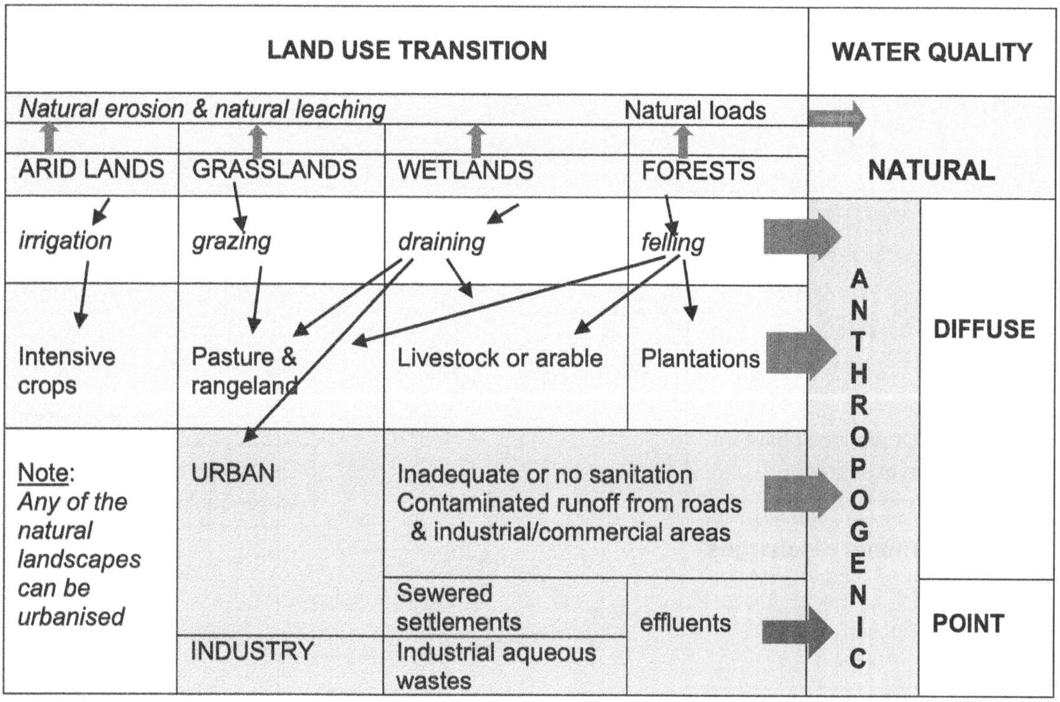

Figure 1.1 Land-use transformation and the generation of pollution (modified from Novotny, 2003).

Novotny and Olem (1994) published a set of characteristics which, with minor modifications and additions, have stood the test of time and are still valid and helpful (see Table 1.1 – modified slightly to avoid misinterpretation across different legislatures with respect to terms such as 'waste').

1.2.2 Evidence of confusion

1.2.2.1 IWA diffuse pollution specialist group survey

The term 'point source' was brought into being to allow control of water pollution in the USA, *but initially without addressing small-scale discharges*, such as urban development below a population threshold and many agricultural activities; these were called simply *nonpoint* sources to indicate their uncontrolled status. Thus, point source referred to sources of pollution which are within the control of section 502(14) of the Clean Water Act in USA. The USEPA website now defines a point source as follows:

> 'The term "point source" means any discernible, confined and discrete conveyance, including but not limited to any pipe, ditch, channel, tunnel, conduit, well, discrete fissure, container, rolling stock, concentrated animal feeding operation, or vessel or other floating craft, from which pollutants are or may be discharged. This term does not include agricultural stormwater discharges and return flows from irrigated agriculture'.

> 'The term "nonpoint source" (NPS) is defined to mean any source of water pollution that does not meet the legal definition of "point source" in section 502(14) of the Clean Water Act' (USEPA, 2020).

The 2020 USEPA definition above merely shows that control of pollution has now been extended to what elsewhere in the world is understood as diffuse pollution. Unfortunately, the term NPS has

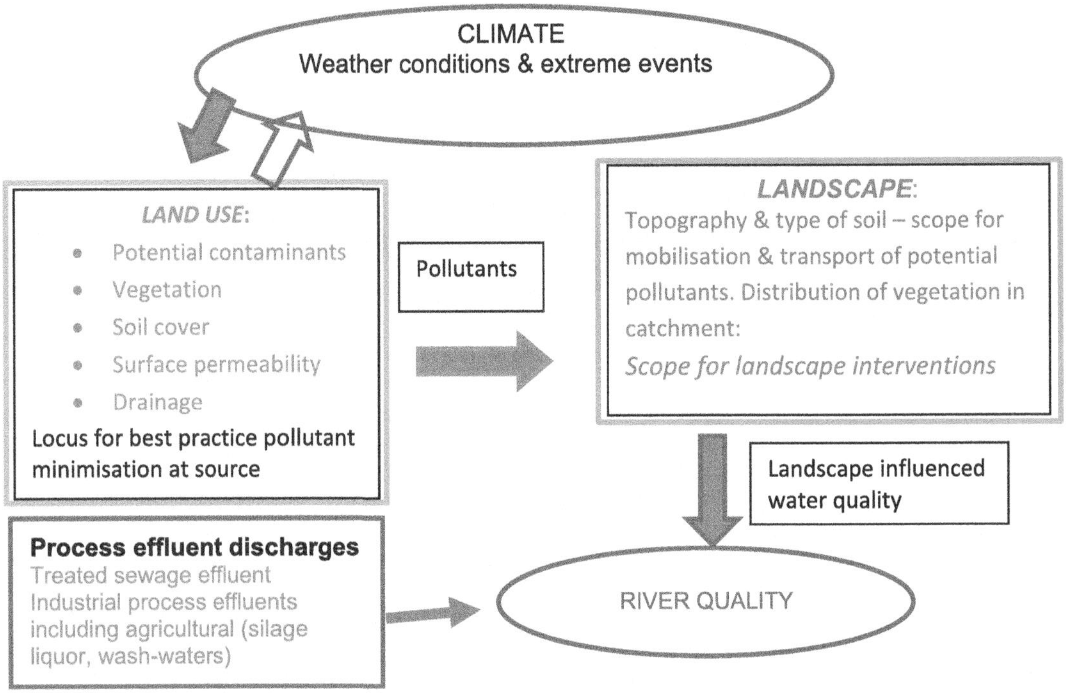

Figure 1.2 Land use and water quality interactions for sources, mobilisation and delivery of pollutants to the water environment.

moved out of its regulatory USA context into international jargon, sometimes resulting in irrational policy arguments, for example that urban runoff cannot be considered as diffuse because it discharges from a lot of 'point' sources (e.g., surface-water drains, SEPA, 1999). Yet in the same appraisal of causes of pollution, where phosphorus (P), nitrate or faecal pathogens drain via field drains (pipes – as is common in agricultural land in the UK), that category of pollution was not identified as a point source. Both of those categories are diffuse in nature. Such inconsistencies diminish the chance of adequate recognition by policy leaders and governments that diffuse pollution, across the spectrum of impacts, needs to be taken seriously. That in turn reduces the resource likely to be given to address it.

Various definitions of diffuse pollution have been published, usually referring to dispersed origins and difficulty in identifying a source (Ellis *et al.*, 2004). In the UK under the auspices of the Chartered Institute for Water and Environmental Management (CIWEM), sixteen organisations came together to produce an agreed definition which tried to describe what was included by specifying categories of sources and sectors (D'Arcy *et al.*, 2000). That was adopted in Novotny (2003) but has failed to meet international needs for a consensus viewpoint.

A survey of members of the Diffuse Pollution and Eutrophication Specialist Group of the International Water Association (IWA) undertaken for this article, produced a revealing picture of confusion and contradiction (Table 1.2). For example 47 out of 62 respondents agreed that 'not from a pipe' was an important or very important characteristic. Yet 44 out of 64 responses also recognised that stormwater (which enters the water environment via pipes) was an important or very important aspect of diffuse pollution.

The perceived importance of not being able to identify a single discrete source was also indicated in the IWA survey (Table 1.2), a widespread perspective in Europe (e.g., Ellis *et al.*, 2004). Does this mean that the relative ease of actually identifying many discrete sources in the field (not the office or

Table 1.1 The key characteristics of diffuse pollution (adapted from Campbell *et al.*, 2004, derived from Novotny and Olem, 1994).

	Characteristic	Comments
1	Pollutants enter the water environment in a diffuse intermittent way, at intervals related to weather events	*Hotspots* occur in a heterogeneous landscape
2	Pollution arises over an extensive area of land and is in transit overland prior to reaching surface waters, or infiltrating into groundwater	Includes wind-blown as well as runoff-induced movement
3	Diffuse sources are difficult or impossible to monitor at point of origin	Often dry unless raining heavily
4	Abatement is focused on land and runoff management practices	The best practice philosophy 'best management practices' (BMPs)
5	Compliance monitoring is carried out on land rather than in water	Assess BMPs
6	Water quality impacts are assessed on a catchment scale	Collectively significant impacts
7	Discharges cannot be measured in relation to effluent limitations	Watercourse quality is the aim for landscape runoff
8	Extent of impacts is related to climatic events and geographic and geological conditions; will differ greatly year on year & place to place	Spot samples year on year are inadequate for management
9	Most important pollutants from diffuse sources are suspended solids, nutrients, faecal pathogens and toxic substances	Now add emerging pollutants, including plastics, micro-plastics, fire retardants, and so on

laboratory) means they are not diffuse pollution issues? (See section 1.2.4 below). In Table 1.2 there is however, strong recognition of the importance of landscape- and weather-driven characteristics.

1.2.2.2 Diffuse or natural pollution?

The final characteristic in Table 1.1 referred to the pollutants most often associated with diffuse source pollution in the water environment. Whilst many pollutants are anthropogenic in origin (manufactured compounds, such as pesticides, solvents, plastic, PCBs etc.), others are natural and are

Table 1.2 Results of a survey of IWA diffuse pollution specialists, asked to indicate importance of various aspects of diffuse pollution.

	Irrelevant	Not Very Important	Neither Important Nor Unimportant	Important	Very Important	No. of Replies
Cannot identify a source	18.8% 12	6.3% 4	12.5% 8	35.9% 23	26.6% 17	64
Not from a pipe	6.5% 4	8.1% 5	9.7% 6	38.7% 24	37.1% 23	62
Weather driven	7.8% 5	15.6% 10	10.9% 7	37.5% 24	28.1% 18	64
Cannot be controlled by legislation	27.4% 17	21% 13	21% 13	21% 13	9.7% 6	62
Landscape related	4.7% 3	1.6% 1	18.8% 12	45.3% 29	29.7% 19	64
Stormwater associated	6.3% 4	12.5% 8	12.5% 8	34.4% 22	34.4% 22	64

Table 1.3 The two super-classes of potential diffuse source pollutants.

Naturally Occurring Substances	Anthropogenic Pollutants (Some Examples)
May dominate during land-use transitions (see Figure 1.1) but not restricted to such circumstances	*Man-made compounds polluting at point of manufacture and/or at points of consumption, including legacy pollutants*
Soil/sediment	Pesticides
Nutrients	PCBs
Natural organic matter (NOM) & associated decomposition products (e.g., geosmin and other odour/taste compounds in raw potable water supplies)	Plastic, including litter, micro- and nanoplastic fibres and particles Solvents
PAHs derived from forest fires	Fire retardants
Cyanobacteria toxins	Nanoparticles, including heavy metals, POPs
Faecal pathogens	Toxic forms of metals in many materials
Geogenic metals	Endocrine-disrupting chemicals
Naturally occurring substances are water pollutants when anthropogenic activities significantly increase the quantities released into the water environment to detrimental effect.	Lubricating oil, hydraulic oil, specialist oils Diesel, gasoline, other fuel oils Timber preservatives, anti-fouling paints And many more… including degradation products

essential to life; for example, nutrients and soil can be pollutants (Hussein & Shahid, 2017; Liu *et al.*, 2011; Sthiannopkao *et al.*, 2006). These natural materials are pollutants when they are mobilised by anthropogenic activities in quantities which have adverse impacts on aquatic systems (Table 1.3). Nutrients of course can be manufactured products and when applied in a landscape may greatly increase the quantities at risk of mobilisation to the water environment, as well being more readily mobilised than natural sources.

Examples of naturally occurring substances causing pollution are the increased mobilisation of sediment and phosphorus from over-grazed and arable farmland, or faecal pathogens associated with high stocking rates for livestock or access to watercourses (Kay *et al.*, 2007), or pigeon roosts and pets contaminating the built environment in urban population centres (Kay *et al.*, 2000; O'Keefe *et al.*, 2005; Reynolds *et al.*, 2020; SEPA, 2020; Wither *et al.*, 2005). For catchment management, how can pollution from natural substances be usefully distinguished from natural pollution? Is it, in some jurisdictions, a blurred issue confused with diffuse pollution? Comparisons between states or countries, or river systems, may be invalid if attribution of causes differs from one place to another.

1.2.3 A new definition

NPS (non-point source) was primarily simply an indication of what was controlled and what was not (Craig & Roberts, 2015; Novotny, 2003; WEF, 1987; USEPA, 2020). This paper aims to derive a practical and readily understood definition of diffuse pollution based on the characteristics of the problem. What are the characteristics which should influence a definition? Ellis *et al.* (2004) for urban contexts defined it as 'pollution that arises from various urban land use activities and which have no obvious discrete source'. The problem with that definition is it precludes recognition of the hotspots which are created by the processes which drive diffuse pollution (see Section 1.2.4) and are therefore important priority targets for management action.

The problem with management by exclusion (e.g., of sewage or industrial effluents only – an old-fashioned point-sources regulatory regime and mindset) is that environment protection agencies working on water pollution became focused on regulating discharges by permits or licences, targeting the largest individual discharges and spending a lot of time and resource on each one. Therefore diffuse pollution was outside the scope of their activities and interests: it would otherwise be a cost

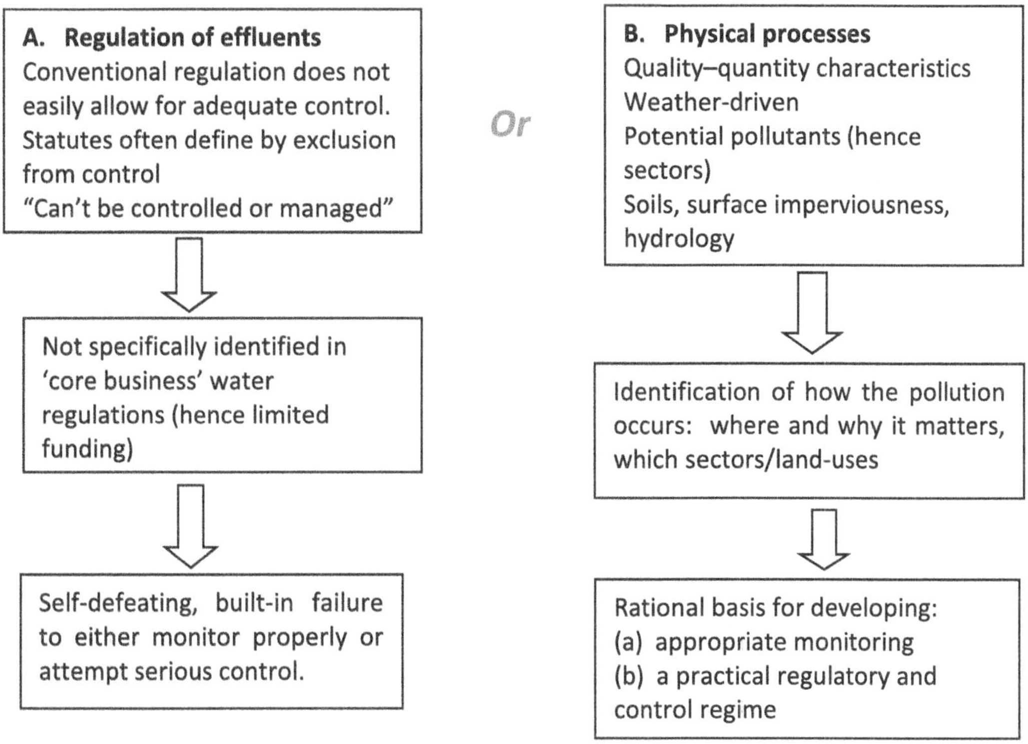

Figure 1.3 Comparison of conceptualisation options for diffuse pollution definitions.

to the agency (far too many sources/discharges) without an obvious cost recovery mechanism. None of those approaches to defining diffuse pollution led to adequate understanding of the issues, nor to sensible, adequately resourced, approaches to controlling them (Figure 1.3).

Figure 1.3 suggests that a definition focused on the physical characteristics of diffuse source pollution would lead to better understanding and hence management of pollution risks. A new definition is clearly necessary, and the following is proposed as a close reflection of the characteristics in Table 1.1, as identified by Novotny and Olem in 1994:

> 'Landscape or atmospheric contamination mobilised and transported into the water environment by weather conditions'

This provides a clear focus on the problem as the starting point for developing appropriate controls including regulations and joining with inter-related issues in a coherent policy framework. On that basis it is of course possible to develop an appropriate and enforceable regulatory regime, for example the Controlled Activities Regulations (Scotland) 2006 and their General Binding Rules (D'Arcy & Hemingway, 2018; D'Arcy *et al.*, 2006a).

1.2.4 Diffuse pollution hotspots

From the perspective of estimating risks of diffuse pollution, landscapes are, at least in the UK, rarely homogeneous. One of the first priorities in a pollution audit of a town or city, or business park or industrial estate, or farmland is to assess the scope for interventions to address specific diffuse pollution hotspots. A diffuse pollution hotspot is a location from which a disproportionately greater (in relation to the area of the local catchment) quantity of pollutants is being mobilised by weather conditions to contaminate the water environment.

Examples of rural hotspots include:

- where topography favours a concentration of flow across a field or hillside, leading to a deep rill and outbreak of runoff into a watercourse;
- where livestock congregate around ring-feeders placed in winter too close to a stream, or beneath sheltering trees along a watercourse, or have open access to a watercourse;
- where livestock congregate at a gateway and contaminated runoff enters the road drainage system and watercourse;
- on a steading (built environment part of a farm) where hotspots can occur in various situations associated with livestock, pesticides, fertiliser (organic and inorganic) and farm machinery and equipment;
- other situations where disproportionately significant risks can occur, sometimes on a temporary (e.g., road and track construction, ditch maintenance) or seasonal basis (e.g., forestry planting and felling activities, sheep dipping etc.).

Example of hotspots in towns and cities include:

- bird roosts involving large numbers in a locality, in contrast to more dispersed but still large numbers of pets (Moriarty *et al.*, 2011; O'Keefe *et al.*, 2005; Wither *et al.*, 2005);
- traffic concentrations – the distribution of pollution risk across the city is heterogeneous, not uniform, and road traffic is a major source of diffuse pollution, affecting air quality (Defra, 2020; Marshall, 2020; SEPA, 2006) as well as contributing to road runoff (Napier *et al.*, 2008);
- 'dirty' premises on industrial estates and business parks, for example, loading and unloading areas, vehicle fleets with *ad hoc* steam cleaning/washdown activities, premises handling large amounts of oil or other materials likely to cause pollution if spilled in transit, storage or use (D'Arcy & Taylor, 2018);
- other impervious areas liable to be heavily contaminated where the pollutants are washed off in wet weather into a surface water drainage system (e.g., extensive areas of exposed soil on construction sites).

1.3 QUANTIFYING DIFFUSE POLLUTION – UK EXAMPLES

1.3.1 High flows and water quality

The characterisation of diffuse pollution as mobilisation of contaminants by weather conditions (especially rainstorms) predicts that generally higher flows will have higher concentrations of those land-use-derived pollutants than during periods of dry weather, in the absence of polluting point sources (effluents in this context). That is not a new observation and is evident in many watercourses studied in many countries (e.g., Ferrier *et al.*, 2005; Fujii *et al.*, 2006; Kay *et al.*, 2007; Mamun *et al.*, 2020; Wyer *et al.*, 2001). Such a clear picture of water quality, however, requires storm event monitoring (integrated flow and chemical contamination assessment) which is not often a routine action by environmental managers. In Scotland, UK, two long-term rural diffuse pollution monitoring stations were established from 2000 by SEPA which provided data to characterise and quantify pollution from example land uses; a livestock-dominated catchment in the wet west of Scotland, and a mixed arable and livestock catchment in the drier east of the country (Greig *et al.*, 2004). High concentrations at high flows were characteristic of the data from all three diffuse pollution monitoring stations. Example data from the farmland sites are published in D'Arcy *et al.* (2006b) and Ferrier *et al.* (2005). Data from a third SEPA diffuse pollution monitoring site, below an industrial estate in the north-east of Scotland also showed higher concentrations at high flows than lower flows (Cundil *et al.*, 2010; Ibanez, 2015). Figure 1.4 shows how concentrations of toxic metals in the East Tullos Burn varied over

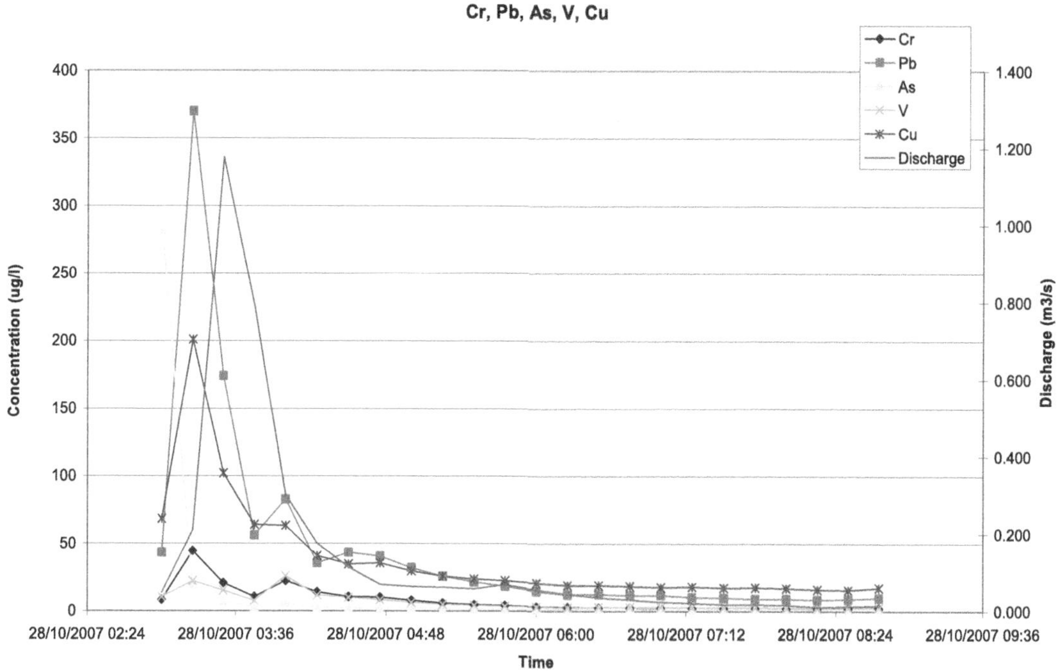

Figure 1.4 Responses of toxic metals to stream flow (example storm event on 28 October 2007), East Tullos Burn (from Cundil *et al.*, 2010).

the duration of a storm event (from Cundil *et al.*, 2010). For the particular storm event presented, the concentrations of each of the toxic metals were higher during high flow parts of the storm event than during lower flow conditions, throughout that particular event.

Elsewhere in the UK, suspended solids and nutrients loads were estimated at two headwaters streams in north-west England, as part of the Demonstration Test Catchments (DTC) programme there (Ockenden *et al.*, 2016). The catchment with the greater prevalence of arable farming (37% of land use) was the Pow Beck, as compared with only 6% in the Newby Beck which has a more upland character. The Pow Beck had double the total phosphorus (TP) load and 3.7 times the total reactive phosphorus (TRP) load of the Newby Beck, in high flows. In both streams the high flow events contributed more than 80% of the annual loads of total suspended solids (TSS) and phosphorus. Summary results for the Pow Beck are reproduced in Table 1.4 below. High flow load was ten times greater than at mid flow for suspended solids. The high flow condition also had the greatest load for the other parameters measured. Full details are given in Ockenden *et al.* (2016).

Also in north-west England, a study at Blackpool identified concentrations of faecal indicator organisms adversely affecting compliance with EU standards for bathing water quality. A primary source was the estuary of the River Ribble, just south of the town (Crowther *et al.*, 2001, 2003).

Highest values for contamination were at the higher flows, and the lower values correspond to the lower, base flows (Figure 1.5). Figure 1.6 shows that the relationship between contamination and flow holds true for many of the tributary rivers potentially contributing faecal pathogens to the EU-designated bathing water at Blackpool. It shows that water quality can be classified as high-flow condition and low-flow condition. The situation is however complicated by the fact that several of the tributaries are draining old towns on combined sewer systems, so the high-flow quality is affected

Table 1.4 Estimated nutrient loads measured for three flow categories in a tributary of the River Eden, north-west England, UK (modified from Ockenden *et al.*, 2016).

Pow Beck Tributary of R. Eden	TSS		TP		TRP		Nitrate-N	
	Tonnes	%	kg	%	Kg	%	Kg	%
High flow ($>Q_{10}$)	530	91	3060	81	1700	76	10 590	52
Mid flow (Q_{90}–Q_{10})	50	9	710	19	520	23	9720	47
Low flow ($<Q_{90}$)	0	0	20	<1	20	<1	160	1
Annual load	580		3790		2240		20 470	
Annual load per ha.	0.55		3.61		2.13		19.5	

Pow Beck: 1 April 2012–31 March 2013; $Q_{10}=0.441$ m³/s; $Q_{90}=0.027$ m³/s; land use mainly improved grassland (dairy and beef cattle plus sheep), but also 37% of land is arable farming.

by combined sewer overflows, as well as runoff from separately sewered districts and from livestock farming in the rural landscapes (e.g., the River Douglas).

The work analysing river flow and water quality relationships for tributaries to coastal bathing waters (Aitken, 2003; Crowther *et al.*, 2001, 2003; Kay *et al.*, 2007; Wyer *et al.*, 2001) led to SEPA and the Environment Agency in the UK developing a 'high flows – poor quality' warning system for bathers (McPhail & Stidson, 2004; Stidson *et al.*, 2012). The research also helped to move forward

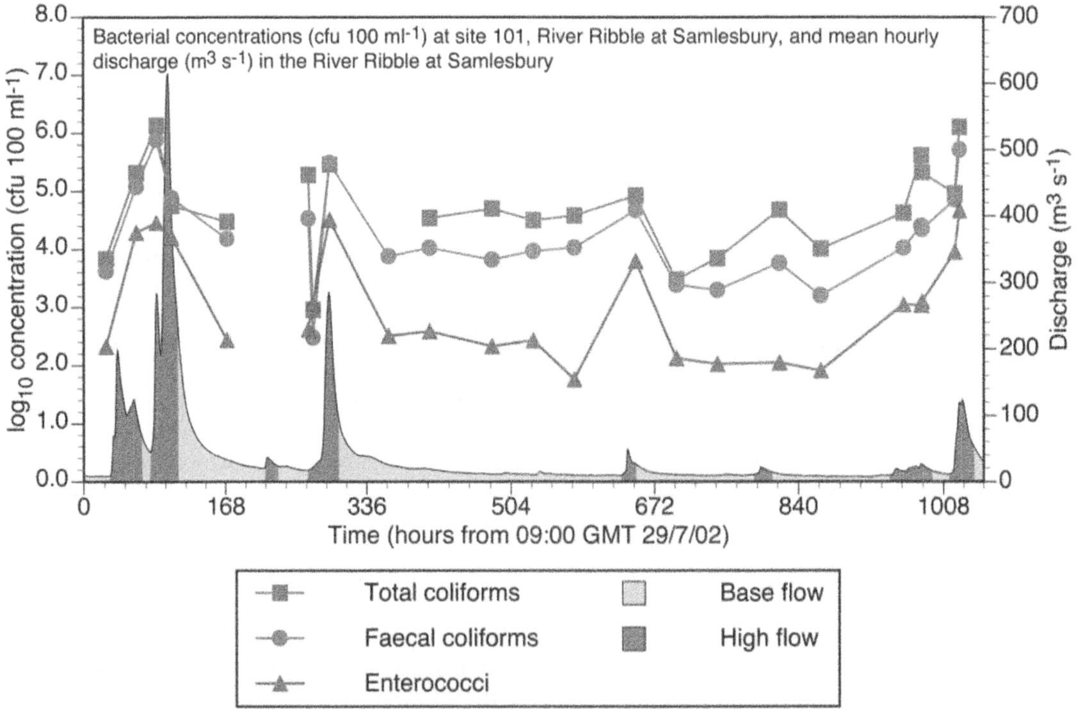

Figure 1.5 Faecal Indicator Organisms measured in the River Ribble (UK) and corresponding river flows (from Crowther *et al.*, 2001, 2003).

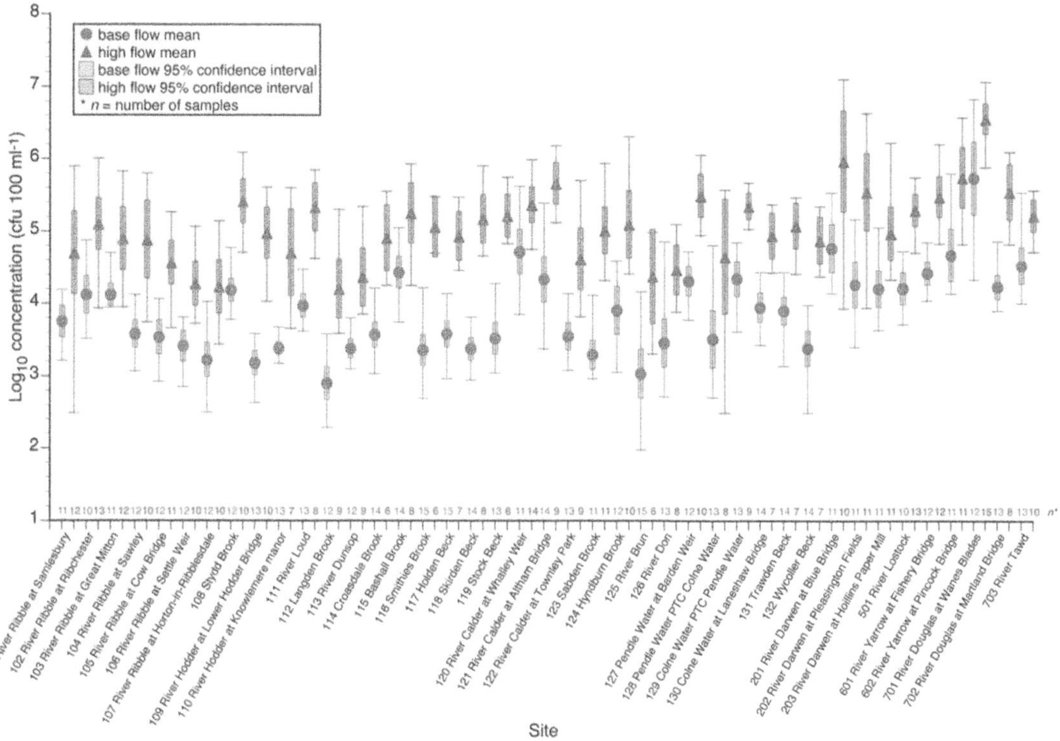

Figure 1.6 Faecal Indicator Organisms in tributaries of the Ribble Estuary in relation to its impact on the recreational waters at Blackpool, UK (from Crowther *et al.*, 2001, 2003).

the business of pollution prevention and control for pollution control agency staff, from a simplistic looking for effluent leaks and spills, to a diffuse pollution aware assessment of whole farm landscapes, not only for bad practice, but also for pathways and practices which deliver a variety of contaminants to the water environment (e.g., Haygarth *et al.* 2005), and how that can be managed, for example with a four-point plan for livestock farming (Audsley, 2021).

1.3.2 Environmental quality standards (EQS) and river concentrations

The classification of a river as polluted is usually determined by comparing water quality measurements for the river or other water body, against agreed environmental quality standards (EQS). For water chemistry, the latter are accepted threshold concentrations of indicator parameters, below which it is hoped there would be no adverse impacts (EU, 2008; Inglezakis *et al.*, 2016). Watercourses impacted by diffuse pollution have high *concentrations* at high flows (Figures 1.4–1.6) and exceedance of an EQS is at least as significant then as at low flow (when the exceedance would normally be expected if the pollution was from large, poorly controlled effluent discharges). Indeed, if exceedance is at higher flows, then the impacts are more severe, since at high flows the *loads* will be massive, likely to deposit suspended contaminants downstream as a storm event declines (e.g., phosphorus deposited in lakes) and the high-flow loads carried by the rivers will adversely affect even coastal waters downstream (Crowther *et al.*, 2001; Kay *et al.*, 2007; Wyer *et al.*, 2001).

EQS values are commonly applied for nutrients as well as toxic substances, and Figure 1.7 shows the importance of rainfall events in the mobilisation of ortho-phosphate from a farming catchment

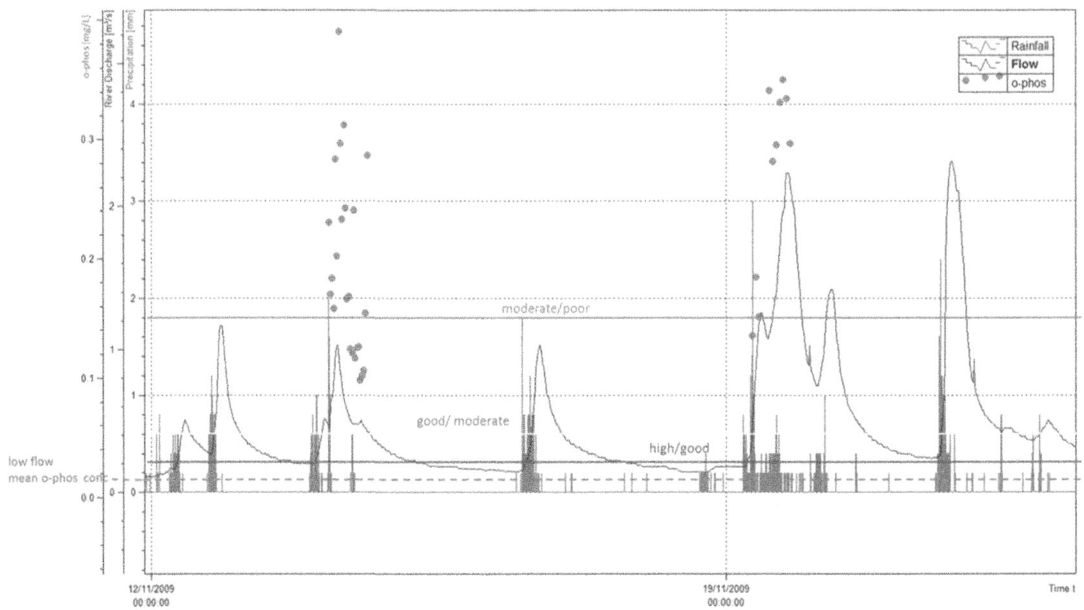

Figure 1.7 Relationship in 2017 between rainfall (mm), river discharge (m^3/s) and ortho-phosphate (mg/L) in a rural stream (Burnside Burn) in Scotland, UK, in relation to good status target for river quality.

in Scotland, indicating concentrations and corresponding water quality targets. The stream quality indicated by the mean low-flow concentration is consistently within the high/good target, but exceedance is observed when rainfall events occur: classic diffuse pollution characteristics.

The storm event monitoring of the Newby Beck and Pow Beck tributaries of the River Eden noted in Section 1.3.1 above (Table 1.4) also allowed assessment of seasonal compliance with quality targets, at different flow regimes. Figure 1.8, from Ockenden *et al.* (2016), shows that the Pow Beck had the poorer water quality across the four seasons, including in low-flow conditions. The low-flow quality may have been associated with septic tanks from the more populous Pow Beck catchment by comparison with the Newby Beck. The authors also suggested it might be associated with wetting and drying and warmer, drier summers favouring P release from sediments. The poor summer condition is therefore likely to be exacerbated by climate change with longer periods of drier warmer weather predicted. Ockenden *et al.* (2016) also predicted increased mobilisation of pollutants driven by more intense rainstorms.

For diffuse pollution impacted watercourses the peak concentration of a pollutant may occur ahead of the peak flow during a rainstorm; the first flush phenomenon, especially evident close to source in small urban catchments (Deletic, 1998; Kim *et al.*, 2006; Mamun *et al.*, 2020; Thames 21, 2019). Although it may then decline during the storm event, often the concentration only reaches a minimum in dry weather (e.g., Figures 1.4 and 1.7). A diffuse pollution impacted watercourse may have good quality as indicated by water chemistry at low flows, but poor ecological quality if diffuse sources are important (Wilson *et al.*, 2005). It may not matter if the concentration has dropped (e.g., after a first flush peak in the hydrograph, before the peak flow is attained) *if the lower value is still in excess of the EQS*. For toxic pollutants the peak concentration in the water body can be a critical value for aquatic life there. In lentic and recreational waters, the *load* of nutrients or persistent pollutants delivered to the sensitive waterbody is important.

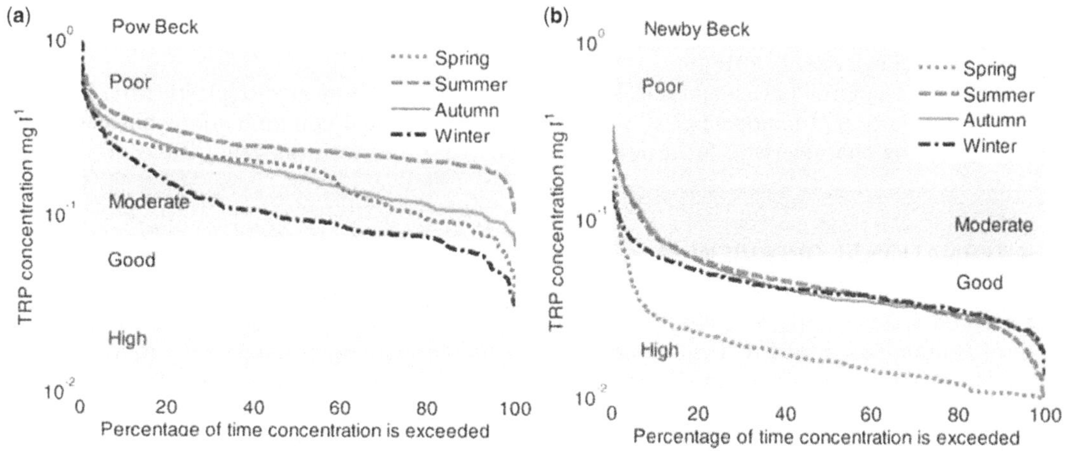

Figure 1.8 The seasonality of exceedance of Water Framework Directive standards for total reactive phosphorus in example study years of storm event monitoring (2012 for Newby Beck and 2013 for Pow Beck). Spring: March, April, May (MAM); summer JJA; autumn SON; winter DJF (Figure from Ockenden *et al.*, 2016).

1.3.3 Diffuse pollutants which are not easily managed by EQS

The EQS system is vital for typical continuous effluent discharges (point sources) since it provides a target pollutant concentration usually achievable by a combination of effluent treatment and water environment dilution. But for some contaminants, for example new chemical substances barely recognised yet for their potential significance in the environment, there are not yet any EQS values. The dilution assumption, of course, requires a larger volume of less-contaminated water into which the concentrated polluting effluent is discharged. That's fine for a process effluent discharged from a factory where the contaminants are industrial chemicals which are not present upstream in the catchment, or for a major sewage effluent discharge which carries a BOD load substantially in excess of naturally low concentrations. But for diffuse sources, especially of the kinds of natural contaminants identified in Table 1.3 (in particular sediment, nutrients and faecal pathogens), the concentrations may only be seen to be problematic after mobilisation by rainfall and hence at higher flows in the watercourses. Then the characteristics of diffuse sources identified by Novotny and Olem (1994) become clear: 'cannot be managed by discharge quality standards at source, cannot be monitored at source'. Consequently the steps to achieve the receiving water standards have to be iterative application of best practices (but increasingly well supported by predictive models for the risk catchments and the measures which might be effective).

Litter or trash is a common problem and familiar sources vary from casual disposal alongside roads, walkways and in public spaces, to deliberate disposal from ships at sea (Nelms *et al.*, 2017). The largest sources are washed from the landscape into watercourses and hence to the seas and oceans (Lebreton *et al.*, 2017). For such highly visible material the dilution factor is questionable, especially floating material. It's unreasonable to apply a 3-D concept to a 2-D one; does it matter how many cubic metres per second of water are passing beneath the mat of floating plastic and other trash on the surface? Breakdown of plastic produces progressively smaller fragments, eventually producing micro- and nanoplastic particles and fibres (Horton *et al.*, 2017; Jambeck *et al.*, 2015; Law & Thompson, 2014). Microplastics are now ubiquitous contaminants of the aquatic environment and a survey of 16 beaches in the Bristol Channel, UK, over a coastal length of 230 km, found microplastics in 15 of them. The contamination was associated with lower-energy beaches (deposition zones); fragments (74%) and industrial plastic pellets (13%) predominated (Wilson *et al.*, 2021). In another UK study, when seals and cetaceans found dead on British beaches were sampled, microplastics were found in every one without exception (Nelms *et al.*, 2019). As well as local beaches, hotspots occur on an oceanic scale (Schuyler *et al.*, 2016).

Plastic is now recognised as a major diffuse pollutant at all scales from macro (plastic litter including fragments) to micro- and nanoparticles. Such is the scale of global contamination that plastic has become a defining characteristic for a new geological era – the Anthropocene (Zalasiewicz *et al.*, 2016).

1.4 INTERNATIONAL OVERVIEW

1.4.1 Inter-connected environmental problems

Land use and water quality are inextricably linked due to the importance of the land use as the source of contaminants, and their mobilisation by weather conditions. Land uses often require or create a landscape which favours rapid runoff and the mobilisation of contaminants. The inter-linked problems of land-use change, diffuse pollution and flood risk management range from resource impairment (including rising costs of water supply and treatment) to habitat destruction and loss of tourism revenue, yet flood risk is still often seen as a separate area of professional expertise to water quality management, as suggested in Figure 1.9 below.

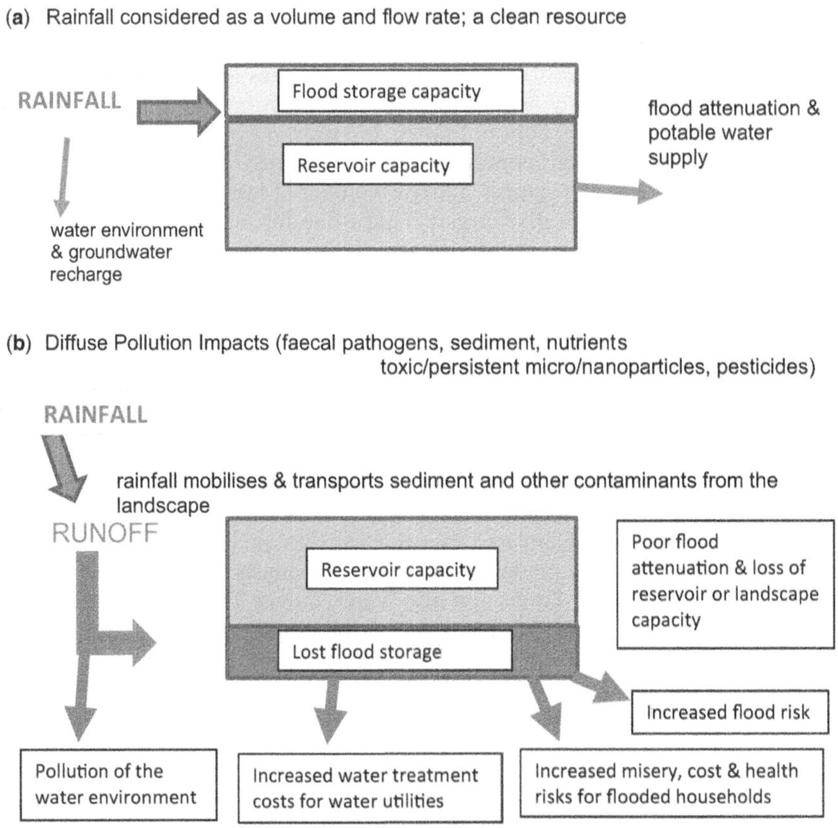

Figure 1.9 Water quantity and quality issues are inter-related; ignoring the water quality aspects is costly.

Reviewing the catastrophic flooding affecting Bangkok, Thailand, in 2011, Supharatid *et al.* (2016) highlighted deforestation, poor land drainage, expanding agriculture and urbanisation as key land use changes which increased both runoff rates and volumes. But those same factors also generate pollution. For flood survivors, it is the contamination which causes lasting concerns and costs after flood waters subside, with faecal pathogens, waterborne disease and contaminated mud in premises affected by the flood waters (World Health Organization, not dated).

The sediment load in high flows also affects the life of reservoirs (Gill, 1979; Wu *et al.*, 2012), and accumulations there can seriously impact the quality of water supplies (Thorne & Fenner, 2011). Sediment load associated with land use change can have a major impact on biodiversity downstream too; for example, significant impacts are highlighted in a review of freshwater biodiversity in Indo-Burma, (Allen *et al.*, 2012); and also for the Great Barrier Reef in Queensland Australia (Great Barrier Reef Water Science Taskforce, 2016). Thus there is considerable scope for common cause presentation to governments of overall environmental damage from land-use change and uncontrolled diffuse pollution. For example, deforestation (with intrinsic loss of habitat) and associated impairment of downriver fisheries, and offshore coral reefs or seagrass beds. Similarly for loss of recruitment habitats for fisheries, as well as tsunami protection zones, when mangroves are cut down and replaced with commercial agriculture or aquaculture, or adversely impacted by commercial/industrial development and their inherent local pollution issues (Chirapart, 2016; Worakhunpiset, 2018). Even when patches of mangrove forest remain, plastic contamination is evident (Ivar do Sul *et al.*, 2014). Impacts can be on recreational resources as well as ecological impacts (Nevers *et al.*, 2018).

1.4.2 Climate change, diffuse pollution and the united nations sustainable development goals

The pressures of climate change are likely to exacerbate the drivers for and impacts of diffuse pollution (Bates *et al.*, 2008):

- Increasing temperatures may increase the probability of phosphorus release from lake and river sediments in summer (hot weather, slow flow) conditions, favouring local sediment anoxia and algal blooms (Ockenden *et al.*, 2016; Thorne & Fenner, 2011).
- Also due to the lower solubility of oxygen in water, higher temperatures are likely to exacerbate marine anoxia, associated with the movement of excess nutrients from intensively farmed landscapes into coastal waters (Mississippi River/Gulf of Mexico Watershed Nutrient Task Force, 2008; Watson *et al.*, 2017).
- Higher temperatures and drier conditions certainly increase the risks of bush and forest fires and the associated exposure of potential pollutants to runoff in subsequent rainfall runoff presents additional risks to water resources (Bladon *et al.*, 2014; Emelko *et al.*, 2011; Wallbrink *et al.*, 2004; Webb, 2012, 2004).
- More extreme rainfall patterns may also exacerbate pollution from runoff associated with storm events, generating increased pollutant loads (Loecke *et al.*, 2017; Ockenden *et al.*, 2017; Sinha *et al.*, 2017).
- The increased intensity of rainstorms has also been implicated in the movement of plastic and other material from the landscape into rivers and coastal waters (Duncan *et al.*, 2020; Lebreton *et al.*, 2017) and lakes (Tanentzap *et al.*, 2021).

Figure 1.10 summarises the above, indicating the importance of climate change phenomena increasing diffuse pollution problems. It also indicates links to the UN Sustainable Development Goals (SDGs), especially SDGs 1, 2, 3, 6, 11 and 13 (United Nations, 2015, 2017). The water–energy–food nexus, whereby the interactions between the factors and SDGs are recognised, is discussed in detail in Heal *et al.* (2021).

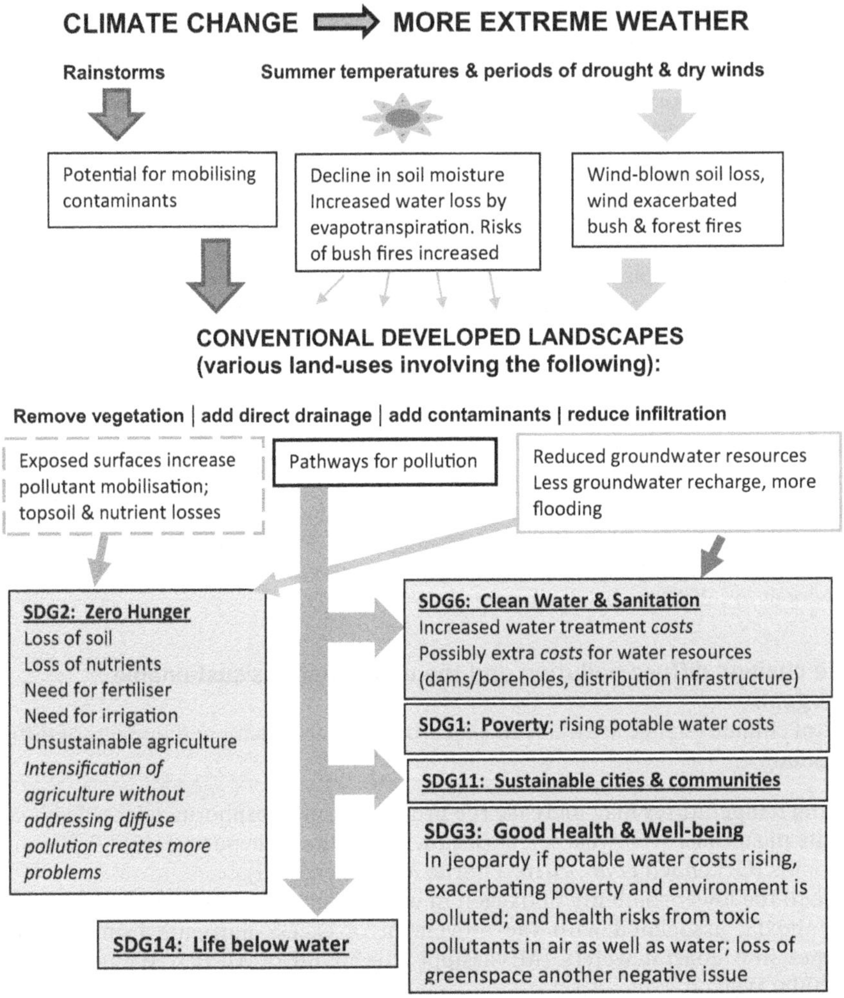

Figure 1.10 Diffuse pollution, climate change and the UN Sustainable Development Goals.

1.5 CONCLUSIONS AND RECOMMENDATIONS

(a) Diffuse pollution impacts are widespread and include serious adverse impacts on water supplies and potential resources, on habitats including globally renowned national parks (Kruger in South Africa and Great Barrier Reef in Australia, e.g.), on human health (e.g. waterborne diseases, faecal pathogens in recreational waters, pesticides), and on local economies (e.g. algal blooms in popular tourism regions).

(b) Diffuse pollution cannot be resolved simply by banning a particular problem pollutant, necessary though that often is. It is only a small part of the challenge.

(c) There is confusion and inconsistency in many countries about precisely what is diffuse pollution and hence how impacts should be source apportioned and reported. That prevents an inclusive presentation of the overall impact of diffuse pollution, which in turn handicaps efforts to achieve a more radical approach to achieving widespread integrated multi-purpose landscape design and additional interventions appropriate for particular land uses.

(d) In making the case for resources to investigate and address diffuse pollution, it's not helpful to fragment the evidence into individual components which alone may not sway governments to provide resources for action.
(e) An inclusive definition of diffuse pollution is offered which is based on key physical environmental characteristics: *contamination of air or landscape, mobilised and transported into the water environment by weather conditions.*
(f) The definition above helps to explain a number of important facts about water quality associated with catchment characteristics, including flood risk links and landscapes.
(g) Important pollutants mobilised in this way include plastic as well as pesticides, nutrients, sediment and toxic substances, including metals and industrial chemicals. Many chemical diffuse pollutants are associated with the point of use as well, occurring more extensively than just at the places of manufacture and production.
(h) The importance of diffuse sources can be inferred from data at high flows compared with low-flow data for the same watercourse; however, environmental monitoring strategies need to be designed to assess that.
(i) High concentrations at high flows result in extremely high loadings (e.g., nutrients and sediment into lakes, reservoirs and coastal waters, and faecal indicator organisms into recreational waters).
(j) The recognition of those characteristics should help drive prevention and catchment management more strongly towards at source actions, including landscape interventions, rather than unrealistic expectations for 'business as usual' plus simple add-ons such as end of system ponds or sediment traps.
(k) The great advances in water quality sensors and telemetry since the 1980s allow for simultaneous monitoring of water quantity and quality at stations along watercourses. In the UK the environment agencies still operate separate flow and quality monitoring with separate, unlinked databases, accessed via separate departments in the agencies (assertion based on Freedom of Information Act enquiries in 2016, 2017, 2019). Integrated monitoring (and reporting) should be routine.
(l) Diffuse pollution sources can certainly be subject to environmental regulation, and there are good examples for reference, but the regulatory regimes must be supported by sector engagement including the general public. Impacts evidence is an essential part of sector engagement. No data – a consequence of inadequate monitoring – does not mean no problem.
(m) Regulation does not, however, necessarily imply adequate control; effectiveness is a function of commitment by the appropriate agency to utilise the statutory powers established by government for that purpose.
(n) Climate change is exacerbating diffuse pollution problems. Unless governments and policy makers are aware that many water pollution issues are associated with weather-driven mobilisation of contaminants from the modified landscapes which dominate inhabited and habitable parts of the planet, the water environment will continue to be degraded, including many important habitats and economically essential resources, with costs for water utilities and consumers. Failure to recognise the importance of the characteristics of diffuse pollution is likely to undermine efforts to achieve important sustainable development goals.

REFERENCES

Aitken M. N. (2003). Impact of agricultural practices and river catchment characteristics on river and bathing water quality. *Water Science and Technology*, **48**(10), 217–224, https://doi.org/10.2166/wst.2003.0577

Allen D. J., Darwall W. R. T. and Smith K. G. (2012). The Status and Distribution of Freshwater Biodiversity in Indo-Burma. IUCN, Cambridge, UK and Gland, Switzerland. ISBN: 9782831714240.

Audsley R. S. (2021). The 4 Point Plan – straightforward guidance for livestock farmers to minimise diffuse pollution risks and benefit your business. Available at https://www.farmingandwaterscotland.org/managing-water-on-your-farm/the-4-point-plan-4pp/ (accessed 26 March 2022).

Bates, B. C., Kundzewicz, Z. W., Wu, S. and Palutikof, J. P. (eds). (2008). Climate Change and Water. Technical Paper of the Intergovernmental Panel on Climate Change. IPCC Secretariat, Geneva. ISBN: 9789291691234.

Bladon K. D., Emelko M. B., Silins U. and Sonte M. (2014). Wildfire and the future of water supply, *Environmental Science and Technology*, **48**, 8936–8943, https://doi.org/10.1021/es500130g

Campbell N., D'Arcy B., Frost C., Novotny V. and Sansom A. (2004). Diffuse Pollution – An Introduction to the Problems and Solutions. IWA Publishing, London, ch. 8, pp. 239–243. ISBN: 1900222531.

Chikozho C., Kadengye D. T., Wamukoya M. and Orindi B. O. (2019). Leaving no one behind? Analysis of trends in access to water and sanitation services in slum areas of Nairobi, 2003–2015. *Journal of Water, Sanitation and Hygiene for Development*, **9**(3), 549–558, https://doi.org/10.2166/washdev.2019.174

Chirapart A. (2016). Transition of species composition, abundance and distribution of the gracilarioid seaweeds (rhodophyta) in coastal areas of the upper gulf of Thailand observed from 2004 to 2007. *Natural History Bulletin of the Siam Society*, **61**(2), 71–88.

Cosgrove P., McInnes N., Dolby S., Gunn D., Shields D., Cosgrove C. and Kortland K. (2017). Forest management and freshwater pearl mussels. *Scottish Forestry*, **71**(1), 14–21.

Cundil A., Asher L., D'Arcy B. J., Napier F., McCreadie B., McLean N. and Clark R. (2010). Industrial estates are diffuse sources of pollution. In: Diffuse Pollution and Eutrophication (DIPCON 2010), E. van Bocove, P. Vanrolleghem, P. Chambers, B. Novotna and G. Theriault (eds), Proceedings and Abstracts CD-ROM, International Water Association, Beaupre, Quebec, Canada.

Craig R. K. and Roberts A. M. (2015). When will governments regulate nonpoint source pollution? A comparative perspective. *Boston College Environmental Affairs Law Review*, **42**(1), 1–64. http://lawdigitalcommons.bc.edu/ealr/vol42/iss1/2 (accessed 27 March 2022).

Crowther J., Kay D. and Wyer M. D. (2001). Relationships between microbial water quality and environmental conditions in coastal recreational water: the fylde coast, UK. *Water Research*, **35**, 4029–4038, https://doi.org/10.1016/S0043-1354(01)00123-3

Crowther J., Wyer M. D., Bradford M., Kay D. and Francis C. A. (2003). Modelling faecal indicator concentrations in large rural catchments using land use and topographic data. *Journal of Applied Microbiology*, **94**(6), 962–973, https://doi.org/10.1046/j.1365-2672.2003.01877.x

D'Arcy B. J. and Hemingway A. (2018). The regulatory regime for bringing SUDS into routine use for industrial estates and business parks in Scotland, UK. In: Wealth Creation Without Pollution – Designing for Industry, Ecobusiness Parks and Industrial Estates, B. J. D'Arcy, L.-H. Kim and M. Maniquiz-Redillas (eds), IWA Publishing, London, ch. 4, pp. 205–218. ISBN: 9781780408330.

D'Arcy B. J. and Taylor V. (2018). An integrated approach for pollution prevention on industrial premises. In: Wealth Creation Without Pollution – Designing for Industry, Ecobusiness Parks and Industrial Estates, B. J. D'Arcy, L.-H. Kim and M. Maniquiz-Redillas (eds), IWA Publishing, London, ch. 17, pp. 237–258. ISBN: 9781780408330.

D'Arcy B. J., Ellis J. B., Ferrier R. C., Jenkins A. and Dils R. (2000). Diffuse Pollution Impacts: the Environmental and Economic Impacts of Diffuse Pollution. Chartered Institution of Water and Environmental Management, London. ISBN: 1870752465.

D'Arcy B. J., Schmulian K. and Wade R. (2006a). Regulatory options for the management of rural diffuse pollution. In: Managing Rural Diffuse Pollution, L. Gairns, K. Crighton and B. Jeffrey (eds), Proceedings of the SAC & SEPA Biennial Conference, SAC, Edinburgh, pp. 192–200.

D'Arcy B. J., May L., Fozzard I. R., Greig S. and Brachet A. (2006b). The restoration of loch leven, Scotland, UK. *Water Science and Technology*, **53**(10), 183–191, https://doi.org/10.2166/wst.2006.311

D'Arcy B. J., Kim L.-H. and Morison P. (2018). Industrial pollution and the water environment: a historical perspective. In: Wealth Creation Without Pollution – Designing for Industry, Ecobusiness Parks and Industrial Estates, B. J. D'Arcy, L.-H. Kim and M. Maniquiz-Redillas (eds), IWA Publishing, London, ch. 1, pp. 1–24. ISBN: 9781780408330.

Defra. (2020). Inquest into the death of Ella Adoo-Kissi-Debrah. Inquest into the death of Ella Adoo-Kissi-Debrah – Defra in the media (blog.gov.uk).

Degerman E., Andersson E. K., Soderberg H., Norrgrann O., Henrikson L., Angelstam P. and Tornnblom J. (2013). Predicting population status of freshwater pearl mussel (*Margaritifera margaritifera*, L.) in central Sweden using instream and riparian zone land-use data. *Aquatic Conservation: Marine and Freshwater Ecosystems*, **23**, 332–342, https://doi.org/10.1002/aqc.2322

Deletic A. (1998). The first flush load of urban surface runoff. *Water Research*, **32**(8), 2462–2470, https://doi.org/10.1016/S0043-1354(97)00470-3

Duncan E. M., Davies A., Brooks A., Chowdhury G. W., Godley B. J., Jambeck J., Maddalene T., Napper I., Nelms S. E., Rackstraw C. and Koldewey H. (2020). Message in a bottle: open source technology to track the movement of plastic pollution. *PLoS ONE*, **15**(12), e0242459, https://doi.org/10.1371/journal.pone.0242459

Edokpayi J. N., Odiyo J. O. and Durowoju O. S. (2017). Impact of wastewater on surface water quality in developing countries: a case study of South Africa. In: Water Quality, H. Tutu (ed.), InTechOpen, London, chap. 18, pp. 401–416. https://doi.org/10.5772/66561; http://www.intechopen.com/books/water-quality (accessed 27 March 2022).

Ellis J. B., Chocat B., Fujita S., Rauch W. and Marsalek J. (2004). Urban Drainage – A Multilingual Glossary. IWA Publishing, London. ISBN: 190022206X.

Emelko M. B., Silins U., Bladon K. D. and Stone M. (2011). Implications of land disturbance on drinking water treatability in a changing climate: demonstrating the need for 'source water supply and protection; strategies. *Water Research*, **45**, 461–472, https://doi.org/10.1016/j.watres.2010.08.051

EU. (2008). Consolidated text: Directive 2008/105/EC of the European Parliament and of the Council of 16 December 2008 on environmental quality standards in the field of water policy, amending and subsequently repealing Council Directives 82/176/EEC, 83/513/EEC, 84/156/EEC, 84/491/EEC, 86/280/EEC and amending Directive 2000/60/EC of the European Parliament and of the Council. http://data.europa.eu/eli/dir/2008/105/2013-09-13 (accessed 27 March 2022).

Ferrier R. and Ellis J. B. (2000). Suspended solids. In: Diffuse Pollution Impacts – The Environmental and Economic Impacts of Diffuse Pollution in the UK, B. J. D'Arcy, J. B. Ellis, R. C. Ferrier, A. Jenkins and R. Dils (eds), Chartered Institution of Water and Environmental Management, London, UK, pp. 33–40. ISBN: 1870752 465. Chapter 4.

Ferrier R., D'Arcy B. J., MacDonald J. and Aitken M. (2005). Diffuse pollution – what is the nature of the problem? *Journal of the Chartered Institution of Water and Environmental Management, WEJ*, **1**, 361–366, https://doi.org/10.1111/j.1747-6593.2005.tb00574.x

Fujii S., Moriya M., Songprasert P. and Ihara H. (2006). Estimation of annual pollutant loadings in two small catchments and examination of their differences caused by regional properties. *Water Science and Technology*, **53**(2), 33–44, https://doi.org/10.2166/wst.2006.036

Gill M. A. (1979). Sedimentation and useful life of reservoirs, *Journal of Hydrology*, **44**(1–2), 89–95, https://doi.org/10.1016/0022-1694(79)90148-3

Great Barrier Reef Water Science Taskforce. (2016). Final Report, May 2016. [online] www.gbr.qld.gov.au (accessed 27 March 2022).

Greig S., Dawson P. and Craig W. (2004). Trends in diffuse pollution: data report to assist with the design and implementation of effective diffuse pollution monitoring programmes. Scottish Environment Protection Agency DPI report no. 21, ref. DPI No. 21/SG/Sept 04, SEPA, Stirling, UK.

Haygarth P. M., Condron L. M., Heathwaite A. L., Turner B. L. and Harris G. P. (2005). The phosphorus transfer continuum: linking source to impact with an interdisciplinary and multi-scaled approach. *Science of the Total Environment*, **344**, 5–14, https://doi.org/10.1016/j.scitotenv.2005.02.001

Heal K. V., Bartosova A., Hipsey M. R., Chen X., Buytaert W., Li H.-Y., McGrane S. J., Gupta A. B. and Cudennec C. (2021). Water quality: the missing dimension of water in the water–energy–food nexus. *Hydrological Sciences Journal*, **66**(5), 745–758, https://doi.org/10.1080/02626667.2020.1859114

Horton A. A., Walton A., Spurgeon D. J., Lahive E. and Svendsen C. (2017). Microplastics in freshwater and terrestrial environments: evaluating the current understanding to identify the knowledge gaps and future research priorities. *Science of the Total Environment*, **586**, 127–141, https://doi.org/10.1016/j.scitotenv.2017.01.190

Hynes H. B. N. (1971). The Biology of Polluted Waters. Liverpool University Press, UK. ISBN: 0853232008.

Hussein A. O. and Shahid S. (2017). Soil erosion susceptibility of johor river basin. *Water and Environment Journal*, **31**, 367–374, https://doi.org/10.1111/wej.12252

Hypoxia Task Force. (2018). Report on Nonpoint Source Progress in Hypoxia Task Force States (PDF) (39 pp, 1 MB, May 2018) Report on Nonpoint Source Progress in Hypoxia Task Force States | Mississippi River/Gulf of Mexico Hypoxia Task Force | US EPA.

Ibanez M. (2015). MSc dissertation, University of Abertay Dundee, Dundee, UK.

Inglezakis V. J., Poulopoulos S. G., Arkhangelsky E., Zorpas A. A. and Menegaki A. N. (2016). Environmental quality standards and emission limit values. In: Environment and Development, S. G. Poulopoulos and V. J. Inglezakis (eds), Chap. 3, pp. 137–212, Elsevier Science & Technology.

Ivar do Sul J. A., Costa M. F., Silva-Cavalcanti J. S. and Araújo M. C. B. (2014). Plastic debris retention and exportation by a mangrove forest patch. *Marine Pollution Bulletin*, **78**(1–2), 252–257. pmid:24321881, https://doi.org/10.1016/j.marpolbul.2013.11.011

Iwugo K., D'Arcy B. J., Heath R. G. M. and Andoh R. (2003). Breaking the pollution cycle in poor African settlements. Paper presented at 29th WEDC Conference, 22–26 September 2003, Abuja, Nigeria.

Jambeck J. R., Geyer R., Wilcox C., Siegler T. R., Perryman M., Andrady A., Narayan R. and Law K. L. (2015). Plastic waste inputs from land into the ocean. *Science*, **347**(6223), 768–771. pmid:25678662, https://doi.org/10.1126/science.1260352

Jones P. D. (2000). The mersey estuary – back from the dead: solving a 150-year old problem. *J. CIWEM*, **14**, 124–130.

Kay D., Wither A. W. and Jenkins A. (2000). Faecal pathogens. In: Diffuse Pollution – the Environmental and Economic Impacts of Diffuse Pollution, B. J. D'Arcy, J. B. Ellis, R. C. Ferrier, A. Jenkins and R. Dils (eds), CIWEM, London, pp. 49–60.

Kay D., Aitken M., Crowther J., Dickson I., Edwards AC. ., Francis C., Hopkins M., Jeffrey W., Kay C., McDonald A. T., McDonald D., Stapleton C. M., Watkins J., Wilkinson J. and Wyer M. (2007). Reducing fluxes of faecal indicator compliance parameters to bathing waters from diffuse agricultural sources, the brighouse Bay study, Scotland. *Environmental Pollution*, **147**, 138–149, https://doi.org/10.1016/j.envpol.2006.08.019

Kim L.-H. and Maniquiz-Redillas M. (2018). Green industry concept and practices. In: Wealth Creation Without Pollution – Designing for Industry, Ecobusiness Parks and Industrial Estates, B. J. D'Arcy, L.-H. Kim and M. Maniquiz-Redillas (eds), IWA Publishing, London, Chap. 5, pp. 65–74. ISBN: 9781780408330.

Kim L.-H., Kang J., Kayhanian M., Gil K.-I., Stenstrom M. K. and Zoh K.-D. (2006). Characteristics of litter waste in highway storm runoff. *Water Science and Technology*, **53**(2), 225–234, https://doi.org/10.2166/wst.2006.056

Kim L.-H., D'Arcy B. J., Ibanez M. and Maniquiz-Redillas M. (2018). Industrial estates as sources of pollution. In: Wealth Creation Without Pollution – Designing for Industry, Ecobusiness Parks and Industrial Estates, B. J. D'Arcy, L.-H. Kim and M. Maniquiz-Redillas (eds), IWA Publishing, London, Chap. 3, pp. 37–51. ISBN: 9781780408330.

Law K. L. and Thompson R. C. (2014). Microplastics in the seas. *Science*, **345**(6193), 144–145. Available from: http://www.ncbi.nlm.nih.gov/pubmed/25013051 https://doi.org/10.1126/science.1254065 (accessed 27 March 2022).

Lebreton L. C. M., Van Der Zwet J., Damsteeg J. W., Slat B., Andrady A. and Reisser J. (2017). River plastic emissions to the world's oceans. *Nature Communications*, **8**(1), 1–10, https://doi.org/10.1038/s41467-016-0009-6

Liu W., Liu W., Lu H., Duan W. and Li H. (2011). Runoff generation in small catchments under a native rainforest and a rubber plantation in Xishuangbanna, southwestern China. *Water and Environment Journal*, **25**, 138–147, https://doi.org/10.1111/j.1747-6593.2009.00211.x

Loecke T. D., Burgin A. J., Riveros-Iregui D. A., Ward A. S., Thomas S. A., Davis C. A. and St. Clair M. A. (2017). Weather whiplash in agricultural regions drives deterioration of water quality. *Biogeochemistry*, **133**, 7–15, https://doi.org/10.1007/s10533-017-0315-z

Mamun A. A., Shams S. and Nuruzzaman M. (2020). Review on uncertainty of the first-flush phenomenon in diffuse pollution control. *Applied Water Science*, **10**, 53, https://doi.org/10.1007/s13201-019-1127-1

Marshall C. (2020). Air pollution death ruling: what comes next? BBC News [online] https://www.bbc.co.uk/news/science-environment-55352247 (accessed 27 March 2022).

McPhail C. and Stidson R. (2004). The scottish signage project. *Water and Environment Magazine*, **9**, 18–19.

Mississippi River/Gulf of Mexico Watershed Nutrient Task Force. (2008). Gulf Hypoxia Action Plan 2008 for Reducing, Mitigating, and Controlling Hypoxia in the Northern Gulf of Mexico and Improving Water Quality in the Mississippi River Basin. USEPA, Washington, DC.

Moriarty E. M., Karki N., Mackenzie M., Sinton L. W., Wood D. R. and Gilpin B. J. (2011). Faecal indicators and pathogens in selected New Zealand waterfowl. *New Zealand Journal of Marine and Freshwater Research*, **45**, 679–688, https://doi.org/10.1080/00288330.2011.578653

Napier F., D'Arcy B. J. and Jefferies C. (2008). A review of vehicle related metals and polycyclic aromatic hydrocarbons (PAHs) in the UK environment. *Desalination*, **226**(1–3), 143–150, https://doi.org/10.1016/j.desal.2007.02.104

Nelms S. E., Coombes C., Foster L. C., Godley B. J., Galloway T. S. G., Lindeque P. K. and Witt M. J. (2017). Marine anthropogenic litter on British beaches: a 10-year nationwide assessment using citizen science data. *Science of the Total Environment*, **579**, 1399–1409, https://doi.org/10.1016/j.scitotenv.2016.11.137

Nelms S. E., Barnett J., Brownlow A., Davison N. J., Deaville R., Galloway T. S., Lindeque P. K., Santillo D. and Godley B. J. (2019). Microplastics in marine mammals stranded around the British coast: ubiquitous but transitory? *Scientific Reports*, **9**, 1075, https://doi.org/10.1038/s41598-018-37428-3

Nevers M. B., Murulee N., Byappanahalli D., Shively P., Buszka I. M., Jackson P. R. and Phanikumar M. S. (2018). Identifying and eliminating sources of recreational water quality degradation along an urban coast, *Journal of Environmental Quality*, **47**, 1042–1050, https://doi.org/10.2134/jeq2017.11.0461

Novotny V. (2003). Water Quality: Diffuse Pollution and Watershed Management. John Wiley, New York. ISBN: 0471396338.

Novotny V. and Olem H. (1994). Water Quality: Prevention, Identification and Management of Diffuse Pollution. Van Nostrand Reinhold, reprinted and distributed by John Wiley, New York.

Ockenden M. C., Deasy C. E., Benskin C. M. W. H., Bevan K. J., Collins A. L., Evans R., Falloon P. D., Forber K. J., Hiscock K. M., Hollaway M. J., Kahana R., Macleod C. J. A., Reaney S. M., Snell M. A., Villamizar M. L., Wearing C., Withers P. J. A., Zhou J. G. and Haygarth P. M. (2016). Changing climate and nutrient transfers: evidence from high temporal resolution concentration–flow dynamics in headwater catchments. *Science of the Total Environment*, **548–549**, 325–339, https://doi.org/10.1016/j.scitotenv.2015.12.086

Ockenden M. C., Hollaway M. J., Bevan K. J., Collins A. L., Evans R., Falloon P. D., Forber K. J., Hiscock K. M., Kahana R., Macleod C. J. A., Tych W., Villamizar M. L., Wearing C., Withers P. J. A., Zhou J. G., Barker P. A., Burke S., Freer J. E., Johnes P. J., Snell M. A., Surridge B. W. J. and Haygarth P. M. (2017). Major agricultural changes required to mitigate phosphorus losses under climate change. *Nature Communications*, **8**, 161, https://doi.org/10.1038/s41467-017-00232-0

OECD. (2010). Eco-innovation in Industry: Enabling Green Growth. OECD, OECD Publishing, Paris.

OECD. (2017). Diffuse Pollution, Degraded Waters: Emerging Policy Solutions. OECD Studies on Water, OECD Publishing, Paris, http://doi.org/10.1787/9789264269064-en

O'Keeffe J. (2018). Eco-innovation opportunities in the waste management sector in Scotland. In: Wealth Creation Without Pollution – Designing for Industry, Ecobusiness Parks and Industrial Estates, B. J. D'Arcy, L.-H. Kim and M. Maniquiz-Redillas (eds), IWA Publishing, London, pp. 89–98. ISBN: 9781780408330 (print).

O'Keefe B., D'Arcy B. J., Davidson J., Barbarito B. and Clelland B. (2005). Urban diffuse sources of faecal indicators. *Water Science and Technology*, **51**(3–4), 183–190, https://doi.org/10.2166/wst.2005.0590

Porter E. (1973). Pollution in Four Industrialised Estuaries. Four Case Studies Undertaken for the Royal Commission on Environmental Pollution. HMSO, London. ISBN 10: 0117300519/ISBN 13: 9780117300514.

Rabalais N. N., Turner E. and Scavia D. (2002). Beyond science into policy: Gulf of Mexico hypoxia and the Mississippi river. *BioScience*, **52**(2), 129–142, https://doi.org/10.1641/0006-3568(2002)052[0129:BSIPGO]2.0.CO;2

Rabotyagov S. S., Campbell T. D., White M., Arnold J. G., Atwood J., Norfleet L., Kling C. L., Gassman P. W., Valcu A., Richardson J., Turner R. E. and Rabalais N. N. (2014). Cost-effective targeting of conservation investments to reduce the northern Gulf of Mexico hypoxic zone. *Proceedings of the National Academy of Sciences of the United States of America*, **111**(52), 18530–18535, https://doi.org/10.1073/pnas.1405837111

Reynolds L., Sala-Comorera L., Martin N. A., Nolan T. M., Stephens J. H., Gitto A., O'Hare G. M. P., O'Sullivan J. J. and Meijera W. G. (2020). Correlation between antimicrobial resistance and faecal contamination in small urban streams and bathing waters. *Science of the Total Environment*, **739**, 140242, https://doi.org/10.1016/j.scitotenv.2020.140242

Schuyler Q., Wilcox C., Townsend K., Wedemeyer-Strombel K. R., Balazs G., van Sebille E. and Hardesty B. D. (2016). Risk analysis reveals global hotspots for marine debris ingestion by sea turtles. *Global Change Biology*, **22**, 567–576, http://www.ncbi.nlm.nih.gov/pubmed/26365568 (accessed 27 March 2022).

SEPA. (1999). Improving Scotland's Water Environment – SEPA State of the Environment Report. Scottish Environment Protection Agency, Stirling, UK.

SEPA. (2006). State of Scotland's Environment 2006. Scottish Environment Protection Agency, Stirling, UK, p. 86.

SEPA. (2020). Bathing Water Profile for Portobello (West) Edinburgh, Scotland. [Online] https://www.sepa.org.uk/media/39970/portobello-west-bathing-water-profile.pdf. (accessed 4 April 2021).

Sinha E., Michalak A. M. and Balaji V. (2017). Eutrophication will increase during the 21st century as a result of precipitation changes. *Science*, **357**, 405–408, https://doi.org/10.1126/science.aan2409

Sthiannopkao S., Takizawa S. and Wirojanagud W. (2006). Effects of soil erosion on water quality and water uses in the upper phong watershed. *Water Science and Technology*, **53**(2), 45–52, https://doi.org/10.2166/wst.2006.037

Stidson R. T., Gray C. A. and McPhail C. D. (2012). Development and use of modelling techniques for real-time bathing water quality predictions. *Water and Environment Journal*, **26**, 7–18, https://doi.org/10.1111/j.1747-6593.2011.00258.x

Supharatid S., Aribarg T. and Supratid S. (2016). Assessing potential flood vulnerability to climate change by CMIP3 and CMIP5 models: case study of the 2011 Thailand great flood. *Journal of Water and Climate Change*, **7**, 52–67, https://doi.org/10.2166/wcc.2015.116

Tanentzap A. J., Cottingham S., Fonvielle J., Riley I., Walker L. M., Woodman S. G., Kontou D., Pichler C. M., Reisner E. and Lebreton L. (2021). Microplastics and anthropogenic fibre concentrations in lakes reflect surrounding landuse. *PLoS Biology*, **19**(9), e3001389, https://doi.org/10.1371/journal.pbio.3001389

Thames 21. (2019). Spatial Quantification of Road Runoff Pollution in Greater London. [Online] https://www.thames21.org.uk/wp-content/uploads/2019/08/Spatial-Quantification-of-Road-Runoff-in-Greater-London-20-08-19.pdf (accessed 27 March 2022).

Thorne O. and Fenner R. A. (2011). The impact of climate change on reservoir water quality and water quality operations: a UK case study. *Water and Environment Journal*, **25**, 74–87, https://doi.org/10.1111/j.1747-6593.2009.00194.x

United Nations. (2015). Resolution adopted by the General Assembly on 25 September 2015, Transforming our world: the 2030 Agenda for Sustainable Development.

United Nations. (2017). Resolution adopted by the General Assembly on 6 July 2017, Work of the Statistical Commission pertaining to the 2030 Agenda for Sustainable Development.

USEPA. (2020). Basic Information about Nonpoint Source (NPS) Pollution | Polluted Runoff: Nonpoint Source (NPS) Pollution. USEPA, [online] www.epa.gov/nps/basic-information-about-nonpoint-source-nps-pollution (accessed 8 March 2021).

Wada K., Nakamura M., Sato K., Tsuno H. and Fukuju S. (2020). GIS Analysis of eighteen-year trend in land use, runoff, and nonpoint load characteristics in the lake biwa–yodo river basin. *Journal of Japan Society on Water Environment*, **43**(5), 141–152, https://doi.org/10.2965/jswe.43.141

Wallbrink P., English P., Chafer C., Humphreys G., Shakesby R., Blake W. and Doerr S. (2004). Impacts of water quality by sediments and nutrients released during extreme bushfires: report 1: a review of the literature pertaining to the effect of fire on erosion and erosion rates, with emphasis on the Nattai catchment, NSW, following the 2001 bushfires. *CSIRO Land and Water Client Report* December 2004, CSIRO, Clayton Victoria, Australia.

Wang X., Junqi L., Yingxia L., Zhenyao S., Xuan W., Zhifeng Y. and Inchio L. (2014). Is urban development an urban river killer? A case study of yongding diversion channel in Beijing, China. *Journal of Environmental Sciences*, **26**, 1232–1237, https://doi.org/10.1016/S1001-0742(13)60593-8

Watson A. J., Lenton T. M. and Mills B. J. W. (2017). Ocean deoxygenation, the global phosphorus cycle and the possibility of human-caused large-scale ocean anoxia. *Philosophical Transactions of the Royal Society A*, **375**, 20160318, https://doi.org/10.1098/rsta.2016.0318

Webb A. (2004). Monitoring the potential impacts of forestry activities on water quality in NSW: preliminary results from the middle brother paired catchment program. In: Erosion in Forests: Proceedings of the Forest Workshop – March 2004, Melbourne, J. Croke, I. Takken and S. Mockler (eds) (Technical Report, Report 04/10, December 2004), Cooperative Research Centre for Catchment Hydrology, Australia, Chap. 30, pp. 149–152. ISBN: 1920813152.

Webb A. A. (2012). Can timber and water resources be sustainably co-developed in south-eastern New south wales, Australia? *Environment, Development and Sustainability*, **14**, 233–252, https://doi.org/10.1007/s10668-011-9319-3

WEF. (1987). The Clean Water Act of 1987. Water Environment Federation, Alexandria, VA, USA.

Wilson C. (2018). The hidden epidemic in the shanty towns of Honiara. [Online] *The Interpreter*, 22 February 2018, The Lowy Institute. https://www.lowyinstitute.org/the-interpreter/hidden-epidemic-shanty-towns-honiara (accessed 27 March 2022).

Wilson C., Clarke R., D'Arcy B. J., Heal K. V. and Wright P. W. (2005). Persistent pollutants urban rivers sediment survey: implications for pollution control. *Water Science and Technology*, **51**(3–4), 217–224, https://doi.org/10.2166/wst.2005.0594

Wilson D. R., Haggar G., Santillo D. and Sheena K. L. (2021). The influence of depositional environment on the abundance of microplastic pollution on beaches in the Bristol channel, UK. *Marine Pollution Bulletin*, **164**, 111997, https://doi.org/10.1016/j.marpolbul.2021.111997

Wither A. W., Rehfisch M. and Austin G. (2005). The impact of bird populations on the microbiological quality of bathing waters. The impact of bird populations on the microbiological quality of bathing waters. *Water Science and Technology*, **51**(3–4), 199–207, https://doi.org/10.2166/wst.2005.0592

Worakhunpiset S. (2018). Trace elements in marine sediment and organisms in the gulf of Thailand. *International Journal of Environmental Research and Public Health*, **15**(4), 810, https://doi.org/10.3390/ijerph15040810

World Economic Forum. (2021). Five Big Bets for the Circular Economy in Africa. [Online] Available from: www.weforum.org/connectedworld/report (accessed 27 March 2022).

World Health Organization. (no date). Flooding and Communicable Diseases Fact Sheet: Risk Assessment and Preventive Measures. World Health Organization, Communicable Disease Working Group on Emergencies, HQ. Microsoft Word -FloodingandCommunicableDiseasesfactsheet.doc (who.int).

Wu C.-H., Chen C.-N., Tsai C.-H. and Tsai C.-T. (2012). Estimating sediment deposition volume in a reservoir using the physiographic soil erosion–deposition model. *International Journal of Sediment Research*, **27**, 362–377, https://doi.org/10.1016/S1001-6279(12)60041-9

Wyer M. D., Kay D., Stapleton C. M., Francis C., Watkins J. and Rushby L. (2001). Faecal Indicator Sources and Budgets for Irvine Bay, Ayrshire. CREH, University of Wales, Aberystwyth, UK.

Zalasiewicz J., Waters C. N., Ivar do Sul J. A., Corcoran P. L., Barnosky A. D., Cearreta A., Edgeworth M., Galuszka A., Jeandel C., Leinfeldere R., McNeill J. R., Steffen W., Summerhayes C., Wagreich M., Williams M., Wolfe A. P. and Yonan Y. (2016). The geological cycle of plastics and their use as a stratigraphic indicator of the anthropocene, *Anthropocene*, **13**, 4–17, https://doi.org/10.1016/j.ancene.2016.01.002

doi: 10.2166/9781789061123_025

Chapter 2

Pesticide, diffuse pollution and food safety

Puangrat Kajitvichyanukul and Thirasant Boonupara

Department of Environmental Engineering, Faculty of Engineering, Chiang Mai University, Chiang Mai 50200, Thailand

Keywords: Emerging pollutants; land use and land-use changes; nature-based solutions; non-point source; point source

2.1 HISTORY AND EMERGING OF PESTICIDE DIFFUSE POLLUTION

Roughly half a century ago, the world was alerted to the need for pesticide control by the very first environmental movement. An environmental science book titled *Silent Spring* by Rachel Carson, published on 27 September 1962, depicted the adverse effects on the environment and humans from the widespread use of pesticides (Bonzi, 2013). Her book brought a tremendous change in environmental concerns for public policy in using toxic chemicals for agriculture. From a starting point in the late 1950s, a decrease in pesticide use was recorded, including the US nationwide ban on DDT, a highly toxic organochlorine insecticide shown to be mutagenic and carcinogenic to human (Shapiro *et al.*, 2016).

The decline in pesticide usage became more pronounced in the 1970s as the use of persistent organic pollutants (POPs) was restricted. The POPs are toxic organic compounds that require a very long period to be degraded under natural conditions in nature and that can accumulate in living organisms and the environment (water and soil). Many pesticides listed as POPs are organochlorine pesticides that have been recognized for their deleterious impacts on the human body and their persistence in the ecosystem (Al-Mulali *et al.*, 2015; Bakirtas & Akpolat, 2018). The reduction and elimination of the release of POPs became an urgent issue, leading to calls for action by global society. Since the 1970s, the use of POPs has been restricted, and the release of these compounds has been prohibited in Europe and the USA. From May 2004, the POP chemicals, including organochlorine pesticides, were listed in Stockholm Convention with the purpose of terminating or restricting the production and use of this toxic chemical group (Xu *et al.*, 2013). These organochlorine pesticides are aldrin, chlordane, chlordecone, dicofol, dieldrin, 1,1′-(2,2,2-Trichloroethane-1,1-diyl)bis(4-chlorobenzene) (or DDT), endosulfan and its related isomers, endrin, heptachlor, hexachlorobenzene (HCB), α-hexa chlorocyclohexane, β-hexachlorocyclohexane, lindane (or γ-hexachlorocyclohexane), mirex, and toxaphene.

The banning of organochlorine pesticides led to the extensive use of organophosphate pesticides such as parathion, malathion, chlorpyrifos, dichlorvos, and other chemicals in this group for insect

Editor(s): Puangrat Kajitvichyanukul and Brian D'Arcy

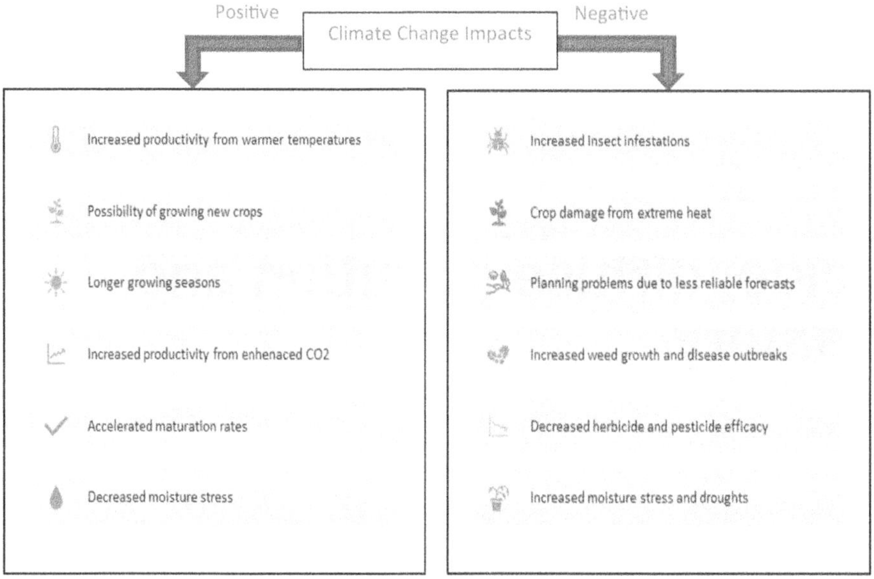

Figure 2.1 The potential impacts of climate change on agriculture (modified from Reilly, 1995).

control. However, these chemicals were soon reported to have bioaccumulation capacity and to remain for an extended period in the ecosystem (Rodrigues *et al.*, 2011). The adverse impact on the central nervous system and respiratory system resulting from human exposure to these pesticides has been recently addressed (Attia *et al.*, 2016). The notorious symptoms of organophosphate pesticides have led to calls for alternative chemical groups such as pyrethroids and carbamates to be used to control insects and pests on farmland. These other pesticides also generate acute intoxication symptoms, including weakness and vomiting, and chronic problems involving retarded neurological effects and liver damage, as well as chromosomal alterations to the exposed human (Carneiro *et al.*, 2013).

Although the application of these toxic pesticides to farmland leads to adverse human health impacts and accumulation in the ecosystem resulting in the contamination of water, soil, and food, these chemicals can increase crop yields considerably. Climate change can have a significant impact on the future use of pesticides. Several entomologists and biologists confirm that global warming will enable some pest species to survive winters and will enhance the development of summer-active species (Patterson *et al.*, 1999; Porter *et al.*, 1991). Subsequently, the use of pesticides may increase, leading to more negative impacts on humans and the environment in the near future. The impact of weather variability and climate change on pesticide applications has been reported by Koleva *et al.* (2011). The pesticide application rate has perhaps already increased due to the significant changes in weather and climate. The potential impacts of climate change on crops are also described by Reilly (1995), as shown in Figure 2.1.

2.2 GROWTH OF PESTICIDE CONSUMPTION

From 1990–2017, the worldwide consumption of pesticides reached 4.1 million tons of active ingredients, of which 52.8% was used in Asia, 30.2% in the Americas (Northern, Central, and South), 13.8% in Europe, 2.1% in Africa, and 1.3% in Oceania (FAOSTAT, 2020). The global consumption of pesticides in each region during 1990–2017 is shown in Figure 2.2, and China is the major country that has the highest pesticide consumption in the world (Figure 2.3). Contamination by pesticides

Figure 2.2 Global pesticide consumption during 1990–2017 (FAOSTAT, 2020).

can pose a significant threat to the environment (water and soil) and, consequently, the food eaten by people.

Asia is a vast land continent and supports approximately 60 percent of the global population on only about 23 percent of the world's agricultural land. In Asia as a whole, food production has incresed faster than the growth of the population. Notably, in the low-income countries of South Asia, *per capita* food availability has risen. Continuously, Asia has made remarkable progress in agricultural productivity. The use of pesticides in increasing food production is also increasing in the developing

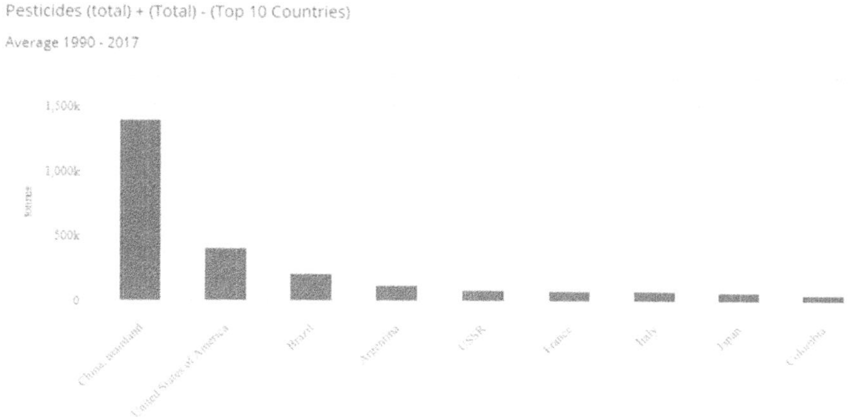

Figure 2.3 Pesticide consumption by country during 1990–2017 (FAOSTAT, 2020).

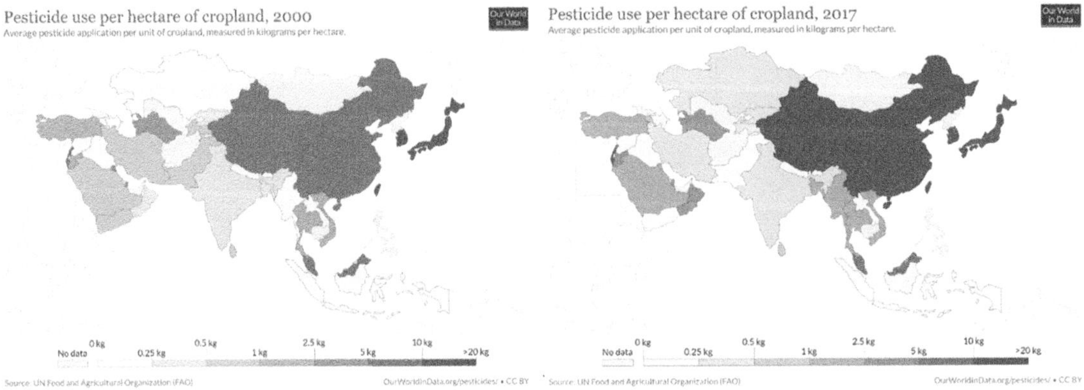

Figure 2.4 Comparison of pesticide use per hectare of cropland in Asia in 2000 and 2017 (modified from Roser, 2019).

countries of Asia, especially in Southeast Asia. Moreover, global warming also plays a significant role as it has a direct impact on pests, especially the summer-active species, and in equatorial regions, including several countries in Asia.

The average pesticide use per hectare of cropland in Asia in 2000 and 2017 is illustrated in Figure 2.4. An increase of pesticide use per hectare is seen in Myanmar, Pakistan, Kazakhstan, and China. The highest pesticide application rate is 13.07 kilograms per hectare in China (Roser, 2019). Table 2.1 shows the annual consumption of pesticides in different Asian countries, as reported in 2017 by FAO. The increase in annual pesticide consumption in Asia is most pronounced in China, Myanmar, Thailand, and Malaysia. However, there has been decreasing pesticide consumption in Japan, India, South Korea, and Indonesia. Overall, the total annual use of pesticides in Asia was 1.10 million tons in 1990, rising to 2.15 million tons in 2017 (Roser, 2019).

For North America, a large amount of pesticide is consumed in the agriculture industry, especially in the USA. The USA accounts for approximately 16–18% of the total world pesticide usage (Atwood & Paisley-Jones, 2017). In the USA during 1990–2012 more than 50% of total pesticide use, almost 200 000 t/year, has been of herbicides followed by insecticides and fungicides (FAOSTAT, 2020). The pesticide use per hectare of cropland in North, Central, and South America in 2000 and 2017 is illustrated in Figure 2.5. The increasing consumption of pesticides during 17 years is apparent in the USA, Canada, and Guatemala for North and Central America, while expanding use is also noticeable in Brazil and Peru. The average pesticide application rates in North America increased from 2.14 kilograms per hectare in 2000 to 2.51 kilograms per hectare in 2017. In comparison, the average pesticide application rate in Brazil was 2.56 kilograms per hectare in 2000, increasing to 5.95 kilograms per hectare in 2017 (Roser, 2019). In Peru, the application increased from 0.41 kilograms per hectare in 2000 to 1.85 kilograms per hectare in 2017 (Roser, 2019). FAO reported that in some areas of South America, the usage rate of pesticides in arable land was as high as 60 kilograms per hectare (Carvalho, 2017). Wagner *et al.* (2017) reported that glyphosate (herbicide) has been the most used pesticide in North America since 2001, followed by atrazine (herbicide) and metolachlor (herbicide). However, for South America, herbicides (such as 2,4-D and paraquat) and insecticides (methamidophos, methomyl, endosulfan, and chlorpyrifos) are the significant most pesticides (Sharma *et al.*, 2019).

In contrast to North America, many countries in Europe have drastically decreased their pesticide consumption. Kristoffersen *et al.* (2008) reported that Denmark, the Netherlands, Germany, and Sweden had a keen political and public interest in lowering the pesticide (especially herbicide) application rate in urban areas. The UK has enhanced awareness and stringent regulations, while

Table 2.1 Annual consumption of pesticide in Asia (source: FAOSTAT, 2020).

Country	Tonnes Pesticide Used			
	1990	2017	Absolute Change	Relative Change
Japan	79 821	52 249	−27 573	−35%
Indonesia	2432	1597	−835	−34%
India	75 000	52 750	−22 250	−30%
South Korea	25 082	20 043	−5039	−20%
Vietnam	20 912	19 154	−1758	−8%
Malaysia	39 406	67 255	+57 882	+71%
Thailand	18 849	35 287	+16 438	+87%
China	775 408	1 770 000	+998 226	+129%
Myanmar	209	17 274	+17 064	+8162%

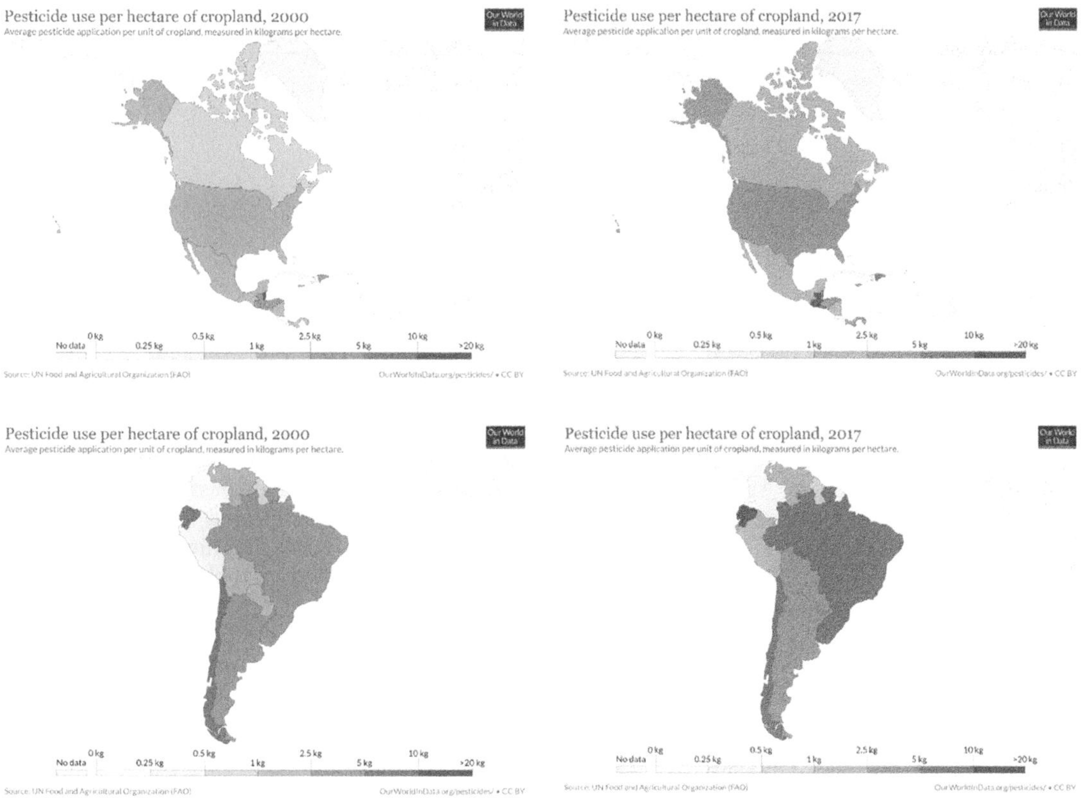

Figure 2.5 Comparison of pesticide use per hectare of cropland in north/central/South America in 2000 and 2017 (modified from Roser, 2019).

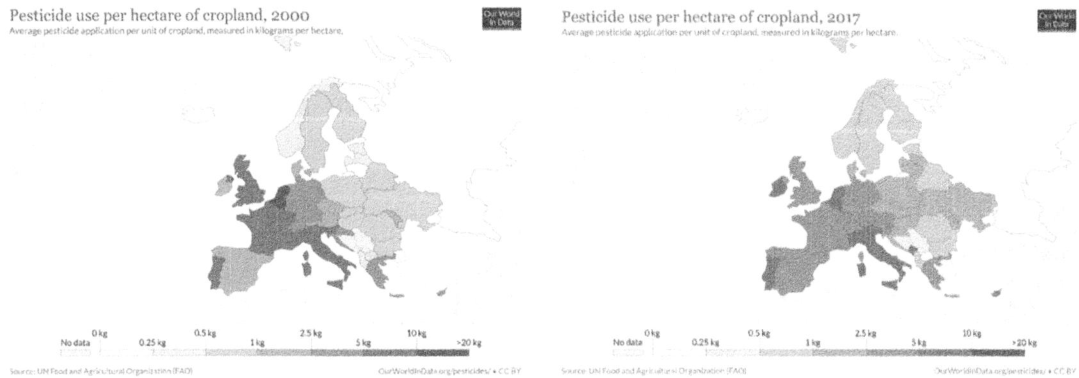

Figure 2.6 Comparison of pesticide use per hectare of cropland in Europe in 2000 and 2017 (modified from Roser, 2019).

Table 2.2 Pesticide use per hectare of cropland in Europe (source: FAOSTAT, 2020).

Country	Pesticide Use Rate (kilogram per hectare)			Relative Change
	2000	2017	Absolute Change	
UK	5.57	3.24	−2.33	−42%
The Netherlands	12.06	7.90	−4.16	−34%
France	5.02	3.63	−1.39	−28%
Denmark	1.39	1.10	−0.29	−21%
Sweden	0.64	0.59	−0.05	−8%
Greece	2.76	2.91	+0.15	+5%
Finland	0.52	0.59	+0.07	+13%
Germany	2.93	4.03	+1.10	+38%
Spain	1.89	3.59	+1.70	+90%
Ukraine	0.82	2.32	+1.50	+183%

Finland and Latvia have no specific regulation for reducing the amount of pesticide application. Figure 2.6 shows pesticide use per hectare of cropland in Europe in 2000 and 2017, as recorded by FAO. Pesticide use per hectare was tremendously lowered from 12.06 kilograms per hectare in 2000 to 7.90 kilograms per hectare in 2017 in the Netherlands, as shown in Table 2.2. The UK also had a significantly reduced amount of pesticide use from 5.57 kilograms per hectare in 2000 to 3.24 kilograms per hectare in 2017, and the annual pesticide consumption in 2017 was 19 844 tons (Roser, 2019). However, there has been some levelling-up, with an increase of pesticide use in Eastern Europe, especially Ukraine, Poland, Slovakia, Lithuania, Latvia, and Hungary, where it had been low in 2000.

2.3 ENVIRONMENTAL CONTAMINATION AND FOOD SAFETY

Pesticides have been used extensively for the purpose of enhancing crop yields and production of agricultural products for humans. However, large amounts of residual pesticides can be accumulated in the agricultural fields and ecosystems. Their persistence and low biodegradability has enhanced

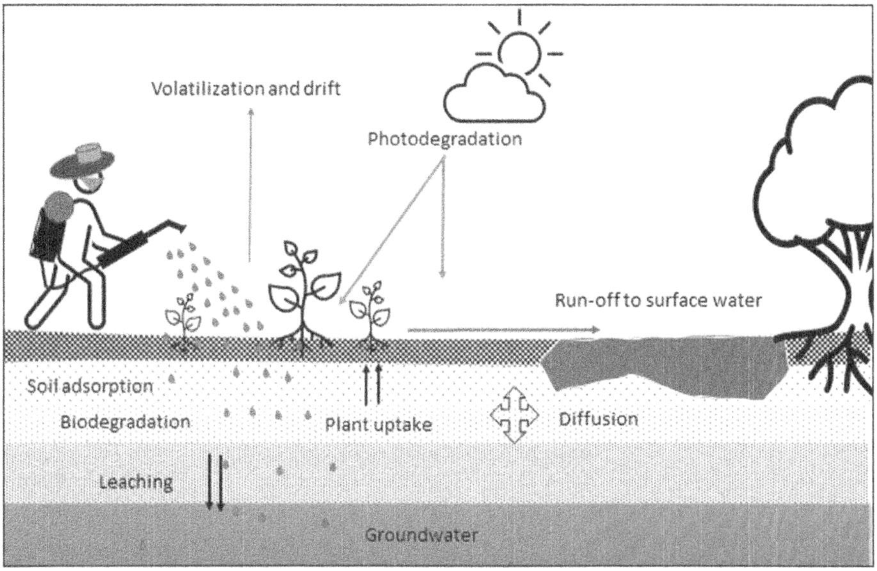

Figure 2.7 Fate and transport of pesticides in the environment.

the problem these toxic chemicals pose. They can cause multiple health problems such as neurological disruption and developmental effects in humans. During application, pesticide contamination of soil and water can occur through eight processes: volatilization; erosion and runoff; photodegradation; soil adsorption; leaching; diffusion; biodegradation by microorganisms; and plant uptake (Figure 2.7). The pesticide may volatilize to the air, be degraded by sunlight, adhere to soil particles (mainly organic matter and clays), be degraded by microorganisms, or be transported through the soil by water.

Volatilization of pesticides from soil surfaces is a crucial process. Physical and chemical properties of pesticides, especially their vapor pressure, are critical factors affecting the degree of pesticide volatilization. For example, the volatilization half-life of endrin is 63 days (Budavari, 1996; De Bruijn & Hermens, 1991), while the half-life value for α-hexachlorocyclohexane is 115 days (Lide, 2000; Schaefer *et al.*, 2015). Soil properties (such as dry or moist soil surface), meteorological conditions, and agricultural practices are also important factors that influence volatilization.

Soil adsorption of pesticides occurs naturally by physical or chemical adsorption. The difference in electrostatic charges, such as the positive charge of pesticide molecules and negative charge of particles on clay and organic matter, leads to electrical attraction and so-called physical adsorption. The formation of strong bonds between the pesticides or their metabolites and organic molecules in the soil causes chemical adsorption. The strong adsorption of pesticides is found in both soil and sediment. The desorption process is also a crucial process affecting the release of pesticides to the environment and the risk of pesticides to human health. The adsorption–desorption phenomena control the behaviour of pesticides in soil, and depend on the type of soil, temperature, pH, and the properties of the pesticide itself.

Table 2.3 shows the fate and environmental degradation of selected pesticides along with their chemical structures. Transformation and degradation of pesticides in soil by microorganisms is possible. The biodegradable pesticides may be transformed over days or weeks by soil microorganisms. However, many pesticides are recalcitrant and can remain in the soil for extended periods (years or decades), such as DDT and many organochlorine pesticides (Larson *et al.*, 1997).

Table 2.3 Environmental degradation of the selected pesticides.

Name and Chemical Structure	Environmental Degradation	References
Endrin $C_{12}H_8Cl_6O$ 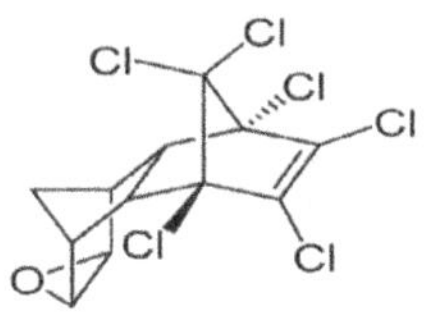	Photodegradation half-life is 7 days. Volatilization half-life is 63 days. It has low mobility in soil with half-life 4–8 years. Volatilization from moist soil surfaces is expected to be an important fate process. If released into water, endrin is expected to adsorb to suspended solids and sediment in water. Bioconcentration in aquatic organisms is very high.	Budavari (1996); De Bruijn and Hermens (1991)
Heptachlor $C_{10}H_5C_{17}$ 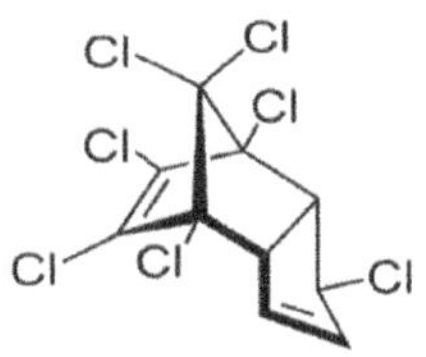	It is stable to daylight, air, moisture, and moderate heat (up to 160°C). Field dissipation half-lives for heptachlor can range from 40 days to 5.5 years. If released into water, heptachlor is expected to adsorb to suspended solids and sediment.	Budavari (1996); Hansch and Leo (1979); Lide (2000); Simpson *et al.* (1995); Tomlin (1994)
Hexachlorobenzene C_6Cl_6 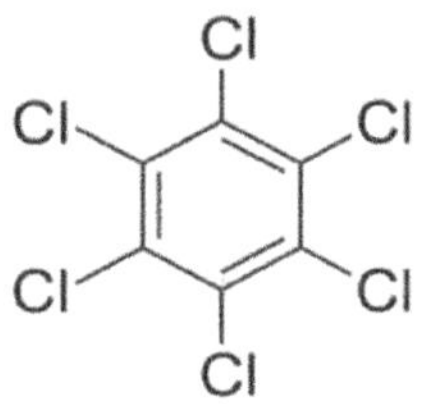	Hexachlorobenzene is not expected to volatilize from dry soil surfaces based upon its vapor pressure. It is persistent to either abiotic or biodegradation processes in soil. If released into water, hexachlorobenzene is expected to adsorb to suspended solids and sediment. The volatilization half-life from a model pond (2 m deep) is estimated as approximately 5 years if adsorption is considered.	O'Neil (2006); MacBean (2008); Hansch *et al.* (1995); Hartley and Kidd (1983)
α-hexachlorocyclohexane $C_6H_6Cl_6$ 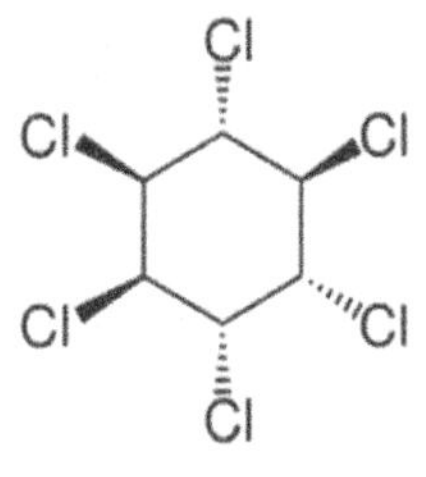	Half-life in air is approximately 115 days. It has low mobility in soil and does not volatilize from dry soil surfaces. The biodegradation half-life was 90 days in moist soil and 63 days in a soil slurry.	Lide (2000); Schaefer *et al.* (2015)
β-hexachlorocyclohexane $C_6H_6Cl_6$ 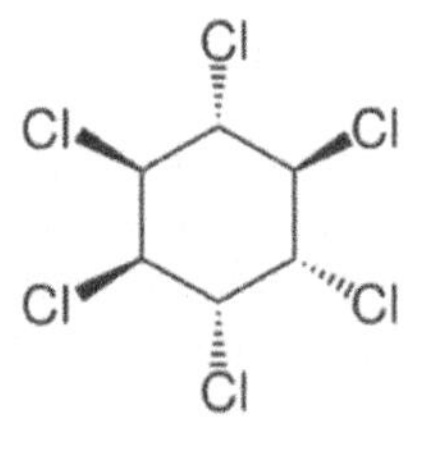	Half-life in air is estimated to be 28 days. It has low to slight mobility in soil and does not volatize from dry soil surface. It has little to no aerobic biodegradation and only slight anaerobic degradation. If released into water, it is expected to adsorb to suspended solids and sediment.	Lide (2000); USEPA (2020)

(Continued)

Table 2.3 Environmental degradation of the selected pesticides (*Continued*).

Name and Chemical Structure	Environmental Degradation	References
γ-hexachlorocyclohexane (or Lindane) $C_6H_6Cl_6$ 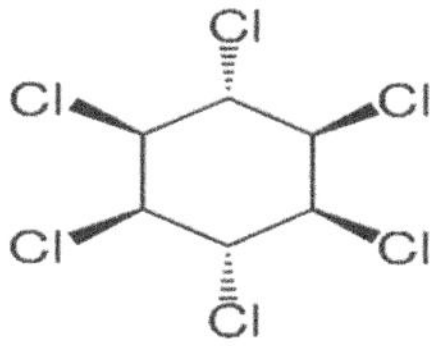	It is resistant to aerobic degradation in water and soil. Slow biodegradation is possible by anaerobic pathways. Its biodegradation in soil is dependent on its desorption from soil particles. It can bioaccumulate in food chains.	Lide (2000); USEPA (2020)
Mirex $C_{10}Cl_{12}$ 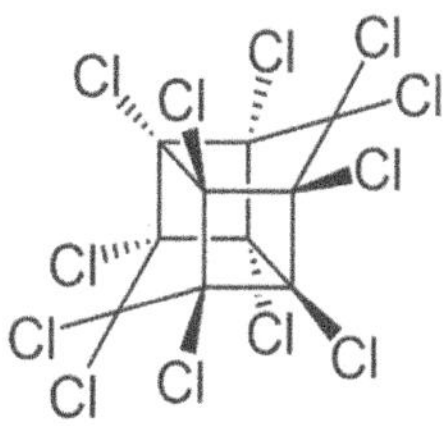	Mirex is slowly degraded by direct photolysis reaction. It is expected to be immobile in soil and may volatilize from moist soil surfaces. Mirex is very slow to biodegrade in the environment and is expected to have a biodegradation half-life of one year or more in most soils.	ATSDR (2011); O'Neil (2006)
Toxaphene 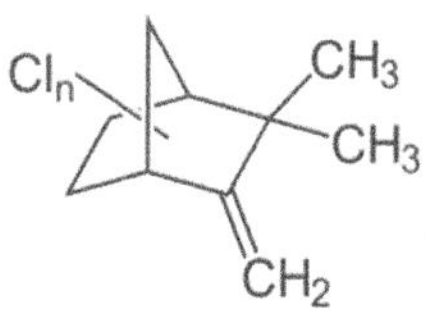	Half-lives in aerobic soil for the toxaphene mixture range from 1 to 11 years. It is not readily aerobic biodegradation for toxaphene components with more than 3 chlorine atoms. Aerobic biodegradation does not occur readily for toxaphene components with more than 3 chlorine atoms.	USEPA (1998); Saleh (1991); Esaac and Matsumura (1980); Sanborn *et al.* (1976)

The degradation of pesticides in soil occurs when pesticides have relatively high solubility, low persistence, and relatively high bioavailability. These factors can enhance the rapid degradation of chemicals (Gianfreda & Rao, 2010). Active degradation of pesticides can also occur when microorganisms can survive in the contaminated environment. Consequently, the rate of degradation depends on pesticide properties, soil characteristics, climate factors (temperature, rainfall, and humidity), and the number of microorganisms capable of decomposing the pesticides (Wołejko *et al.*, 2020). Where these factors do not apply, pesticides are persistent and remain in the soil. They can accumulate in the soil for extended periods, as shown in Table 2.4, which illustrates the types of residual pesticides in the soil of different countries.

Recently, a large study carried out in many regions of the world reported contamination by pesticides of not only soil and sediment but also rivers, lakes, and groundwater. Soils are the primary contaminant sources for all types of pesticides. However, soil erosion by surface runoff, and river discharges carry considerable amounts of various pesticides to rivers and lakes. Many events of pesticide contamination in surface water, carried by soil or sediment particles, have been reported in many regions.

Table 2.5 illustrates the residual pesticides in rivers, lakes, and groundwater in different countries. Three groups of insecticides – organochlorine, organophosphate ester, and carbamate – are frequently found in rivers and lakes in China. The triazine group such as atrazine and its derivative, simazine, and hydroxyatrazine are the most widely used herbicides and have been detected as residual chemicals in surface water in China (Brauns *et al.*, 2018; Sun *et al.*, 2018). Environmental monitoring data in Asia

Table 2.4 Types of residual pesticides in soil from selected countries.

Country	Pesticides	References
India	HCHs, DDTs, heptachlor, chlordanes, endosulfan, aldrin, dieldrin, endrin, endrin aldehyde, endrin ketone, methoxychlor	Kata *et al.* (2015)
Nepal	DDT, DDD, DDE	Bhandari *et al.* (2020)
Pakistan	3-Phenoxybenzaldehyde (3-PBA), fenitothion, chlorpyrifos, MCPA (methyl ester), profenofos, triazophos, bifenthrin	Rafique *et al.* (2016)
India	HCH, DDT	Mishra *et al.* (2011)
China	HCH, DDT, HCB, chlordane, endosulfan, aldrin, dieldrin, endrin	Wang *et al.* (2011)

Table 2.5 Types of residual pesticides in river, lake, and groundwater in selected countries.

Country	Environmental Media	Insecticide	Herbicides	References
China	river	Carbofuran, Acephate, Methamidophos, Isoprocarb, Fenobucarb, Acetamiprid, Imidacloprid	Atrazine, Acetochlor, Hydroxyatrazine, Desethylatrazine, Deisopropylatrazine, Terbuthylazine, 2,4-D, Bentazone, Simazine, 2,6-dichloro-benzamide, Metolachor, Isoproturon	Brauns *et al.* (2018); Sun *et al.* (2018); Xu *et al.* (2019)
	lake	HCH, Heptachlor, Aldrin, Dieldrin, Endrin, Endosulfan, Methoxychlor DDE, DDD, DDT	–	Xu *et al.* (2007)
	groundwater	DDT, Endosulfan,HCH, Monocrotophos, Chlorpyrifos	–	He *et al.* (2020)
India	river	α-HCH, β-HCH, γ-HCH, δ-HCH, Total HCH o,p′-DDE, p,p′-DDE, p,p′-DDD, o,p′-DDT, p,p′-DDT, Endosulfan	–	Khuman and Chakraborty (2019)
	groundwater	α-HCH, β-HCH, γ-HCH, δ-HCH, Total HCH, o,p′-DDE, p,p′-DDE, p,p′-DDD, o,p′-DDT, p,p′-DDT	–	
Malaysia	river	Lindane, Heptachlor, Endosulfan, Dieldrin, DDT, DDE	–	Leong *et al.* (2007)
Vietnam	lake	HCHs, DDTs, Endrin, Heptachlor, Aldrin	–	Hung and Thiemann (2002)
Japan	lake	Isoprothiolane, Diazinon, Fenitrothion	Molinate, Simetryn, Oxadiazon, Thiobencarb,	Sudo *et al.* (2002)
South Korea	groundwater	Aldicarb, Carbaryl	2,4-D	Lee *et al.* (2019)
Turkey	lake	BHC, Lindane, Aldrin, Dieldrin, Heptachlor, Heptachlor, DDT, DDE, DDD	–	Ozmen *et al.* (2008)

from many works illustrate that organochlorine insecticides such as DDTs, hexachlorocyclohexane, and endosulfan are the significant pesticides found in surface water in many countries including China, Turkey, Vietnam, Malaysia, and India (Hung & Thiemann, 2002; Ozmen *et al.*, 2008; Xu *et al.*, 2007); 2,4-dichlorophenoxyacetic acid or 2,4-D is also a well used herbicide and frequently contaminates surface waters in China and Southeast Asia (Brauns *et al.*, 2018). However, different types of herbicides such as molinate, simetryn, oxadiazon, and thiobencarb are found in lakes in Japan (Sudo *et al.*, 2002).

Vammen *et al.* reported many types of pesticide contamination in surface water in Latin America (Vammen *et al.*, 2019). Glyphosate concentrations in the range of 0.10–0.7 mg/L were detected at sampling points of streams and wetlands to the north of Buenos Aires, Argentina (Peruzzo *et al.*, 2008). Atrazine, metolachlor, and flutriafol were found in surface water in the drainage area from the agriculture area of Mato Grosso, Brazil (Casara *et al.*, 2012). Carbendazim (fungicide), simazine, hexazinone, 2,4-D, and picloram were reported as residual pesticides in surface waters near the agricultural-forestry area in Southern Chile (Palma *et al.*, 2004).

In Europe, routine monitoring is conducted. In the Netherlands, of 408 pesticides and 52 metabolites, 63 pesticides and 6 metabolites were categorised as high-priority pesticides according to their presence in groundwater, surface water, and drinking water (Sjerps *et al.*, 2019). Concentrations of many pesticides, including thiamethoxam, acetamiprid, fluopyram, mandipropamid, fluxapyroxadand, and fluxastrobin, were detected above the drinking water standard of 0.1 μg/L in surface water (Sjerps *et al.*, 2019). The most frequently detected pesticides in drinking water in Germany and France are bentazone, carbendazim, glyphosate, and the metabolite AMPA (Loos *et al.*, 2010; Schreiner *et al.*, 2016).

The plant uptake of pesticides is the crucial pathway for food security concerns. Tables 2.6 and 2.7 illustrate concentrations of the residual pesticides detected in several types of fruits and edible plants in many countries. Food safety is a globally important issue nowadays. Contamination of the environment by pesticides leads to a residue of toxic chemicals in the food chain and, consequently, a potential negative impact on human health. Many pesticides are endocrine disruptors that can disturb the functioning of various hormones and many adverse effects on the human (Mandrich, 2014). The undesirable failure of the human body, including the central nervous system, respiratory system, and neurological effects as well as chromosomal alterations are pressing issues for food security around the world. A call for food safety through a reduction of pesticide usage in agriculture activities is rising in both developing and developed countries today. This article evidences the situation of pesticide usage and pesticide contamination as the new emergent diffuse pollution, which requires further monitoring to obtain information for regulation and decision-making to protect the environment now and in the future.

2.4 CHALLENGES AND RECOMMENDATIONS

From past to present, pesticides have played an important role in increasing agricultural production, and they are continuously employed in a high amounts in many countries. The application of insecticides, herbicides, and fungicides leads to their occurrence in surface water, groundwater and, eventually, drinking water. Pesticides are accumulating in soil to significant amounts in many regions. Pesticides in soil are transported to water bodies and can impact the whole ecological food chain, through uptake by plants, and concentration in aquatic animals. Residues of these persistent chemicals in vegetables and plants have been reported from around the world, as well as their adverse effects on human health.

Recently, many publications have reported mitigation strategies to reduce pesticide loading into water bodies, including surface water and groundwater (Reichenberger *et al.*, 2007). Practical and feasible mitigation measures are recommended for implementation at both farm and catchment scales. The performance of grassed buffer strips located at the lower edges of fields has been evaluated and

Table 2.6 Concentrations of residual pesticides in fruits in selected countries.

Country	Pesticides	Type of Fruits	Concentration Range	References
India	Acephate, Benfuracarb, Profenofos, Carbendazim, Dimethoate, Ethion, Monocrotophos	Apple	14–1325 μg/kg	Sivaperumal *et al.* (2015)
	Acephate, Dimethoate, Difenoconazole, Monocrotophos	Banana	49–904 μg/kg	
	Acephate, Monocrotophos	Grape	9–1559 μg/kg	
	Acephate, Monocrotophos, Carbendazim	Guava	11–1455 μg/kg	
	Acephate	Mango	23–100 μg/kg	
	Acephate, Monocrotophos	Orange	141–294 μg/kg	
	Acephate, Benfuracarb, Chlorpyrifos, Ethion, Monocrotophos	Pomegranate	8–1517 μg/kg	
Cambodia	DDTs, HCHs, Mirex, HCB	Banana	3.09 ± 1.04 ng/g	Wang *et al.* (2011)
Kuwait	Oxamyl, Cypermethrin, Profenofos, Deltamethrin, Fenpropathrin, Malathion, Monocrotophos, Aldrin, Primiphosmethyl, Diazinon, Chlorpyrifosmethyl, Imidacloprid, Acetamiprid, Thiophanatemethyl, Metalaxyl, Difenoconazole	Watermelon Apple Grapes Strawberry	nd–0.29 mg/kg nd–0.65 mg/kg nd–1.32 mg/kg nd–1.2 mg/kg	Jallow *et al.* (2017)
China	Chlorpyrifos, Methamidophos, Cyfluthrin, Triadimefon, Fenpropathrin, Chlorothalonil, Omethoate, Cyhalothrin, Fenvalerate, Dichlorvos, Acephate, Parathion-methyl, Fimethoate, Parathion, Bifenthrin, Isocarbophos, Phorate, Deltamethrin, Triazophos, Fenitrothion, Permethrin	Apple Peach Orange Pear Grape	0.0001–0.0041 mg/kg 0.0002–0.002 mg/kg nd 0.0001–0.0078 g/kg 0.0001–0.032 mg/kg	Chen *et al.* (2011)
Japan	Imidacloprid, Acetamiprid	Apple	134–1670 ppb, 45.2–802 ppb	Wanatabe *et al.* (2001)
	Imidacloprid	Apple juices Grape juices Orange juices	8.2–96 μg/L 10.9–128.7 μg/L 54.7–383.3 μg/L	Watanabe *et al.* (2007)

has demonstrated generally reduced pesticide runoff and erosion losses. Buffer strip widths of at least 8 m can reduce the pesticide load with high efficiency, especially where a significant portion of pesticide is transported in the sediment phase (Reichenberger *et al.*, 2007). From the EXPOSIT 1.1 model (Winkler, 2001), 50% pesticide load reduction was achieved with a 5 m buffer strip width, and 97.5% pesticide load reduction was reported for a 20 m width (Winkler, 2001). The research from Gril and Lacas (2006) and Reichenberger *et al.* (2007) suggested that the riparian buffer strips are most probably much less effective than edge-of-field buffer strips for pesticide load reduction (Gril & Lacas, 2006; Reichenberger *et al.*, 2007). The significance of riparian buffer strips is higher for drift control than for run-off control.

Table 2.7 Concentrations of residual pesticides in edible plants in selected countries.

Country	Pesticides	Type of Vegetable	Concentration Range	References
India	Acephate	Brinjal	124–1198 μg/kg	Sivaperumal *et al.* (2015)
	Acephate, Chlorpyrifos, Monocrotophos	Cauliflower	10–1556 μg/kg	
	Acephate, Chlorpyrifos, Ethion, Monocrotophos, Profenofos	Cabbage	19–1170 μg/kg	
	Acephate, Ethion, Monocrotophos	Okra	10–2774 μg/kg	
	Propoxur, Dimethoate	Onion	37–147 μg/kg	
	Acephate, Ethion, Monocrotophos, Profenofos	Potato	41–2634 μg/kg	
	Acephate, Dimethoate, Ethion, Monocrotophos, Profenofos	Tomato	16–1219 μg/kg	
Cambodia	DDTs, HCHs, CHLs, DRINs, Mirex, HCB	Bitter guard	5.39 ± 2.28 ng/g	Wang *et al.* (2011)
		Cabbage	1.79 ± 0.63 ng/g	
		Carrot	2.58 ± 0.57 ng/g	
		Chinese Radish	2.35 ± 0.63 ng/g	
		Cucumber	2.82 ± 1.41 ng/g	
		Eggplant	4.51 ± 2.34 ng/g	
		Long Bean	2.20 ± 0.99 ng/g	
		Morning glory	2.32 ± 0.66 ng/g	
		Mustard Green	1.28 ± 0.28 ng/g	
		Petsai	2.20 ± 0.50 ng/g	
		Sponge gourd	2.14 ± 1.29 ng/g	
		Wintermelon	4.95 ± 2.86 ng/g	
Kuwait	Oxamyl, Cypermethrin, Deltamethrin, Fenpropathrin, Malathion, Profenofos, Monocrotophos, Primiphos methyl, Diazinon, Chlorpyrifos methyl, Imidacloprid, Acetamiprid, Thiophanate methyl, Metalaxyl, Difenoconazole, Aldrin	Tomato	nd–0.51 mg/kg	Jallow *et al.* (2017)
		Carrot	nd–0.03 mg/kg	
		Bell pepper	nd–0.05 mg/kg	
		Eggplant	nd–0.13 mg/kg	
		Cucumber	nd–1.2 mg/kg	
		Zucchini	nd–0.08 mg/kg	
		Cabbage	nd–0.1 mg/kg	
		Potato	nd	
Indonesia	RHCHs (aHCH, b-HCH, c-HCH, d-HCH), HCB, RDrins (Aldrin and Dieldrin), Endrin, RHeptachlors (heptachlor, cis-heptachlor epoxide, transheptachlor epoxide), RChlordanes (oxychlordane, cis-chlordane, trans-chlordane, trans-nonachlor, cis-nonachlor), RDDTs (o,p0 - DDE, p,p0 -DDE, o,p0 -DDD, p,p0 -DDD, o,p0 -DDT, p,p0 -DDT), Mirex	Carrot	<0.7 ng/g	Shoiful *et al.* (2013)
		Potato	<0.7 ng/g	
		Cucumber	<0.7 ng/g	
		Corn	<0.7 ng/g	
		Onion	<0.7 ng/g	

(Continued)

Table 2.7 Concentrations of residual pesticides in edible plants in selected countries (*Continued*).

Country	Pesticides	Type of Vegetable		References
Bangladesh	Malathion, Malathion, Carbaryl, Cartap, Carbofuran, Chlorpyrifos, Cypermethrin, Diazinon, Fenitrothion	Eggplant	0.005–0.700 mg/kg	Chowdhury *et al.* (2013)
	Chlorpyrifos, Diazinon, Malathion, Carbosulfan, Phenthoate, Dimethoate, Carbofuran, Carbaryl	Tomato	0.004–0.7 mg/kg	
	Cypermethrin, Chlorpyrifos, Acephate, Cypermethrin, Carbaryl	Cauliflower	0.02–0.460 mg/kg	
	Endosulfan, Malathion, Carbofuran, Chlorpyrifos, Cypermethrin	Cabbage	0.013–1.00 mg/kg	
	Diazinon, Propiconazole, Carbofuran, Diazinon, Chlorpyrifos, Endosulfan, Cartap, Carbosulfan, Carbaryl, Acephate, Dimethoate, Fipronil	Potatoes	0.008–0.3 mg/kg	
	Chlorpyrifos, Cypermethrin, Dicofol Deltamethrin, Carbaryl, Diazinon	Cucumbers	0.007–0.3 mg/kg	
	Fenvalerate, Carbofuran, Propiconazole, Chlorpyrifos, 2,4-D	Carrot	0.011–0.4 mg/kg	
	Malathion, Chlorpyrifos, 2,4-D, diazinon, Dimethoate, Propiconazole	Onion	0.01–0.5 mg/kg	
China	Chlorpyrifos, Methamidophos, Cyfluthrin, Triadimefon, Fenpropathrin, Chlorothalonil, Omethoate, Cyhalothrin, Fenvalerate, Dichlorvos, Acephate, Parathion-methyl, Dimethoate, Parathion, Bifenthrin, Isocarbophos, Phorate, Deltamethrin, Triazophos, Fenitrothion, Permethrin	Chinese cabbage	0.0004–0.074 mg/kg	Chen *et al.* (2011)
		Spinach	0.0003–0.156 mg/kg	
		Legumes	0.0019–0.0144 mk/kg	
		Cabbage	0.0005–0.035 mg/kg	
		Tomato	0.0002–0.03 mg/kg	
		Radish	0.0002–0.313 mg/kg	
		Cauliflower	0.0001–0.036 mg/kg	
		Cucumber	0.0001–0.0055 mg/kg	
		Leaf mustard	0.0003–0.148 mg/kg	
		Lettuce	0.0003–0.076 mg/kg	
		Capsicum	0.0002–0.027 mg/kg	
		Eggplant	0.0001–0.029 mg/kg	
		Broccoli	0.0008–0.27 mg/kg	
		Pakchoi cabbage	0.0009–0.22 mg/kg	
		Celery	0.0001–0.054 mg/kg	
Japan	Imidacloprid, Acetamiprid	Cucumber	0.2–0.50 mg/kg in dry seasons	Wanatabe *et al.* (2001)
		Green pepper	2.04–2.84 mg/kg in dry seasons	
		Tomato	140–1420 ppb, 45.6–938 ppb	
Turkey	DDVP	Cucumber	0.16–13.92 mg/l	Cengiz *et al.* (2006)
	Diazinon		0.02–6.56 mg/kg	

(*Continued*)

Table 2.7 Concentrations of residual pesticides in edible plants in selected countries (*Continued*).

Country	Pesticides	Type of Vegetable		References
India	Lambda-cyhalothrin	Tea	<0.001–0.137 mg/kg	Seenivasan and Muraleedharan (2009)
Saudi Arabia	Bioallethrin, Carbaryl, Chlorpyrifos, Dieldrin, Ethiofencarb, Heptenophos, Malathion, Methoxychlor, Lindane, Propoxur, Pyrethrin I, Tefluthrin	Tomato	below detection limit (0.08 mg/kg)–17.27 mg/kg	Osman *et al.* (2010)
		Squash	0.0002–0.0211 mg/kg	
		Cucumber	0.25–7.443 mg/kg	
		Egg-plant	0.588–2.844 mg/kg	
		Green pepper	0.388–2.85 mg/kg	
		Lettuce	1.505–3.811 mg/kg	
		Carrot	0.878–3.518 mg/kg	
		Cabbage	0.675–12.206 mg/kg	
Thailand	Acetochlor, Alachlor, Clomazone, Isoxaflutole, Metribuzin, Oxadiazon, Pendimethalin,	Soybean	0.072–6.747 mg/kg	Pornprom *et al.* (2010)

Many works (Schulz & Peall, 2001) suggest that constructed wetlands are very effective in reducing pesticide loading into surface water. A small wetland 50 m long and 1.5 m wide was effective in removing pesticides such as chlorpyrifos from the water passing through it (Moore *et al.*, 2001). However, for atrazine herbicide removal, a travel distance of 100–280 m through the wetland is required (Moore *et al.*, 2000). Grassed waterways with a flow length of 24.4 m were reported (Rohde *et al.*, 1980) to reduce trifluralin herbicide by 96% (dry) and 86% (wet). Elsewhere, 72% (dry) and 69% (wet) reduction of 2,4-D was also reported using grassed waterways (Asmussen *et al.*, 1977).

In-field cleaning and on-farm handling are reported as effective methods to reduce the amount of leftover taken back to the farmyard. A biobed consisting of a hole in the ground filled with a mixture of chopped straw, peat, and topsoil is suggested as a practical approach for the proposed treatment method for pesticide-contaminated water (Torstensson & Castillo, 1997). Separately, Felgentreu and Bischoff (2006) reported that the recycling of biobed leachates could increase the removal efficiency of the biobeds (Felgentreu & Bischoff, 2006). A biomass bed, modified biobed, using four mixtures from farm organic filter materials, and recycling of the leachate, can enhance pesticide degradation up to five times faster than standard soil.

Farmers are already aware of the harm caused by residual pesticides in the environment and food chain, but this awareness needs to be reinforced. Best management practice by farmers in respect of pesticide handling and application, and the adoption of mitigation measures are essential measures to reduce damaging diffuse pesticide pollution. Routine monitoring of pesticides throughout the ecosystem is critical in clearly demonstrating the extent of pollution and the effectiveness of the control measures being taken.

ACKNOWLEDGEMENT

Authors would like to express their gratitude to Tom Leatherland, former senior water quality scientist, Scottish Environment Protection Agency (SEPA) for his contribution in improving the quality of this article. This work is supported by National Higher Education Science Research and Innovation Policy Council (Contract No.C16F630255) and Thailand Research Fund (Grant No.IRN62W0005).

REFERENCES

Al-Mulali U., Ozturk I. and Lean H. H. (2015). The influence of economic growth, urbanization, trade openness, financial development, and renewable energy on pollution in Europe. *Natural Hazards*, **79**(1), 621–644. https://doi.org/10.1007/s11069-015-1865-9d

Asmussen L. E., White A. W., Jr., Hauser E. W. and Sheridan J. M. (1977). Reduction of 2,4-D load in surface runoff down a grassed waterway. *Journal of Environmental Quality*, **6**, 159–162, https://doi.org/10.2134/jeq1977.00472425000600020011x

ATSDR, Vaalgamaa S., Vähätalo A. V., Perkola N. and Huhtala S. (2011). Photochemical reactivity of perfluorooctanoic acid (PFOA) in conditions representing surface water. *Science of The Total Environment*, **409**(16), 3043–3048. https://doi.org/10.1016/j.scitotenv.2011.04.036

Attia M. Y., Abdelbary A. A., Khaled M. M. and Abdelfattah A. H. (2016). Veno-venous extracorporeal membrane oxygenation in a case of organophosphorus poisoning. *The Egyptian Journal of Critical Care Medicine*, **4**(1), 43–46. https://doi.org/10.1016/j.ejccm.2016.02.006

Atwood D. and Paisley-Jones C. (2017). Pesticides industry sales and usage: 2008–2012 Market Estimates, US Environmental Protection Agency, Washington, DC.

Bakirtas T. and Akpolat A. G. (2018). The relationship between energy consumption, urbanization, and economic growth in new emerging-market countries. *Energy*, **147**, 110–121. https://doi.org/10.1016/j.energy.2018.01.011

Bhandari G., Atreya K., Scheepers P. T. J. and Geissen V. (2020). Concentration and distribution of pesticide residues in soil: non-dietary human health risk assessment. *Chemosphere*, 126594. https://doi.org/10.1016/j.chemosphere.2020.126594

Bonzi R. S. (2013). Meio século de Primavera silenciosa: um livro que mudou o mundo. *Desenvolvimento E Meio Ambiente*, **28**, 207–215, https://doi.org/10.5380/dma.v28i0.31007

Budavari, S. (ed.) (1996). The Merck Index – An Encyclopedia of Chemicals, Drugs, and Biologicals. Whitehouse Station, NJ Merck Co., New Jersey.

Brauns B., Jakobsen R., Song X. and Bjerg P. L. (2018). Pesticide use in the wheat-maize double cropping systems of the north China Plain: assessment, field study, and implications. *Science of The Total Environment*, **616–617**, 1307–1316, https://doi.org/10.1016/j.scitotenv.2017.10.187

Carneiro R. P., Oliveira F. A. S., Madureira F. D., Silva G., de Souza W. R. and Lopes R. P. (2013). Development and method validation for determination of 128 pesticides in bananas by modified QuEChERS and UHPLC-MS/MS analysis. *Food Control* **33**, 413–423. https://doi.org/10.1016/j.foodcont.2013.02.027

Carvalho F. P. (2017). Pesticides, environment, and food safety. *Food and Energy Security*, **6**(2), 48–60, https://doi.org/10.1002/fes3.108

Casara K. P., Vecchiato A. B., Lourencetti C., Pinto A. A. and Dores E. F. G. C. (2012). Environmental dynamics of pesticides in the drainage area of the São Lourenço River headwaters, Mato Grosso State, Brazil. *Journal of the Brazilian Chemical Society*, **23**, 1719–1731, https://doi.org/10.1590/S0103-50532012005000037

Cengiz M. F., Certel M. and Gocmen H. (2006). Residue contents of DDVP (Dichlorvos) and diazinon applied on cucumbers grown in greenhouses and their reduction by duration of a pre-harvest interval and post-harvest culinary applications. *Food Chemistry*, **98**(1):127–135. https://doi.org/10.1016/j.foodchem.2005.05.064

Chen C., Qian Y., Chen Q., Tao C., Li C. and Li Y. (2011). Evaluation of pesticide residues in fruits and vegetables from Xiamen, China. *Food Control*, **22**(7), 1114–1120, https://doi.org/10.1016/j.foodcont.2011.01.007

Chowdhury A. Z. M., Fakhruddin A. N. M., Nazrul Islam M., Moniruzzaman M., Gan S. H. and Khorshed Alam M. (2013). Detection of the residues of nineteen pesticides in fresh vegetable samples using gas chromatography–mass spectrometry. *Food Control*, **34**(2), 457–465. https://doi.org/10.1016/j.foodcont.2013.05.006

De Bruijn J. and Hermens J. (1991). Uptake and elimination kinetics of organophosphorous pesticides in the guppy (poecilia reticulata): correlations with the octanol/water partition coefficient. *Environmental Toxicology and Chemistry*, **10**(6), 791–804. https://doi.org/10.1002/and so on.5620100610

Esaac E. G. and Matsumura F. (1980). Metabolism of insecticides by reductive systems. *Pharmacology & Therapeutics*, **9**(1), 1–26. https://doi.org/10.1016/0163-7258(80)90014-5

FAOSTAT (2020). Pesticide Use, Food and agriculture organization of the United Nations. Retrieved from: http://www.fao.org/faostat/en/?fbclid=IwAR3VqLCbBolO4139FiTlLah6ssYD28LLbstB4DoETH-ot6zUKijowANB8RQ#data/RP/visualize (accessed 28 July 2020)

Felgentreu D. and Bischoff G. (2006). Studies on Inactivating Wastewater and Residual Liquids Containing Plant Protection Products by 'biobeds'. Abstracts of the SCI Meeting 'Pesticide behaviour in soils, water and air', Warwick, UK, 27–29 March, vol. **A 10**.

Gianfreda L. and Rao M. A. (2010). The influence of pesticides on soil enzymes. In: Soil Enzymology. (Soil Biology Series, **22**), G. Shukla and A. Varma (eds.), Springer Verlag, Berlin/Heidelberg, Germany, pp. 293–312.

Gril J. J. and Lacas J. G. (2006). Practical Aspects of the Use of Grassed or Wooded Buffer Zones to Control Water Pollution by Pesticides. Abstracts of the SCI Meeting 'Pesticide behaviour in soils, water and air', Warwick, UK, 27–29 March, vol. **B 35**. Abstract.

Hansch C. and Leo A. (1979). Substituent Constants for Correlation Analysis in Chemistry and Biology. John Wiley and Sons, New York.

Hansch C., Leo A. and Hoekman D. (1995). Exploring QSAR. Hydrophobic, Electronic, and Steric Constants. ACS Professional Reference Book. ACS, Washington. https://doi.org/10.1021/jm950902o

Hartley D. and Kidd H. (eds.) (1983). The Agrochemicals Handbook. Royal Society of Chemistry, Surrey, UK.

He X. S., Zhang Y. L. and Liu Z. H. (2020). Interaction and coexistence characteristics of dissolved organic matter with toxic metals and pesticides in shallow groundwater. *Environmental Pollution*, **258**, 113736. https://doi.org/10.1016/j.envpol.2019.113736

Hung D. Q. and Thiemann W. (2002). Contamination by selected chlorinated pesticides in surface waters in Hanoi, Vietnam. *Chemosphere*, **47**, 357–367. https://doi.org/10.1016/s0045-6535(01)00342-3

Jallow M. F. A., Awadh D. G., Albaho M. S., Devi V. Y. and Ahmad N. (2017). Monitoring of pesticide residues in commonly used fruits and vegetables in Kuwait. *International Journal of Environmental Research and Public Health*, **14**, 833. https://doi.org/10.3390/ijerph14080833

Kata M., Rao S. S. and Mohan K. R. (2015). Spatial distribution, ecological risk evaluation and potential sources of organochlorine pesticides from soils in India. *Environmental Earth Sciences*, **74**(5), 4031–4038 https://doi.org/10.1007/s12665-014-3189-6

Khuman S. N. and Chakraborty P. (2019). Air–water exchange of pesticidal persistent organic pollutants in the lower stretch of the transboundary river Ganga, India. *Chemosphere*, **233**, 966–974 https://doi.org/10.1016/j.chemosphere.2019.05.223

Koleva N. G., Schneider U. and McCarl B. A. (2011). Pesticide externalities from the US agricultural sector – the impact of internalization, reduced pesticide application rates, and climate change. *Biophysical Chemistry*, **6**, 153–161.

Kristoffersen P., Rask A. M., Grundy A., Franzen I., Kempenaar C., Raisio J., Schroeder H., Spijker J., Verschwele A. and Zarina L. (2008). A review of pesticide policies and regulations for urban amenity areas in seven European countries. *Weed Research*, **48**, 201–214. https://doi.org/10.1111/j.1365-3180.2008.00619.x

Larson S. J., Capel P. D. and Majewski M. S. (1997). Pesticides in surface waters–distribution, trends, and governing factors. In: Series of Pesticides in Hydrologic System, R. J. Gilliom (ed.), Ann Arbor Press, Chelsea, Michigan, US, p. 3.

Lee H. J., Kim K. Y., Hamm S. Y., Kim M., Kim H. K. and Oh J. E. (2019).Occurrence and distribution of pharmaceutical and personal care products, artificial sweeteners, and pesticides in groundwater from an agricultural area in Korea. *Science of The Total Environment*, **659**, 168–176. https://doi.org/10.1016/j.scitotenv.2018.12.258

Leong K. H., Benjamin Tan L. L. and Mustafa A. M. (2007). Contamination levels of selected organochlorine and organophosphate pesticides in the Selangor River, Malaysia between 2002 and 2003. *Chemosphere*, **66**(6), 1153–1159. https://doi.org/10.1016/j.chemosphere.2006.06.009

Lide D. R. (2000). 81st edn, CRC Handbook of Chemistry and Physics. CRC Press, Florida.

Loos R., Locoro G., Comero S., Contini S., Schwesig D., Werres F., Balsaa P., Gans O., Weiss S., Blaha L., Bolchi M., and Gawlik B. M. (2010). Pan-European survey on the occurrence of selected polar organic persistent pollutants in ground water, *Water Resources*, **44**(14), 4115–4126.

MacBean, C. (ed.) (2008). 15th edn, e-Pesticide Manual – Hexachlorobenzene (117-84-1). British Crop Protection Council, Alton, UK.

Mandrich L. (2014). Endocrine disrupters: the hazards for human health. *Cloning & Transgenesis* **3**: e110. https://doi.org/10.4172/2168-9849.1000e110

Mishra K., Sharma R. C. and Kumar S. (2011). Contamination levels and spatial distribution of organochlorine pesticides in soils from India. *Ecotoxicology and Environmental Safety*. **76**, 215–225, https://doi.org/10.1016/j.ecoenv.2011.09.014

Moore M. T., Rodgers J. H., Jr., Cooper C. M. and Smith S., Jr. (2000). Constructed wetlands for mitigation of atrazine-associated agricultural runoff. *Environmental Pollution*, **110**, 393–399, https://doi.org/10.1016/S0269-7491(00)00034-8

Moore M. T., Bennett E. R., Cooper C. M., Smith S., Jr., Shields F. D., Jr. and Milam C. D. (2001). Transport and fate of atrazine and lambda-cyhalothrin in agricultural drainage ditches: a case study for mitigation. *Agriculture Ecosystems & Environment*, **87**, 309–314. https://doi.org/10.1016/S0167-8809(01)00148-7

O'Neil, M. J. (ed.) (2006). The Merck Index – An Encyclopedia of Chemicals, Drugs, and Biologicals. Merck and Co., Inc., New Jersey.

Osman K. A., Al-Humaid A. M., Al-Rehiayani S. M. and Al-Redhaiman K. N. (2010). Monitoring of pesticide residues in vegetables marketed in Al-Qassim region, Saudi Arabia. *Ecotoxicology and Environmental Safety*, **73**, 1433–1439. https://doi.org/10.1016/j.ecoenv.2010.05.020

Ozmen M., Ayas Z., Güngördü A., Ekmekci G. F. and Yerli S. (2008). Ecotoxicological assessment of water pollution in Sariyar Dam Lake, Turkey. *Ecotoxicology and Environmental Safety*, **70**, 163–173. https://doi.org/10.1016/j.ecoenv.2007.05.011

Palma G., Sánchez A., Olave Y., Encina F., Palma R. and Barra R. (2004). Pesticide levels in surface waters in an agricultural-forestry basin in Southern Chile. *Chemosphere*, **57**, 763–770, https://doi.org/10.1016/j.chemosphere.2004.08.047

Patterson D. T., Westbrook J. K., Joyce R. J. V., Lingren P. D. and Rogasik J. (1999). Weeds, insects, and diseases. *Climatic Change*, **43**, 711–727, https://doi.org/10.1023/A:1005549400875

Peruzzo P. J., Porta A. A. and Ronco A. E. (2008). Levels of glyphosate in surface waters, sediments and soils associated with direct sowing soybean cultivation in north pampasic region of Argentina. *Environmental Pollution*, **156**, 61–66. https://doi.org/10.1016/j.envpol.2008.01.015

Pornprom T., Sukcharoenvipharat W. and Sansiriphun D. (2010). Weed control with pre-emergence herbicides in vegetable soybean (Glycine max L. Merrill). *Crop Protection*, **29**, 684–690, https://doi.org/10.1016/j.cropro.2010.02.003

Porter J. H., Parry M. L. and Carter T. R. (1991). The potential effects of climatic-change on agricultural insect pests. *Agricultural and Forest Meteorology*, **57**, 221–240, https://doi.org/10.1016/0168-1923(91)90088-8

Rafique N., Tariq S. R. and Ahmed D. (2016). Monitoring and distribution patterns of pesticide residues in soil from cotton/wheat fields of Pakistan. *Environmental Monitoring and Assessment*, **188**(12), 695. https://doi.org/10.1007/s10661-016-5668-6

Reichenberger S., Bach M., Skitschak A. and Frede H.-G. (2007). Mitigation strategies to reduce pesticide inputs into ground- and surface water and their effectiveness: a review, *Science of the Total Environment*, **384**, 1–35, https://doi.org/10.1016/j.scitotenv.2007.04.046

Reilly J. (1995). Climate change and global agriculture: recent findings and issues. *American Journal of Agricultural Economics*, **77**, 243–250, https://doi.org/10.2307/1243242

Rodrigues F. de M., Mesquita P. R. R., de Oliveira L. S., de Oliveira F. S., Menezes Filho A., de P. Pereira P. A. and de Andrade J. B. (2011). Development of a headspace solid phase microextraction/gas chromatography-mass spectrometry method for determination of organophosphorus pesticide residues in cow milk. *Microchemical Journal*, **98**, 56–61, https://doi.org/10.1016/j.microc.2010.11.002

Rohde W. A., Asmussen L. E., Hauser E. W., Wauchope R. D. and Allison H. D. (1980). Trifluralin movement in runoff from a small agricultural watershed. *Journal of Environmental Quality*, **9**, 37–42, https://doi.org/10.2134/jeq1980.00472425000900010011x.

Roser M. (2019). *Pesticides*. Published online at OurWorldInData.org. Retrieved from: https://ourworldindata.org/pesticides (accessed 22 February 2021).

Saleh M. A. (1991). Toxaphene: chemistry, biochemistry, toxicity and environmental fate. In: Reviews of Environmental Contamination and Toxicology, G. W. Ware (ed.), Springer, New York, **118**, 1–85.

Sanborn J. R., Metcalf R. L. and Bruce W. N. (1976). The fate of chlordane and toxaphene in a terrestrial aquatic model ecosystem. *Environmental Entomology*, **5**, 533–538, https://doi.org/10.1093/ee/5.3.533

Schaefer C., Peters P. and Miller R. K. (eds.) (2015). Drugs During Pregnancy and Lactation: Treatment Options and Risk Assessment. Academic Press, London.

Schreiner V. C., Szöcs E., Bhowmik A. K., Vijver M. G. and Schäfer R. B. (2016). Pesticide mixtures in streams of several European countries and the USA. *Science of the Total Environment*, **573**, 680–689. https://doi.org/10.1016/j.scitotenv.2016.08.163

Schulz R. and Peall S. K. C. (2001). Effectiveness of a constructed wetland for retention of nonpoint-source pesticide pollution in the Lourens River Catchment, South Africa. *Environmental Science and Technology*, **35**, 422–426, https://doi.org/10.1021/es0001198

Seenivasan S. and Muraleedharan N. N. (2009). Residues of lambda-cyhalothrin in tea. *Food and Chemical Toxicology*, **47**(2), 502–505. https://doi.org/10.1016/j.fct.2008.12.010

Shapiro G. D., Dodds L., Arbuckle T. E., Ashley-Martin J., Ettinger A. S., Fisher M., Taback S., Bouchard M. F., Monnier P., Dallaire R., Morisset A. S. and Fraser W. (2016). Exposure to organophosphorus and organochlorine pesticides, perfluoroalkyl substances, and polychlorinated biphenyls in pregnancy and the association with impaired glucose tolerance and gestational diabetes mellitus: the MIREC study. *Environmental Research*, **147**, 71–81, https://doi.org/10.1016/j.envres.2016.01.040

Sharma A., Kumar V., Shahzad B., Tanveer M., Sidhu G. P. S., Handa N., Kholi S. K., Yadav P., Bali A. S., Parihar R. D., Dar O. W., Singh K., Jasrotia S., Bakshi P., Ramakrishnan M., Kumar S., Bhardwaj R. and Thukral A. K. (2019). Worldwide pesticide usage and its impacts on ecosystem, *SN Applied Sciences*, **1**, 1446, https://doi.org/10.1007/s42452-019-1485-1

Shoiful A., Fujita H., Watanabe I. and Honda K. (2013). Concentrations of organochlorine pesticides (OCPs) residues in foodstuffs collected from traditional markets in Indonesia. *Chemosphere*, **90**, 1742–1750, https://doi.org/10.1016/j.chemosphere.2012.10.022

Simpson C. D., Wilcock R. J., Smith T. J., Wilkins A. L. and Langdon A. G. (1995). Determination of octanol-water partition coefficients for the major components of technical chlordane. *Bulletin of Environmental Contamination and Toxicology*, **55**, 149–153, https://doi.org/10.1007/BF00212402

Sivaperumal P., Anand P. and Riddhi L. (2015). Rapid determination of pesticide residues in fruits and vegetables, using ultra-high-performance liquid chromatography/time-of-flight mass spectrometry. *Food Chemistry*, **168**, 356–365, https://doi.org/10.1016/j.foodchem.2014.07.072

Sjerps R. M. A., Kooij P. J. F., Loon A. V. and Wezel A. P. V. (2019). Occurrence of pesticides in Dutch drinking water sources, *Chemosphere*, **235**, 510–518, https://doi.org/10.1016/j.chemosphere.2019.06.207

Sudo M., Kunimatsu T. and Okubo T. (2002). Concentration and loading of pesticide residues in Lake Biwa basin (Japan). *Water Research*, **36**, 315–329, https://doi.org/10.1016/S0043-1354(01)00196-8

Sun S., Chen Y., Lin Y. and An D. (2018). Occurrence, spatial distribution, and seasonal variation of emerging trace organic pollutants in source water for Shanghai, China. *Science of The Total Environment*, **639**, 1–7. https://doi.org/10.1016/j.scitotenv.2018.05.089

Tomlin C. D. S. (ed.) (1994). 10th edn, The Pesticide Manual – World Compendium. The British Crop Protection Council, Surrey.

Torstensson L. and Castillo M. P. (1997). Use of biobeds in Sweden to minimize environmental spillage from agricultural spraying equipment. *Pestic Outlook*, **8**, 24–27.

United States Environmental Protection Agency (1998). Status of Pesticides in Registration, Reregistration, and Special Review, USEPA 738-R-98-002, Washington, DC, p. 44.

United States Environmental Protection Agency (2020). Integrated Risk Information System (IRIS): Summary on beta-Hexachlorocyclohexane (beta-HCH) (319-85-7). Available from: http://www.epa.gov/iris/ (accessed 13 April 2020).

Vammen K., Vaux H., Roldán G., González E., Tundisi J., Izurieta R. and Fabrega J. (2019). Water Quality in the Americas. Risks and Opportunities. Inter-American Network of Academies of Sciences, México.

Wagner V., Antunes P. M., Irvine M. and Nelson C. R. (2017). Herbicide usage for invasive non-native plant management in wildlandareas of North America. *Journal of Applied Ecology*, **54**(1), 198–204, https://doi.org/10.1111/1365-2664.12711

Wanatabe S., Ito S., Kamata Y., Omoda N., Yamazaki T., Munakata H., Kaneko T. and Yuasa Y. (2001). Development of competitive enzyme-linked immunosorbent assays (ELISAs) based on monoclonal antibodies for chloronicotinoid insecticides imidacloprid and acetamiprid. *Analytica Chimica Acta*, **427**, 211–219, https://doi.org/10.1016/S0003-2670(00)01126-0

Wang H. S., Sthiannopkao S., Du J., Chen Z. J., Kim K. W., Yasin M. S. M., Hashim J. H., Wong C. K. C. and Wong M. H. (2011). Daily intake and human risk assessment of organochlorine pesticides (OCPs) based on Cambodian market basket data. *Journal of Hazardous Materials*, **192**(3), 1441–1449, https://doi.org/10.1016/j.jhazmat.2011.06.062

Watanabe E., Baba K., Eun H. and Miyake S. (2007). Application of a commercial immunoassay to the direct determination of insecticide imidacloprid in fruit juices. *Food Chemistry*, **102**, 745–750, https://doi.org/10.1016/j.foodchem.2006.06.031

Winkler R. (2001). Konzept zur Bewertung des Eintrags von Pflanzenschutzmitteln in Oberflächen-und Grundwasser unter besonderer Berücksichtigung des Oberflächenabflusses (Dokumentation zum Modell EXPOSIT). Umweltbundesamt; [in German. www.bvl.bund.de/pflanzenschutz/Zulassung/Umwelt EXPOSITDok.pdf] (accessed 1 March 2021).

Wolejko E., Jablonska-Trypuc A., Wydro U., Butarewicz A. and Lozowicka B. (2020). Soil biological activity as an indicator of soil pollution with pesticides – A review. *Applied Soil Ecology*, **147**,103356.

Xu X., Yang H., Li Q., Yang B., Wang X. and Lee F. S. C. (2007). Residues of organochlorine pesticides in near shore waters of LaiZhou Bay and JiaoZhou Bay, shandong Peninsula, China. *Chemosphere*, **68**, 126–139, https://doi.org/10.1016/j.chemosphere.2006.12.021

Xu W., Wang X. and Cai Z. (2013). Analytical chemistry of the persistent organic pollutants identified in the Stockholm convention: a review. *Analytica Chimica Acta*, **790**, 1–13, https://doi.org/10.1016/j.aca.2013.04.026

Xu M., Huang H., Li N., Li F., Wang D. and Luo Q. (2019). Occurrence and ecological risk of pharmaceuticals and personal care products (PPCPs) and pesticides in typical surface watersheds, China. *Ecotoxicology and Environmental Safety*, **175**, 289–298, https://doi.org/10.1016/j.ecoenv.2019.01.131

doi: 10.2166/9781789061123_045

Chapter 3

Long term studies in Lake Pátzcuaro, eutrophication and recovery efforts during one decade

L. Bravo-Inclán[1,*], J. J. Sánchez-Chávez[1], A. C. Tomasini-Ortiz[2], R. González-Villela[1], M. Mijangos-Carro[1] and F. Bernal-Brooks[3]

[1]Department of Hydrobiology and Environmental Assessment, Mexican Institute of Water Technology (IMTA), Jiutepec, Morelos 62550, Mexico
[2]Department of Waste Water Treatment, IMTA, Jiutepec, Morelos, Mexico
[3]Instituto de Investigaciones sobre los Recursos Naturales, Universidad Michoacana de San Nicolás de Hidalgo, Morelia, Michoacán 58830, Mexico
*lubravo@tlaloc.imta.mx

Keywords: Eutrophication; Lake Pátzcuaro; trophic state; tropical lake; water quality

3.1 INTRODUCTION

In Mexico there are only 17 water bodies with areas greater than 100 km^2 and just eight of them are natural. Some remarkable lakes are: Chapala in Jalisco, and Cuitzeo and Pátzcuaro in Michoacán which are located in the same region; they all form part of the western zone of the Trans-Mexican Volcanic Belt. National limnological studies have mainly focused on water bodies like Chapala and Pátzcuaro, the first and third largest lakes in Mexico (Alcocer & Bernal-Brooks, 2010).

The constant nutrient enrichment of the water bodies, driven by increased inputs of phosphorus and nitrogen, has in recent years been recognized as a major issue, with the consequent problems associated with eutrophication, excessive growths of algal communities and aquatic weeds, which reduce water quality, affect aquatic communities and human health, and limit the uses of the lake (Zhang *et al.*, 2018). Lakes and reservoirs are coming under increasing pressure through development for uses like agriculture, industry, drinking water, fishing habitats and hydroelectricity generation. This is particularly common in tropical countries. The increased use of fertilizers and intensive animal husbandry has resulted in order-of-magnitude increases in nutrient inputs into surface and coastal waters (Novotny, 2003).

The classification of lakes and reservoirs according to water quality and trophic state has been an important issue in applied limnology. This lake classification is useful for regional comparison of water bodies, evaluation of management projects and setting of future goals (Cooke *et al.*, 2005; Nürnberg, 1996).

Between 2003 and 2007, the Lake Pátzcuaro (LP) recovery program included various projects in a multidisciplinary and integrated basin/lake environmental recovery plan (García-Villanueva, 2007).

Editor(s): Puangrat Kajitvichyanukul and Brian D'Arcy

Bearing these goals in mind, regular monitoring of the lake is particularly important, with the objective of diagnosing its problems and studying the feasibility of recovery measures, and by implementing ecologically sound basin/lake management methods preventing or slowing its eutrophication and gradual deterioration.

The main objective of this study was to characterize the water quality of LP, with results from 37 samplings (6 per year from September 2006 to December 2010, taken about every two months, and two more in October and November 2016), with a particular emphasis on the trophic state changes in the lake. Other limnological characteristics were also included, such as dissolved oxygen profiles.

3.1.1 Study area

LP is located at a high altitude (2035 m a.s.l.), west of the central plateau of Mexico, and lies between 19° 32′ and 19° 42′ N as part of the Trans-Mexican Volcanic Belt. The drainage basin is endorheic and tecto-volcanic, with an estimated surface area of 929 km^2. The lake is warm tropical and the mean evaporation is significantly greater than the precipitation (1393 and 775 mm, respectively). During the study, it had a maximum depth of 10.9 m and a mean depth of 4.7 m, which classifies it as a shallow lake (mean depth <5 m), and it had an almost constant daily mixing (Cooke *et al.*, 2005; Sánchez-Chávez *et al.*, 2011).

The climate in this region is temperate subhumid, with a mean annual temperature of 16.3°C and pronounced dry (November–May) and wet seasons (June–October). The basin population was estimated at between 120 000 and 140 000 (Chacon-Torres, 1993; García-Villanueva, 2007; Quintas *et al.*, 2016). Despite LP's high altitude, its geographical location, south of the Tropic of Cancer (below 23° 26′ latitude N), makes this water body a warm tropical one.

Literature on limnological studies is often more associated with temperate water bodies, which are located above 24° or 25° of north latitude. However, greater parts of the central and south Mexican territory are located in a warm climate. The basic differences between warm (or 'tropical') lakes and reservoirs and the temperate ones are as follows: (a) the criterion to define a warm lake is based on a minimum temperature of 10°C and a minimum annual average of 15°C; (b) in general, the warm lakes' annual cycles are limited to just dry and rainy periods; (c) they exhibit biological processes accelerated by the high temperatures; (d) they display high rates of evaporation and precipitation; and (e) occasionally, nitrogen is the limiting factor (Salas & Martino, 1991). Némery *et al.* (2016) commented that tropical regions are characterized by contrasted wet and dry seasons, and that these water bodies are highly vulnerable to eutrophication, especially during the long dry and warm seasons, when nutrients accumulate and algal blooms increase.

The release of untreated or inefficiently treated wastewater is a common practice in Mexican lakes. In LP, high nutrient pollution levels within the basin are caused by growing domestic-touristic sources. The development of wastewater treatment infrastructure in our country remains insufficient or lacks a permanent program to sustain the operation and maintenance expenses in a permanent and sustainable way (Doan *et al.*, 2015; Némery *et al.*, 2016; Tomasini-Ortiz *et al.*, 2016).

The economy of the LP basin depends mainly on tourism, forestry, fishing and hand-craft activities, which makes this zone very important at the national level (Chacon-Torres, 1993; García-Villanueva, 2007). In the last fifty years, the drainage basin and its lake have undergone serious environmental deterioration, such as: soil erosion and runoff from deforested uplands (85% of the basin soils display some degree of degradation); reduction of the surface and depth of the lake; municipal wastewater effluents, originated from fifteen towns, including Pátzcuaro and Quiroga cities; diffuse pollution concerns, caused by population, agriculture and livestock increases; and lack of environmental culture and a population-shared vision about the problems of the drainage basin (García-Villanueva, 2007; Mijangos-Carro *et al.*, 2008). Also, LP has an overexploited multi-species fishery. Fishery yields from the lake peaked in 1988 but have drastically declined to near record lows since the last decade of the previous century, as a consequence of a decrease in water quality, the introduction of non-native species and overfishing (Ramírez-Herrejón *et al.*, 2014).

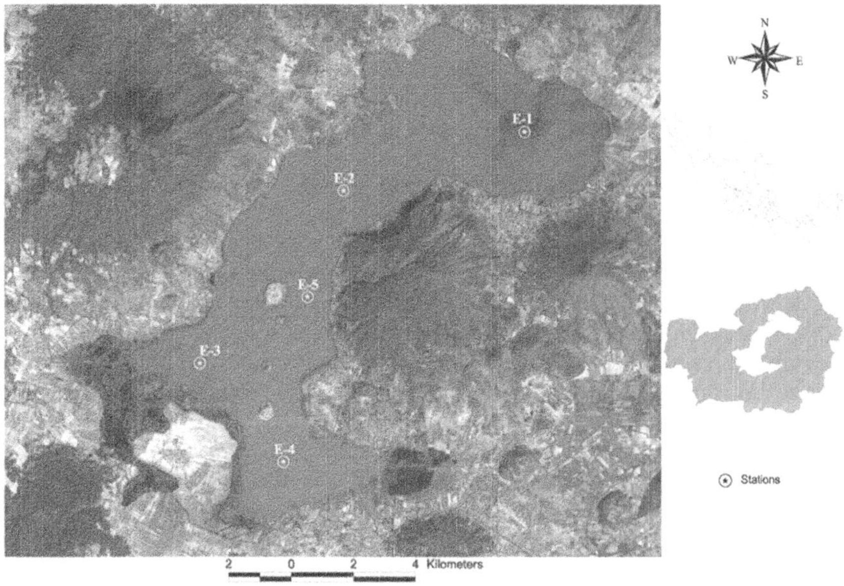

Figure 3.1 Distribution of sampling points in Lake Pátzcuaro.

3.2 MATERIALS AND METHODS

Based on the lake morphology and previous studies, five surface sampling stations were established (Figure 3.1). The sampling methods used in this study were described by Salas and Martino (1991) with modifications proposed by Olvera-Viascán *et al.* (1998). These methods are based on the methodology of eutrophication assessment of warm water tropical lakes. Most of the samples were taken in plastic containers that had been rinsed with the water to be sampled, and were held at 4°C during transport to the laboratory.

In every station, field determinations and Secchi disk (SD) readings were made and two water samples were taken: for chlorophyll *a* (Chl *a*), an integrated one using the hose method (or a Van Dorn bottle) from the surface to three times the Secchi disk, and, for the rest of the parameters, a second using a Van Dorn bottle from four depths (e.g., for a 3 m depth, water was collected at subsurface, 0.5, 1.5, and 2 m).

Analyses of physical and chemical variables – total phosphorus (TP), total nitrogen (TN), Chl *a*, and hardness – were conducted in the laboratory using standard methods (APHA, 2012). Profiles were performed for temperature, dissolved oxygen, pH and conductivity, using a field multi-parameter YSI models 6000 and 6600, from the surface to the bottom of every station and at 0.5 m intervals and at 0.5 m above the depth. Chl *a* pigment was determined using the spectrophotometric method (APHA, 2005). Trophic State Index (TSI, Carlson, 1977) for Secchi disk, chlorophyll *a* and TP were used, and the TN index calculation was taken from Kratzer and Brezonik (1981).

3.3 RESULTS AND DISCUSSION

The lake was classified as warm tropical polymictic, with frequent mixing of the water column; the mixing action is almost daily-continuous due to wind action and it presents the same water quality through the whole water column, as observed in 1983–1990 (Chacon-Torres, 1993). The temperature profile shows no change with depth, and during January and February, the values fluctuated between 17.0 to 15.4°C, with a slight increase in the first metre of depth, to a maximum temperature of 22.2 to 20.0°C in the summer months (June to August). Also, LP shows a largely well mixed and highly oxygenated water body, with an average surface value of 6.77 ± 1.77 mg L^{-1}; near the sediment water interface (0.5 to ~0.15 m above) and during October and November, there is some degree of hypoxia, with values between 4 and 1 mg L^{-1}.

The lake presented a hardwater quality (mean total hardness = 175.2 ± 17.8 mg L^{-1} $CaCO_3$) alkaline (median pH = 8.72 ± 0.43 units; total alkalinity = 472 ± 42 mg L^{-1}) and, like the Lake Chapala, it presents remarkable turbidity, caused by sediments constantly kept in suspension by wind. High nutrient concentrations in the lake have led to degraded water quality, resulting in growths of cyanobacteria, as well as floating aquatic macrophytes such as water hyacinth, and marsh-like plants such as cattail (*Typha* sp.) and reed (*Phragmites australis*) in the south of the lake. From 1930 to 2011, there was a downward trend in the depth of the lake (~6.5 m, equal to 38.2% reduction), leading to a loss of the aquatic zone and a steady increment of the terrestrial zone, which has been more acute in the south of the lake (Sánchez-Chávez *et al.*, 2011).

During the study period the phytoplankton was dominated by blue–green algae such as *Aphanizomenon gracile, Anabaena planctonica* and *Microcystis aeuroginosa*; these species could lead to the presence of toxins (e.g., Microcystin-LR), as well as odor, fisheries impacts and aesthetic problems in the lake (Berry *et al.*, 2011; Tomasini-Ortiz *et al.*, 2012). During the years from 2009 to 2011, Tomasini-Ortiz *et al.* (2012) found steady increases in the number of cells of cyanobacteria blooms in the lake (>10 000 000 cells/mL, which is regarded by the WHO as 'very high' in terms of human health risk). Berry *et al.* (2011) wrote that the dense bloom of cyanobacteria is particularly notorious on the northeastern side of the lake, persisting during the morning as a visible scum on the surface of the water, until it is dissipated by the afternoon wind-induced mixing.

Table 3.1 shows the trophic lake limits reported by different authors and based on several water bodies around the world. For LP, the Chl *a* concentration with a mean annual level of 33.5 μg L^{-1} surpasses the eutrophic–hypereutrophic (25 μg L^{-1}) (Nürnberg, 1996). In addition, the Secchi disk values, with a mean of 0.26 m, confirm the hypereutrophic (HE) state of the lake. Mean values of TP and TN were at 171 and 2213 μg L^{-1}, respectively. The TP values render the lake as hypereutrophic, but the TN data suggest a eutrophic condition for the waterbody (Table 3.2). Secchi disk and TP values do show a high probability of the lake being HE, although there are still some indications of being

Table 3.1 Trophic limits for lakes and reservoirs.

Variable	Limits			Comments	References
	O-M	M-E	E-H		
Total phosphorus (μg L^{-1})	10	30	100	1	Nürnberg (1996)
	30	**70**	–	2,3	Salas and Martino (1991)
Total nitrogen (μg L^{-1})	310	920	2940	2	Kratzer and Brezonik (1981)
Chlorophyll *a* (μg L^{-1})	3.5	9	25	1	Nürnberg (1996)
Secchi disk transparency (m)	4	2	1	1	Nürnberg (1996)
	6	3	1.5	2	Nürnberg (1996)

O-M, transition between oligo- and mesotrophic; M-E, transition between meso- and eutrophic; E-H, transition between eu- and hypereutrophic.

[1]Average epilimnetic summer values.

[2]Annual average values.

[3]Data derived from American tropical warm water lakes.

Table 3.2 Results of four variables and trophic state classification according to the proposed limits.

Parameter	Mean ± Standard Deviation	Range	Trophic State Classification[1]	
			Probability A	Probability B
Secchi disk (m)	0.26±0.06	0.10–0.41	HE – 100%	–
Total phosphorus (μg L^{-1})	171±98	40–540	HE – 88%	E – 12%
Chlorophyll *a* (μg L^{-1})	33.5±40.2	1.8–246	HE – 53%	E – 47%
Total nitrogen (μg L^{-1})	2213±755	658–4517	E – 90%	HE – 8%

[1]Classification of trophic state limits according to Salas and Martino (1991) (for TP); Nürnberg (1996) (for Chl *a* and SD) and Kratzer and Brezonik (1981) (for TN).

eutrophic (E). For instance, the Chl *a* and TN results still imply a 47 and 90% possibility, respectively, that the lake is E.

The relatively low depth of this lake (mean depth of 4.7 m), and its long water residence time (about 10 years according to Chacon-Torres (1993); Bravo-Inclán *et al.* (2011) suggested LP had a longer T_w, between 10 years to $\approx\infty$), together with the closed nature of the basin, the polymictic status, the wind action and the increasing point and diffuse pollution, explains the extremely low transparency values, and the habitat degradation caused by the lower flushing capability of a body of water with longer water renovation and hence, more solids and nutrient input. During the 1970s Vollenweider showed the relevance of water renewal, which is an important modifier of the nutrient loading which affects nutrient water concentrations and sediment accumulation in a water body; LP is naturally affected by the high deficiency of water renewal process (Doan *et al.*, 2015; Némery *et al.*, 2016).

Table 3.3 contains data of four parameters (Chl *a*, SD, TP and TN) evaluated with the Trophic State Index (TSI) (Carlson, 1977; Kratzer & Brezonik, 1981, used only for the TSI of N). The TSI using the four parameters gave an annual average of 70.2 ± 9.1 units (eutrophic state), which is higher than the trophic state found between 1983 and Chacon-Torres (1993); this author estimated a meso-eutrophic condition for LP, with a mean annual TP value of 64 ± 19 μg L^{-1}. On closer inspection it was found that the values of TP and SD were in higher and similar ranges of 62–90 and 75–84 units, respectively; but the Chl *a* values were consistently lower, within a range of 48–69 units. Figure 3.2 shows the monthly TSI values of the three previously mentioned parameters. The TP trend in the period was increasing, with the highest values in the summer and autumn of 2010, as well as in October 2016. Figure 3.3 displays the trend of TN and TP monthly averages at the lake surface. For TN, there is an increment of values in the first four years, but 2010 and 2016 do not follow this tendency, as they show lower values. On the other hand, except for 2010, there is a steady increment, and more extended variations of TP values (and Secchi disk values, Figure 3.2) and of the trophic state condition over time; the TP increment also suggested a change to an N limitation. It is worth noting that, with a slight

Table 3.3 Results of the trophic state Index (TSI) for four parameters (Chl *a*, SD, TP and TN), parameters are arranged in order.

Parameter	Mean	Range
TSI–SD	79.4 ± 2.0	75.0–83.8
TSI–TP	76.0 ± 6.0	62.4–90.9
TSI–TN	65.3 ± 3.2	59.7–73.3
TSI–Chl *a*	60.0 ± 5.8	48.3–69.5
Average total value:	**70.2 ± 9.1**	

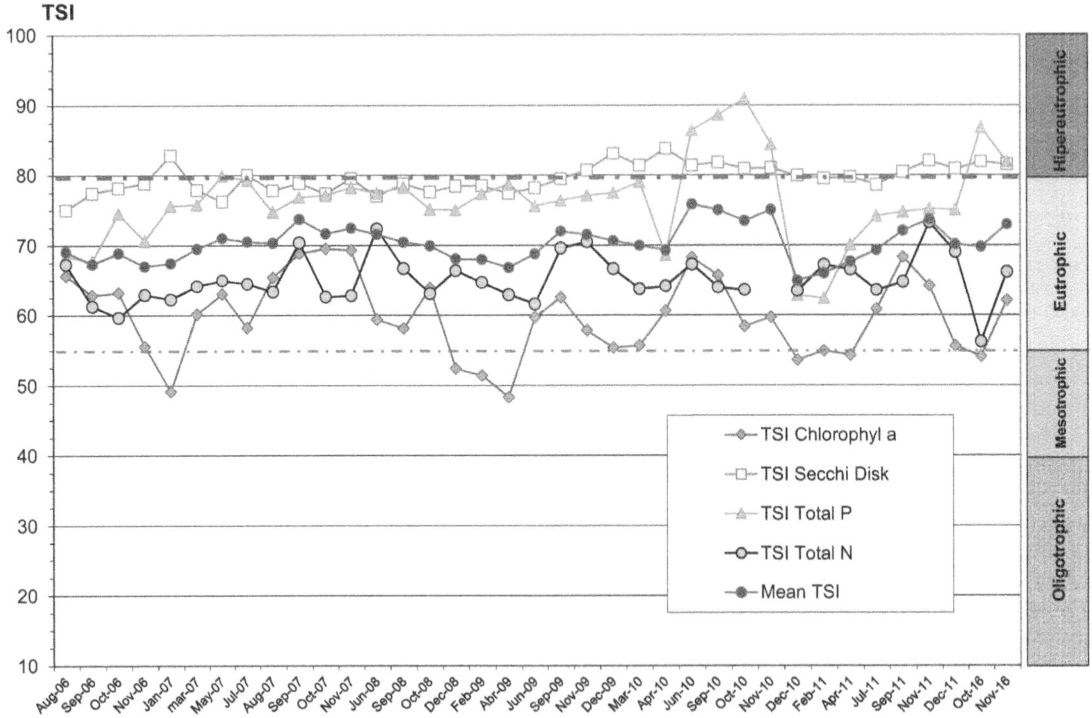

Figure 3.2 Plot of Trophic State Index (TSI), including month average values of chlorophyll a, Secchi disk and total phosphorus, data from August 2006 to November 2016.

displacement in first place in favour of Chl *a*, TSI values for Chl *a* and TP more or less follow with the same up and down trend.

To assess the interplay of nutrients and trophic state, the TN to TP 9:1 ratio was utilized (proposed by Vollenweider in 1983 according to Salas and Martino (1991)). Therefore, lakes with TN to TP ratios lower than 9 units are potentially nitrogen limited (Bravo-Inclán *et al.*, 2011). It is worth saying that Bernal-Brooks *et al.* (2016) suggested that $N:P < 16$ would be evidence of N limitation. Figure 3.4 presents the TN to TP ratio in this seven-year study. Except for the unexpected increment shown in the year 2011, the trend in LP shows a clear decrease in the ratio, which favors a potential probability for N limitation with time (values were 9.7 in 2010, and 7.3 units in 2016 and a mean global value of 14.5 units). These findings are in agreement with the increment of cyanobacteria blooms and trophic level increase in the water body, and also in agreement with González-Villela *et al.* (2012) who, through a correlation and regression analysis for the years of 2006–2011, found significant positive trends for turbidity, conductivity, TN and TP in LP.

Bernal-Brooks *et al.* (2016) conducted a study to reveal the limiting nutrient of five water bodies (four lakes and one reservoir) located in Michoacán state with two different methodologies and data from 1998 and 2006–2010 (numbers from this study): (a) the N:P ratio and (b) bioassays with the use of algal growth potential assay. For LP, they found that there is a change from N and P limitation to an N limitation (three out of four stations); also, they found an increase in P loading associated with this N limitation change and a scenario of habitat stress for the local species.

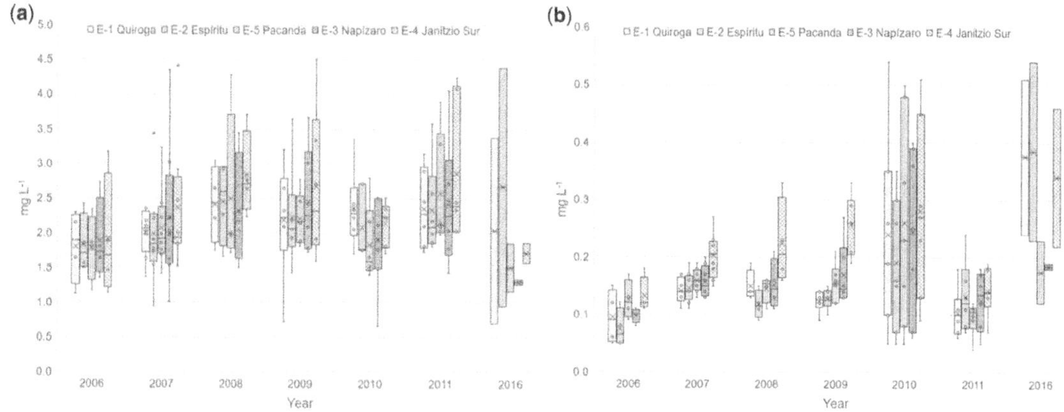

Figure 3.3 Month average values of total nitrogen (a) and total phosphorous (b), trends in Lake Pátzcuaro.

Zhang *et al.* (2018) commented that TN:TP ratios play an important role in aquatic ecosystems; also, they added that when eutrophication increases and the TN:TP ratio is low, cyanobacteria become dominant, as in LP. They further suggested that, with eutrophication increase, the ratio of TN/TP could be correlated with N loss due to denitrification induced by P accumulation and P increase in lakes, due to strong P release from anoxic sediments.

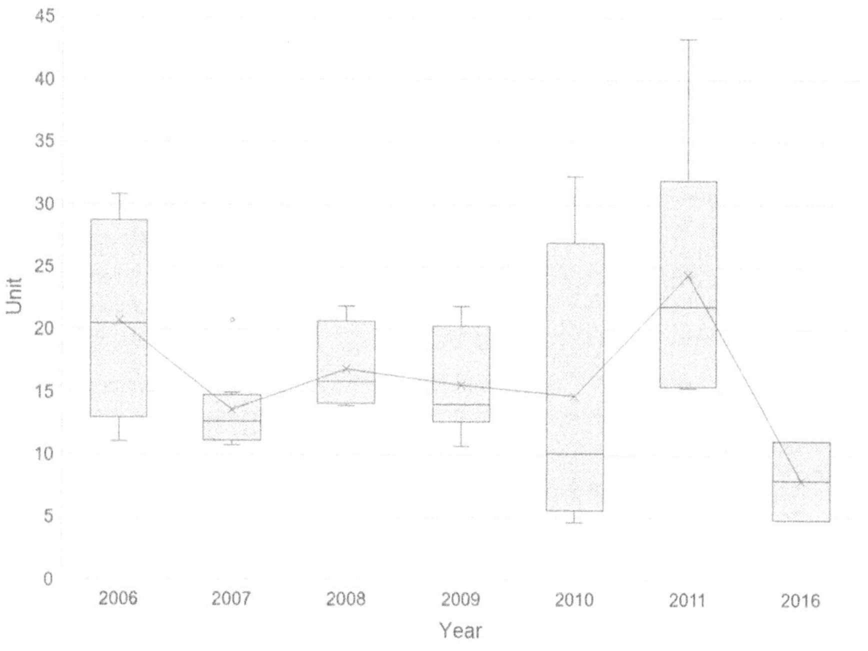

Figure 3.4 Lake Pátzcuaro TN:TP annual ratio change.

3.4 CONCLUSIONS

TSI–Chl a results are one of the most reliable and simple water productivity indexes; and in the case of LP this shows a eutrophic to hypereutrophic system. Chl *a* used with the results of Secchi disk values, and TP and TN concentrations, aid the identification of medium- and long-term trends in the recovery of a water body. Other conclusions are:

- Lake Pátzcuaro is a high-altitude, warm and polymictic waterbody (with almost daily mixing), it does not present a thermocline, and had the presence of an almost oxic layer.
- The lake has a remarkable turbidity caused by sediments constantly kept in suspension by wind-driven forces, besides of the fact of being a shallow lake.
- The high values of total phosphorus and chlorophyll, and relatively low values of total nitrogen and Secchi disk, as well as the presence of cyanobacteria, but without bottom anoxia, suggest that the lake is eutrophic, but with a high probability of a hypereutrophic state.
- The greatest pollution in the lake was found during summer and autumn, and it has a greater affect on the southern part of the lake; also, the wind force pushes cyanobacteria blooms up to the northeastern side of the lake.

The classification of lakes, according to water quality and trophic state, is useful for the evaluation of management projects and the setting of future rehabilitation goals. In the case of LP nutrients, the mean annual TN (as with other water bodies in Mexico) seems to be a better trophic state predictor variable compared with the TP values. The information on the trends in water quality parameters in the lake will serve as feedback in the assessment of watershed nutrient load control, including point and diffuse pollution.

ACKNOWLEDGEMENTS

We are grateful with the Gonzalo Rio Arronte Foundation (FGRA), for the financial support for this study, report no. TC-0632.4; this formed part of the 'Environmental Recovery Plan of LP, 2003–2007' (FGRA–IMTA, described in García-Villanueva, 2007).

REFERENCES

Alcocer J. and Bernal-Brooks F. (2010). Limnology in Mexico. *Hydrobiologia*, **644**, 15–68, https://doi.org/10.1007/s10750-010-0211-1

APHA. (2005). Standard Methods for the Examination of Water and Wastewater, 21st edn. American Public Health Association, American Water Works Association, Water Environment Federation, Washington, DC, USA.

APHA. (2012). Standard Methods for the Examination of Water and Wastewater. 22nd edn. American Public Health Association, American Water Works Association, Water Environment Federation, Washington, DC, USA.

Berry J. P., Lee E., Walton K., Wilson A. E. and Bernal-Brooks F. (2011). Bioaccumulation of microcystins by fish associated with a persistent cyanobacterial bloom in lago de pátzcuaro (michoacán, Mexico). *Environmental Toxicology and Chemistry*, **30**(7), 1621–1628, https://doi.org/10.1002/etc.548

Bernal-Brooks F., Sánchez-Chávez J. J., Bravo-Inclán L., Hernández R., Martínez A. K., Lind O. and Dávalos-Lind L. (2016). The algal growth-limiting nutrient of lakes located at Mexico's mesa central. *Journal of Limnology*, **75**(s1), 169–178, https://doi.org/10.4081/jlimnol.2016.1439

Bravo-Inclán L., Olvera-Viascán V., Saldaña-Fabela P., Sánchez-Chávez J. J. and Tomasini-Ortíz A. C. (2011). Trophic state assessment in warm-water tropical lakes and reservoirs of the central region of Mexico. Selected Papers of 14th International Conference of the IWA Diffuse Pollution and Eutrophication Specialist Group (IWA DIPCON 2010), 12–17 September 2010, Québec, Canada, pp. 194–203.

Carlson R. E. (1977). A trophic state index for lakes. *Limnology and Oceanography*, **22**(2), 361–368, https://doi.org/10.4319/lo.1977.22.2.0361

Chacon-Torres A. (1993). Lake pátzcuaro, Mexico: watershed and water quality deterioration in a tropical high-altitude latin American lake. *Lake and Reservoir Management*, **8**(1), 37–47, https://doi.org/10.1080/07438149309354457

Cooke G. D., Welch E. B., Peterson S. A. and Nichols S. A. (2005). Restoration and Management of Lakes and Reservoirs. 3rd edn. CRC/Lewis, Boca Raton.

Doan P. T. K., Némery J., Schmid M. and Gratiot N. (2015). Eutrophication of turbid tropical reservoirs: scenarios of evolution of the reservoir of cointzio, Mexico. *Ecological Informatics*, **29**, 192–205, https://doi.org/10.1016/j.ecoinf.2015.01.006

García-Villanueva N. H. (2007). Memoria ilustrada del programa para la recuperación ambiental de la cuenca del lago de pátzcuaro. Avances 2003–2007. (ilustrated memoirs of the environmental recovery plan of lake pátzcuaro basin. Advances 2003–2007). In: Gonzalo Río Arronte Foundation/Mexican Institute of Water Tech (FGRA-IMTA), N. H. García-Villanueva (ed.), Mexico (In Spanish), 144pp.

González-Villela R., Sánchez-Chávez J., Bravo-Inclán L., Tomasini-Ortiz A. C., Sánchez-Chávez L., Córdova-Rodríguez M. A. and García-Villanueva N. (2012). Multivariate analysis in the climate change studies and water quality of lake pátzcuaro (Mexico). *Journal of Mathematics and System Science*, **2**(8), 469–481.

Kratzer C. R. and Brezonik P. L. (1981). A carlson-type trophic state index for nitrogen in Florida lakes. *Water Resources Bulletin*, **17**(4), 713–715, https://doi.org/10.1111/j.1752-1688.1981.tb01282.x

Mijangos-Carro M., Izurieta-Dávila J., Gómez-Balandra A., Hernández-López R., Huerto-Delgadillo R., Sánchez-Chávez J. and Bravo-Inclán L. (2008). Importance of diffuse pollution control in the pátzcuaro lake basin in Mexico. *Water Science & Technology*, **58**(11), 2179–2186, https://doi.org/10.2166/wst.2008.820

Némery J., Gratiot N., Doan P. T. K., Duvert C., Alvarado-Villanueva R. and Duwig C. (2016). Carbon, nitrogen, phosphorus and sediment sources and retention in a small eutrophic tropical reservoir. *Aquatic Sciences*, **78**,171–189, https://doi.org/10.1007/s00027-015-0416-5

Novotny V. (2003). Water Quality: Diffuse Pollution and Watershed Management. 2nd edn. John Wiley, New York, USA.

Nürnberg G. K. (1996). Trophic state of clear and colored, soft- and hardwater lakes with special consideration of nutrients, anoxia, phytoplankton and fish. *Lake and Reservoir Management*, **12**(4), 432–447, https://doi.org/10.1080/07438149609354283

Olvera-Viascán V., Bravo-Inclán L. and Sánchez-Chávez J. (1998). Aquatic ecology and management assessment in valle de bravo reservoir and its watershed. *Aquatic Ecosystem Health & Management*, **1**(3–4), 277–290, https://doi.org/10.1080/14634989808656924

Quintas I., Gómez-Balandra M. A. and Vervoort W. (2016). Rain variability and population growth to explain historical levels of the pátzcuaro lake in Mexico. *Journal of Water Resource and Protection*, **8**, 168–182, https://doi.org/10.4236/jwarp.2016.82014

Ramírez-Herrejón J. P., Zambrano L., Mercado-Silva N., Torrez-Téllez A., Pineda-García F., Caraveo J. and Balart E. (2014). Long term changes in the fish fauna of lago de pátzcuaro in central méxico. *Latin American Journal of Aquatic Research*, **42**(1), 137–149, https://doi.org/10.3856/vol42-issue1-fulltext-11

Salas H. J. and Martino P. (1991). A simplified phosphorus trophic state model for warm-water tropical lakes. *Water Research*, **25**, 341–350, https://doi.org/10.1016/0043-1354(91)90015-I

Sánchez-Chávez J., Bravo-Inclán L., Tomasini-Ortiz C. and Bernal-Brooks F. (2011). Calidad del agua del lago de pátzcuaro (water quality of lake pátzcuaro). In: Estudio ecosistémico del lago de Pátzcuaro: aportes en gestión ambiental para el fomento del desarrollo sustentable, R. I. Huerto-Delgadillo, S. Vargas-Velázquez, and C. F. Ortíz-Paniagua (eds), IMTA, Mexico (In Spanish), 29–48.

Tomasini-Ortiz A. C., Moeller-Chávez G., Sánchez-Chávez J. J. and Bravo-Inclán L. (2012). Cianobacterias y cianotoxinas en el lago de pátzcuaro, michoacán, méxico (Cyanobacteria and cyanotoxins in lake pátzcuaro, michoacán, Mexico). *Revista AIDIS*, **5**(2), 93–101 (In Spanish).

Tomasini-Ortiz A. C., Bravo-Inclán L., Sánchez-Chávez J. J. and Moeller-Chávez G. (2016). Monitoreo de descargas de aguas residuales y su impacto en el lago de pátzcuaro, méxico (2006–2011) (waste water discharge sampling and its impact in lake pátzcuaro, Mexico). *Revista AIDIS*, **9**(1), 61–74 (In Spanish).

Zhang Y., Song C., Ji L., Liu Y., Xiao J., Cao X. and Zhou Y. (2018). Cause and effect of N/P ratio decline with eutrophication aggravation in shallow lakes. *Science of the Total Environment*, **627**, 1294–1302, https://doi.org/10.1016/j.scitotenv.2018.01.327

doi: 10.2166/9781789061123_055

Chapter 4

Impacts of oil spill incidences on levels of polycyclic aromatic hydrocarbons (PAHs) in coastal areas of Asian Countries

S. Pongpiachan

School of Social & Environmental Development, National Institute of Development Administration (NIDA), 118 Sereethai Road, Klongchan, Bangkapi, Bangkok 10240, Thailand

Keywords: Polycyclic aromatic compounds; Coastal areas; Risk assessment; Terrestrial soils; GC-MS-MS

4.1 INTRODUCTION

Polycyclic Aromatic Hydrocarbons (PAHs) can be classified as Persistent Organic Pollutants (POPs), which are toxic pollutants that can be preserved in various environmental compartments (ChooChuay *et al.*, 2020a, 2020b; Deelaman *et al.*, 2020; Pongpiachan, 2015; Pongpiachan *et al.*, 2010, 2013, 2018, 2019). After they are emitted into the atmosphere, they will rapidly distribute between gaseous and particulate phases (Dachs & Eisenreich, 2000; Kaupp & McLachlan, 1999; Simcik *et al.*, 1998). This gas–particle partitioning mechanism plays a crucial role in governing atmospheric fate of PAHs (Akyüz & Çabuk, 2010; Degrendele *et al.*, 2016; Fernández *et al.*, 2002). Subsequently, PAHs continue to be transported from warmer to colder regions of the planet through the geochemical process called 'global distillation' or 'the grasshopper effect' (Degrendele *et al.*, 2016; Gouin *et al.*, 2004; Wang *et al.*, 2007). Previous studies highlight its adverse health impacts such as lung cancer (Liao *et al.*, 2011; Sarigiannis *et al.*, 2015), breast cancer (Bonner *et al.*, 2005; Parikh & Wei, 2016), impaired lung function in asthmatics and thrombotic effects (ACGIH, 2005), kidney dysfunctions (Singh *et al.*, 2016), redness and inflammation of the skin (Khairy *et al.*, 2009), and other acute symptoms including eye irritation, nausea, vomiting, diarrhea, and confusion (Bølling *et al.*, 2009). In consequence the International Agency for Research on Cancer (IARC) categorizes some PAHs as acknowledged, potentially, or probably carcinogenic to humans (Group 1, 2A or 2B) (IARC, 2010). Among these congeners are benzo[a]pyrene (Group 1), naphthalene, chrysene, benzo[a]anthracene, benzo[k]fluoranthene and benzo[b]fluoranthene (Group 2B) (IARC, 2010). Since they pose a risk to the environment and human health globally, the Stockholm Convention on Persistent Organic Pollutants commits to take appropriate measures to prevent the emissions of PAHs into the environment or at least reduce such emissions as far as is technically feasible and economically acceptable (Lallas, 2001).

It is well known that primary sources of PAHs result from both anthropogenic (e.g., vehicle releases, industrial emissions, domestic heating) and natural emissions (e.g., forest fires, volcano eruptions) (Kim *et al.*, 2003; Kozak *et al.*, 2017; Lohmann *et al.*, 2000; Pongpiachan *et al.*, 2017; Slezakova *et al.*, 2013; Yang *et al.*, 1998, 2002). In contrast, secondary sources occur from the generation of a pollutant in the ambient air from both the homogeneous and heterogeneous chemical reactions of precursors emitted from air pollution contributors, for example nitro-PAH formation from heterogeneous reactions of PAHs with NO_2, NO_3/N_2O_5, and OH radicals (Jariyasopit *et al.*, 2014; Zhang *et al.*, 2010, 2011). Conclusively, a re-volatilization source results from both primary and secondary pollutants depositing on terrestrial soils (e.g., resuspended particles) or lake, riverine, and marine surfaces, followed by re-volatilization to the ambient air (Cabrerizo *et al.*, 2014; Ma *et al.*, 2011).

Over the past few decades, the prominence of oil spill incidents has raised public concerns about the accumulation of PAHs in coastal and marine deposits around the world (Nwaichi *et al.*, 2016; Onojake *et al.*, 2013; Pongpiachan *et al.*, 2018; Seegar *et al.*, 2015; Veiga *et al.*, 2010). This process is a result of several unique features and physicochemical properties of PAHs. Firstly, PAHs are comparatively persistent in the environment; this is a consequence of their relatively strong aromatic ring structure which means they are not particularly vulnerable to photochemical decomposition and homogeneous/heterogeneous chemical reactions, enabling them to possess long environmental lifetimes (Arey, 1998; Malloci *et al.*, 2003; Miller & Olejnik, 2001). Secondly, PAHs are hydrophobic organic compounds, indicating that convective precipitation is not effective in transferring these toxic congeners from atmosphere to terrestrial surface (Lim *et al.*, 2007; Ozaki *et al.*, 2006; Park *et al.*, 2002). Thirdly, PAHs can undergo both bioaccumulation and biomagnification, suggesting that animals near the top of the food chain are most affected because of these processes (D'adamo *et al.*, 1997; Froehner *et al.*, 2011; Tao *et al.*, 2018). Although many investigations have focused on clarifying the aquatic fate of PAHs in riverine sediments, marine deposits and coastal soils, little is known about the chemical characterization, diagnostic binary ratios, and ecotoxicological studies associated with PAH congeners. Overall, the major aims of this chapter are (i) to comprehensively review both spatial and temporal distribution of PAH congeners in oil spills contaminating coastal areas around the world, (ii) to investigate the diagnostic binary ratios and source apportionment of PAHs in the polluted sites, and (iii) to both quantitatively and qualitatively evaluate the ecotoxicity of PAHs in the oil spill areas with the assistance of risk assessment models.

4.2 MATERIALS AND METHODS

4.2.1 Sampling equipment

4.2.1.1 Terrestrial soils

In order to perform geotechnical surveys, it is crucial to select proper soil sampling equipment, which is suitable for the required investigations. Generally, a geotechnical investigation can be categorized in two ways: *surface survey* or *subsurface examination* of a target observatory site. Surface exploration can be defined as any geotechnical activity which will involve geophysical methods, geologic mapping, and photogrammetry. Subsurface survey can be considered as any geological activities, which normally includes soil sampling and laboratory tests of the collected soil samples (Figure 4.1). To understand the physicochemical properties of subsurface soils and rocks, the techniques employed include a shovel test pit, trenching, boring, Hand- or Machine-Driven Auger, Continuous Flight Auger, Split-spoon/SPT Sampler, Modified California Sampler, Shelby Tube Sampler and Piston samplers (Pongpiachan, 2011).

4.2.1.2 Riverine and marine deposits

Previous studies have used many techniques for collecting riverine and marine sediments, including AMS Shallow Water Bottom Dredge, Deep Water Bottom Dredge, Ekman Dredge, LaMotte Bottom Sampling Dredge, and Integrating Sediment Sampler (Figure 4.1) (Pongpiachan, 2011). As illustrated in Figure 4.2, Ekman Dredges were used for the collection of marine sediment samples as a part of the Thai–German Cooperative Research and Capacity Building: Ocean Geoscience, Marine Ecology and Engineering for Natural Hazard Management and Sustainable Use of Marine Resources: Tsunami Deposits in Near-Shore and Coastal Waters of Thailand (TUNWAT) (Pongpiachan *et al.*, 2013).

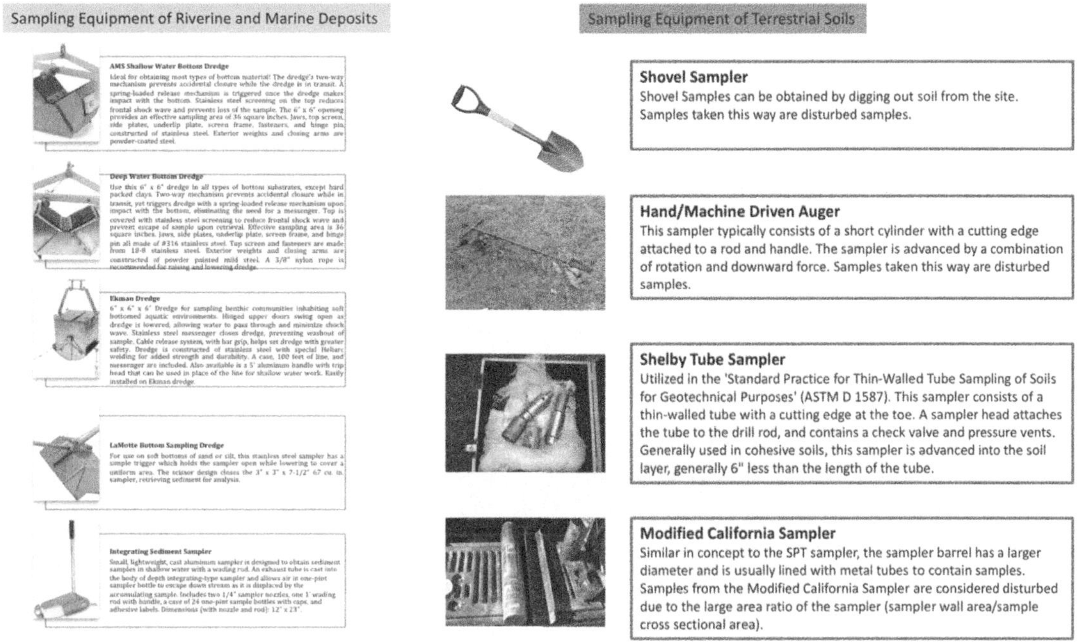

Figure 4.1 Various types of terrestrial soil and sediment samplers (Pongpiachan, 2011).

4.2.2 Chemical extraction of PAHs

All organic solvents (e.g., DCM, hexane, cyclohexane, nonane and toluene) should theoretically be HPLC grade. Many studies associated with chemical characterization of PAHs in environmental samples (Pongpiachan *et al.*, 2013, 2017, 2018, 2019) have used a cocktail of 15 PAHs Norwegian Standard (NS 9815: S-4008-100-T) (phenanthrene (Phe), anthracene (An), fluoranthene (Fluo), pyrene (Pyr), 11h-benzo[a]fluorene (11H-B[a]F), 11h-benzo[b]fluorene (11H-B[b]F), benz[a]anthracene (B[a]A), chrysene (Chry), benzo[b]fluoranthene (B[b]F), benzo[k]fluoranthene (B[k]F), benzo[a]pyrene (B[a]P), benzo[e]pyrene (B[e]P), indeno[12,3-cd]pyrene (Ind), dibenz[a,h]anthracene (D[a,h]A), benzo[g,h,i]perylene (B[g,h,i]P); each 100 μg mL^{-1} in toluene: unit: 1 × 1 mL) and a mix of recovery Internal Standard PAHs (d_{12}-perylene (d_{12}-Per), d_{10}-fluorene (d_{10}-Fl); each 100 μg mL^{-1} in xylene: unit: 1 × 1 mL) in Standard stock solutions of 4 μg mL^{-1} of deuterated PAHs (used as internal standard) and 100 μg mL^{-1} of native PAHs usually prepared in nonane. Working solutions were obtained by appropriate dilution in *n*-cyclohexane. All solutions should be stored in amber coloured vials at −20°C. All materials used (e.g., silica gel, glass and cotton wool etc.) should be Soxhlet extracted with DCM for 24 h and kept dry (in desiccator) until use. All glassware should be cleaned by washing with decanted water before drying at 55°C and rinsed with DCM just before use (Pongpiachan *et al.*, 2009, 2012).

A certain mass of freeze-dried sediment or terrestrial soil samples (e.g., 5 g or 10 g) will generally be transferred to a Teflon-capped bottle and spiked with an internal standard (e.g., d_{10}-phenanthrene) and treated in 20 mL of dichloromethane (DCM) for 15 min in ultrasonic bath (Tian *et al.*, 2008). This ultrasonic extraction was widely used in various studies related with chemical extraction of PAH congeners in sediments and terrestrial soils (Banjoo & Nelson, 2005; Marvin *et al.*, 1992). Alternatively, all terrestrial soils and marine sediments can be Soxhlet extracted, cleaned-up, concentrated, and quantitatively analysed by using Gas Chromatography–Mass Spectrometry (GC–MS) (Chen *et al.*, 2013; Pongpiachan *et al.*, 2013, 2018, 2019). In other recent research sediment samples been partially

Figure 4.2 Tsunami sediment sampling under Thai–German Cooperative Research and Capacity Building: Ocean Geoscience, Marine Ecology and Engineering for Natural Hazard Management and Sustainable Use of Marine Resources: Tsunami Deposits in Near-Shore and Coastal Waters of Thailand (TUNWAT), November 2009 (Project collaboration between National Institute of Development Administration, Chulalongkorn University and The University of Bremen) (Pongpiachan, 2011).

oven-dried at 40°C and mixed with anhydrous sodium sulfate to remove any remaining water, then extracted with methylene chloride by Automated Solvent Extractor (Soliman *et al.*, 2014). In addition, a reciprocal shaker coupled with DCM can also be employed for chemical extraction of PAH compounds in sediment samples (Smith *et al.*, 2008). Further details associated with fractionization, GC–MS analysis and quality control/quality assurance (QA/QC) have previously been mentioned in other studies and will not be discussed here (Pongpiachan *et al.*, 2009, 2012).

4.3 RESULTS AND DISCUSSION

4.3.1 Chemical characterization of PAHs

Table 4.1 displays the statistical descriptions of PAHs in the terrestrial soils, riverine and marine sediments of oil spill affected areas in Asian countries (ACS) (see Figure 4.3) coupled with World Marine Sediments (WMS) reported in previous studies. ACS comprises terrestrial soils, riverine and marine sediments collected from Liaohe Estuary (Lang *et al.*, 2012), Bohai Bay (Jiao *et al.*, 2010), Yellow River Estuary (Hui *et al.*, 2009), Jiaozhou Bay Wetland (Yang *et al.*, 2014), Guan River Estuary (He *et al.*, 2014), Chongming (Wang *et al.*, 2012), Quanzhou Bay (Zhuang *et al.*, 2011), Baguang Wetland (Zeng *et al.*, 2013), Ramsar Samp of Hong Kong (Cai *et al.*, 2013), Dalian, China (Hong *et al.*, 2018), Jahor Strait, Malaysia (Keshavarzifard *et al.*, 2019), West Port, Malaysia (Sany *et al.*, 2014), Bangladesh (Zuloaga *et al.*, 2013), India (Zuloaga *et al.*, 2013), Koh Samed Island, Thailand

Table 4.1 Statistical descriptions of PAH contents collected from oil spill affected coastal areas in Asian Coastal Soils & Sediments (ACS) (this study) and World Marine Sediments (WMS) (Pongpiachan *et al.*, 2013).

	ACS		WMS	
	Average ($n = 5106$)	Stdev ($n = 106$)	Average ($n = 26$)	Stdev ($n = 26$)
Phe	148	254	673	1621
An	30.7	64.4	180	447
Fluo	169	322	802	2285
Pyr	142	277	776	2095
B[a]A	89.6	175	483	1092
Chry	102	203	407	840
B[b]F	226	818	663	1911
B[k]F	149	488	391	623
B[a]P	119	332	376	794
Ind	80.6	244	427	841
D[a,h]A	65.2	172	225	832
B[g,h,i]P	569	3294	347	730
ΣPAHs12	1891	3504	5749	4576

(Pongpiachan *et al.*, 2018), and Sori Island, Korea (Yim *et al.*, 2002). For comparison WMS represents sediments collected from marine and riverine sediments in USA, UK, Hong Kong, China, Italy, Benin, France, Spain, Brazil, and Russia (Pongpiachan *et al.*, 2013). The total concentrations of twelve probably carcinogenic PAHs in ACS ranged from 32 ng g^{-1} to 32 571 ng g^{-1} with an arithmetic mean of 1871 $\pm$ 5533 ng g^{-1}. In this chapter, the $\sum_{12}$PAHs represents to the total of chemically characterized Phe, An, Fluo, Pyr, B[a]A, Chry, B[b]F, B[k]F, B[a]P, Ind, D[a,h]A and B[g,h,i]P.

As displayed in Table 4.1, the ACS $\sum_{12}$PAHs was almost three times lower than those values reported in the WMS $\sum_{12}$PAHs. It is also interesting to note that the decreasing order of PAH congener detected in ACS was B[g,h,i]P > B[b]F > Fluo>B[k]F > Phe>Pyr>B[a]P > Chry>B[a]A > Ind>D[a,h]A > An. In contrast, the decreasing order of PAH content observed in WMS was Fluo>Pyr>Phe>B[b]F > B[a]A > Ind>Chry>B[k]F > B[a]P > B[g,h,i]P > D[a,h]A > An. Since Fluo and Pyr are two PAH congeners that are widely acknowledged as diesel and gasoline markers (Ho *et al.*, 2009), it appears reasonable to interpret that traffic emissions still play a major role in governing PAH contents in global marine deposits. It is also interesting to note that D[a,h]A, which has 4% contribution for both sediments (see Figure 4.4), is typically found in biomass burning particles (de Oliveira Alves *et al.*, 2015; Jenkins *et al.*, 1996). This indicates that the biomass combustion derived PAH congeners are minor of importance in the case of both coastal soils, riverine and marine sediments. The relatively high contribution of B[g,h,i]P detected in ACS can be described by its comparatively low photolysis rate constants (Niu *et al.*, 2007). Under simulated sunlight containing UV-A (315–400 nm) + UV-B (280–315 nm) and UV-A (315–400 nm), the rate constants for the photolysis of B[g,h,i]P were 2.3×10^{3} h^{-1} and 1.8×10^{3} h^{-1}, respectively (Niu *et al.*, 2007). These rate constants are much lower than those of other low molecular weight PAHs (e.g., acenaphthylene, acenaphthene). It is also crucial to underline that sunlight can inhibit biodegradation of pristine and phytane (Bacosa *et al.*, 2015) and thus complicate the PAH profiles related with oil spill incidence in ACS. Although biodegradation can play an important part in altering PAH profiles, photolysis induced by UV is generally acknowledged as one of the most crucial processes of long-term weathering especially for crude oil (Shankar *et al.*, 2015). The fact that the percentage contribution of ΣPAH(5,6) in ACS (i.e., 64%) was comparatively higher than that of WMS (i.e., 42%) also supports this explanation (Figure 4.5).

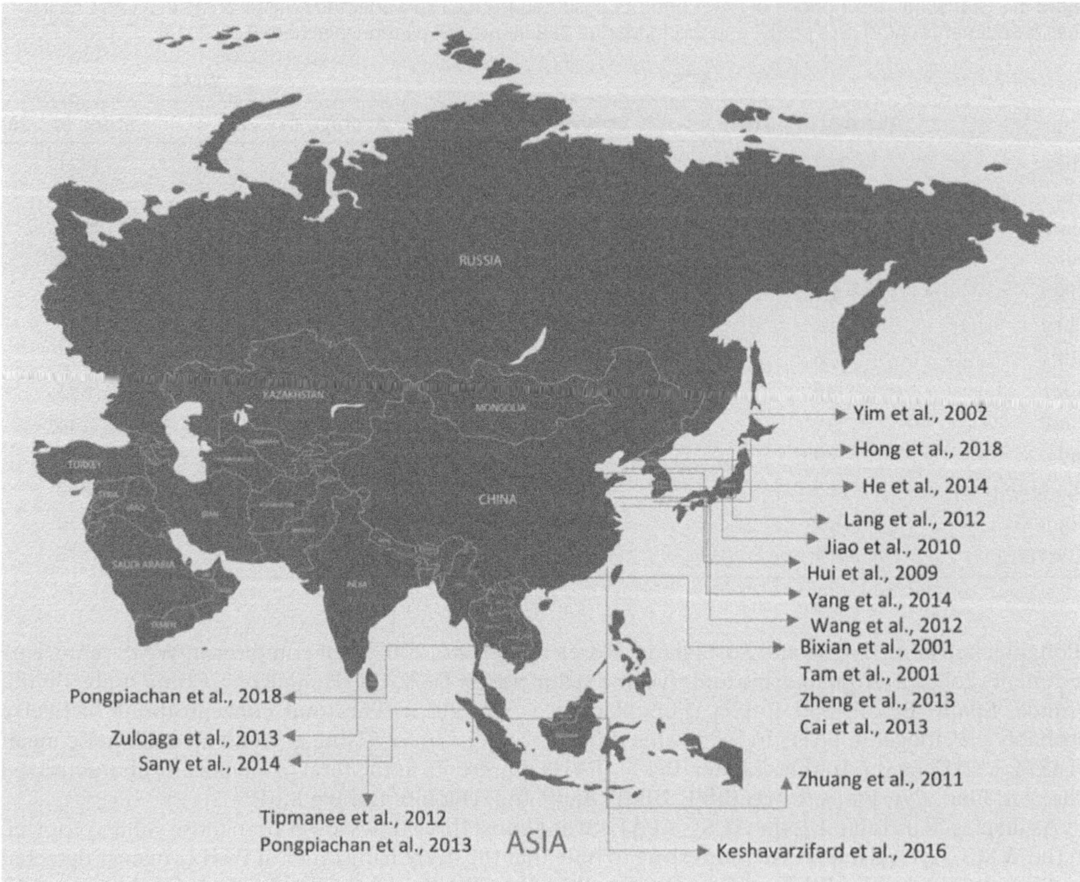

Figure 4.3 A map of sampling sites of PAHs in riverine and marine deposits coupled with coastal soils in Asia.

4.3.2 Diagnostic binary ratios of PAHs

Over the past few decades, numerous studies have used diagnostic binary ratios for identifying potential emission sources of PAH congeners in ambient air, terrestrial soils, riverine sediments and marine deposits (ChooChuay *et al.*, 2020a, 2020b; Deelaman *et al.*, 2020; Pongpiachan *et al.*, 2013, 2018). In spite of its capability in categorizing source contributors, some vagueness related with gas-particle partitioning, photodecomposition, homogeneous/heterogeneous chemical reactions and biodegradation of PAH congeners still remains regarding the accuracy of using this method for source characterization (Pongpiachan, 2016). It appears that the photodecomposition mechanism and traffic emissions have provided small contributions in changing PAH diagnostic binary ratios. In using the PAH ratios it is crucial to pay close attention to details for the data interpretation since NO_3/N_2O_5, OH radicals, and O_3 from traffic emissions can cause significant alterations of the binary ratios (Pongpiachan, 2016). In this study, four different types of diagnostic binary ratios of PAH congeners were applied to assess the impacts of source contributions (see Figure 4.6).

During the past few years, there are copious studies on applications of diagnostic binary ratios for categorizing potential emission sources of PAH congeners such as petroleum, grass, wood, and coal

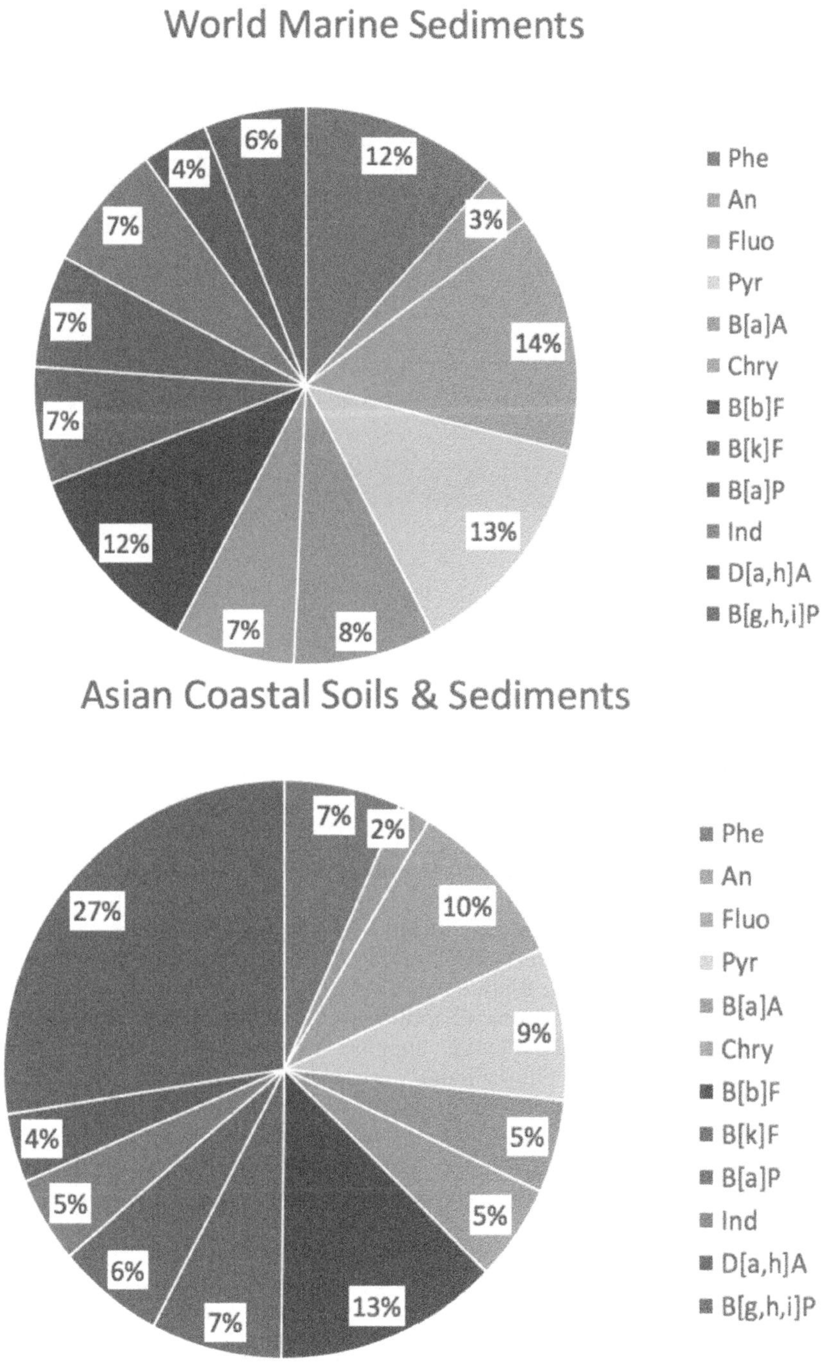

Figure 4.4 Percentage contributions of PAH congeners collected from World Marine Sediments and Asian Coastal Soils and Sediments.

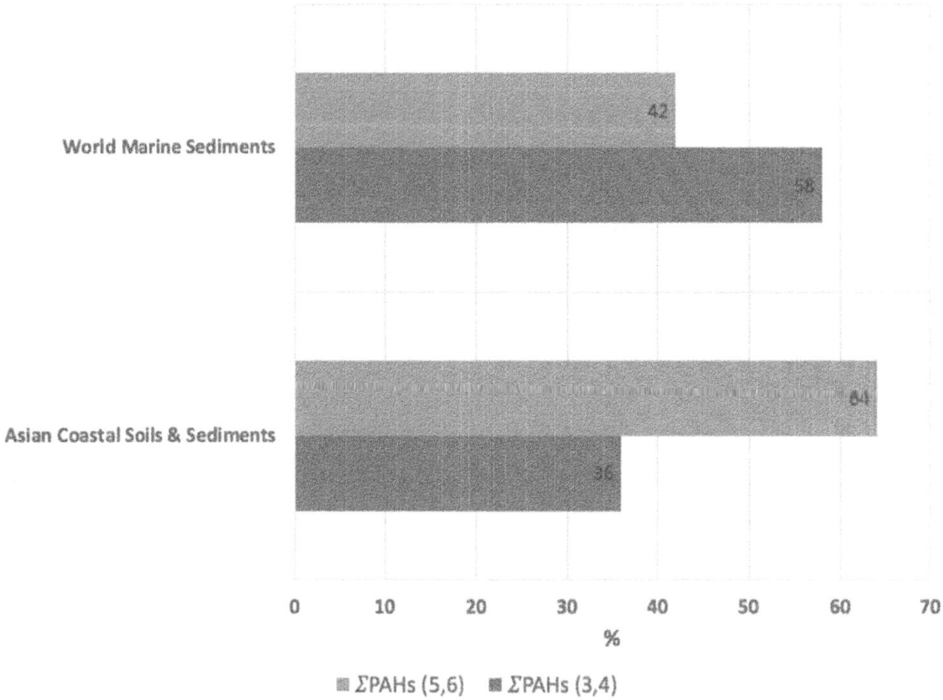

Figure 4.5 Percentage contributions of ΣPAH(5,6) and ΣPAH(3,4) collected from World Marine Sediments and Asian Coastal Soils & Sediments.

combustions in different environmental compartments (ChooChuay *et al.*, 2020a, 2020b; Deelaman *et al.*, 2020; Pongpiachan *et al.*, 2013, 2015, 2017, 2018, 2019). In this chapter, five different types of two-dimensional plots of diagnostic binary ratios were employed for source categorization (e.g., petrogenic and pyrolytic potential sources). As illustrated in Figure 4.6, the diagnostic binary ratios of An/(An + Phe), Fluo/(Fluo + Pyr), B[a]A/(B[a]A + Chry), and Ind/(Ind + B[g,h,i]P) in coastal soils and marine sediments collected from oil spilled area in Asian countries were plotted in two dimensions and compared with those of other chamber experiments (Pongpiachan *et al.*, 2018). Here, the diagnostic binary ratios of Fluo/(Fluo + Pyr) *vs.* B[a]A/(B[a]A + Chry) from ACS samples are almost entirely situated in the areas of pyrolytic source. Equivalently, the two-dimensional plots of Fluo/(Fluo + Pyr) *vs.* Ind/(Ind + B[g,h,i]P) underlined the impacts of biomass and coal combustions in ACS samples. In contrast, the two-dimensional plots of An/(An + Phe) *vs.* B[a]A/(B[a]A + Chry) and An/(An + Phe) *vs.* Fluo/(Fluo + Pyr) highlighted that almost half of ACS samples were potentially influenced by oil spill incidences. This discrepancy can be explained by the fact that there are many possible confounding factors. Firstly, there are some ambiguities in the application of diagnostic binary ratios as previously mentioned. Secondly, numerous potential sources can play an important role in governing PAH profile in terrestrial soils, riverine and marine sediments in oil spill affected coastal areas. In addition, it appears plausible to summarize that oil spill incidence is not the only potential source of PAH congeners in ACS samples but also other pyrogenic combustion categories, highlighting complexity of releasing source strengths in the coastal areas of Asian countries.

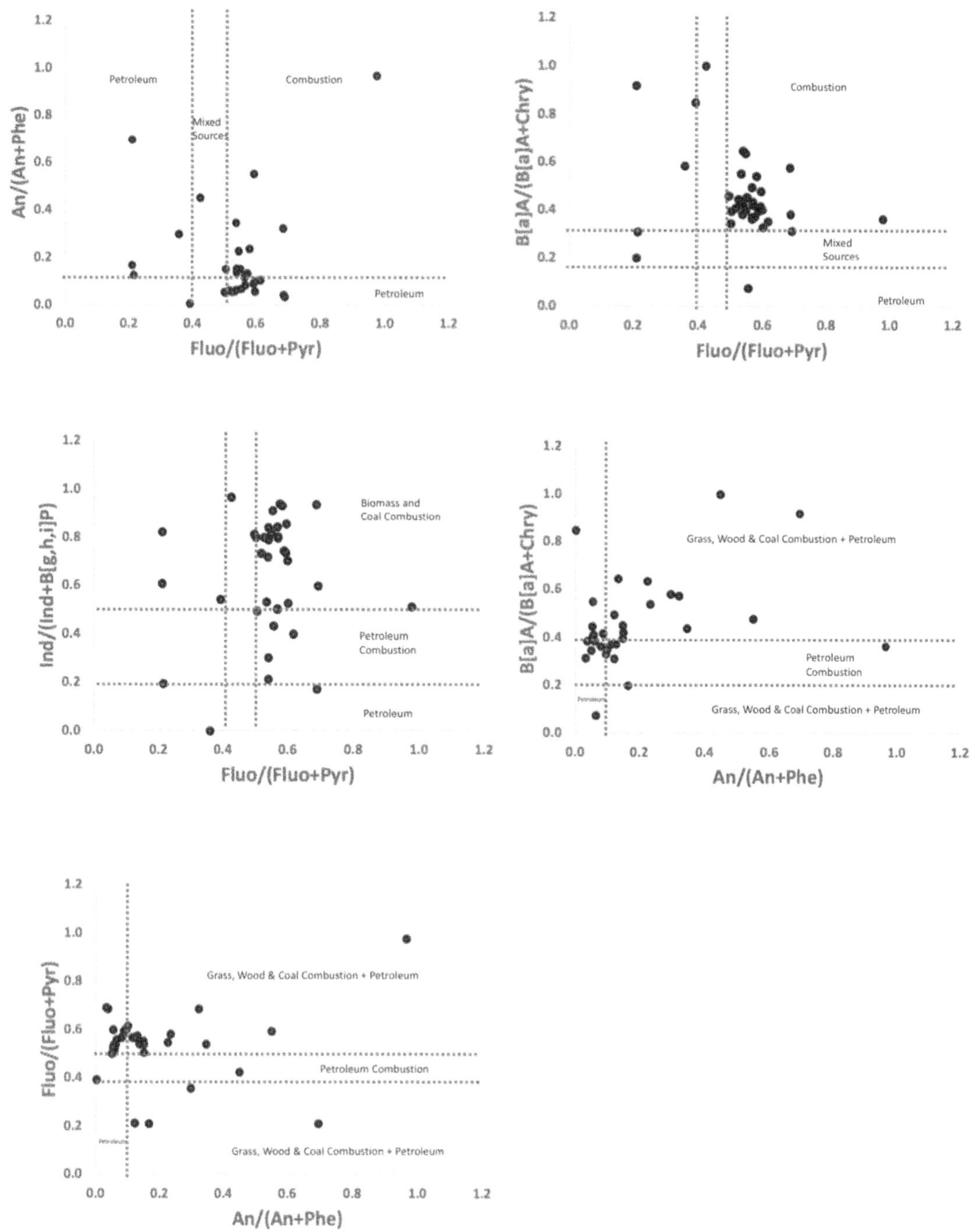

Figure 4.6 Diagnostic binary ratios of PAH congeners obtained from this study in comparison with those of previous studies associated with source characterization.

4.3.3 Risk assessment of PAHs

Ecotoxicological risk assessment has been globally conducted on aerosols, marine sediments, riverine deposits and terrestrial soils over the past few decades (Apiratikul *et al.*, 2020; Bandowe *et al.*, 2014; Chen *et al.*, 2016; Goswami *et al.*, 2016; Pongpiachan *et al.*, 2013). In this chapter, an evaluation of both ecotoxicological cancer and non-cancer risks was performed by applying the following formulas (i.e., Equations 4.1 and 4.2) based on the framework reported by the Agency for Toxic Substances and Disease Registry (ATSDR), which is a federal public health agency of the US Department of Health and Human Services. It is crucial to underline that an assessment of ecotoxicological cancer risks was conducted to evaluate presence of negative human health impact (carcinogenic and non-carcinogenic) as a result of ingestion and/or dermal contact with PAH-contaminated soils and/or deposits.

$$\text{Cancer Risk} = SF \times \left(\frac{CS \times IR \times EF \times ED \times CF}{BW \times AT_c} + \frac{CS \times SA \times AF \times EF \times ED \times DAF \times CF}{BW \times AT_c} \right) \quad (4.1)$$

where the abbreviations can be described as follows (Pongpiachan *et al.*, 2013):

- SF: Slope Factor (1.5 mg kg^{-1} body weight $day^{-1})^{-1}$
- CS: Chemical Concentration (ng g^{-1})
- CF: Conversion Factor of 10^{-6} kg mg^{-1}
- EF: Exposure Frequency = 40 days $year^{-1}$
- ED: Exposure Duration = 10 years
- IR: Ingestion Rate = 50 mg sediment day^{-1}
- BW: Body Weight = 53 kg
- AT_c: Averaging Time for cancer risk = 25 550 days
- AF: Skin Adherence Factor = 0.3 mg sediment cm^{-2}
- DAF: Dermal Absorption Factor = 0.03 (unitless)
- SA: Skin Surface Area (available for contact) = 4700 cm^2

The ecotoxicological assessment of non-cancer risk can be conducted by applying Equation 4.2, where the Hazard Index (HI) reflects the non-cancer risk, which can be calculated by dividing the sum of Average Daily Dose Ingestion (ADD_i: mg kg^{-1} body weight day^{-1}) and Average Daily Dose Dermal (ADD_d: mg kg^{-1} body weight day^{-1}) by the oral reference dose (RfD) (ATSDR, 2005).

$$HI = \frac{((CS \times IR \times EF \times ED \times CF)/(BW \times AT_{NC})) + ((CS \times SA \times AF \times EF \times ED \times DAF \times CF)/(BW \times AT_{NC}))}{RfD} \quad (4.2)$$

where the further abbreviations of incidental ingestion formula (Eq. 4.2) can be written as follows:

- HI: Hazard Index
- RfD: Reference Dose (3.0×10^{-4} mg kg^{-1} body weight day^{-1})
- AT_{NC}: Averaging Time for non-cancer effects = 3650 days.

As shown in Table 4.2, the average cancer risk levels of PAH congeners in ACS varied from 1.25×10^{-6} to 2.33×10^{-5} with an average of 6.44×10^{-6}, which is three times lower than that of WMS. Although the degree of cancer risk that is of interest is a function of body weight, exposure frequency and duration, ingestion rate, and PAH concentration, the US EPA declares excess cancer risks that are below about 1 chance in 1 000 000 (1×10^{-6} or 1E-06) to be so small as to be negligible, and risks above 1×10^{-4} to be sufficiently large that some sort of remediation is desirable (Pongpiachan *et al.*, 2013). Excess cancer risks that vary between 1×10^{-6} and 1×10^{-4} are usually regarded as acceptable based on the Role of the Baseline Risk Assessment in Superfund Remedy Selection Decisions (Clay,

Table 4.2 Statistical descriptions of cancer risk (Eq. (4).(1)) by using PAH congeners detected in oil spill affected coastal areas in Asian Coastal Soils & Sediments (ACS) (this study) and World Marine Sediments (WMS).

	ACS		WMS	
	Average	**Stdev**	**Average**	**Stdev**
Phe	6.06×10^{-6}	1.04×10^{-5}	2.75×10^{-5}	6.63×10^{-5}
An	1.25×10^{-6}	2.64×10^{-6}	7.37×10^{-6}	1.83×10^{-5}
Fluo	6.90×10^{-6}	1.32×10^{-5}	3.28×10^{-5}	9.35×10^{-5}
Pyr	5.81×10^{-6}	1.13×10^{-5}	3.18×10^{-5}	8.57×10^{-5}
B[a]A	3.66×10^{-6}	7.16×10^{-6}	1.97×10^{-5}	4.47×10^{-5}
Chry	4.15×10^{-6}	8.30×10^{-6}	1.66×10^{-5}	3.44×10^{-5}
B[b]F	9.26×10^{-6}	3.35×10^{-5}	2.71×10^{-5}	7.82×10^{-5}
B[k]F	6.11×10^{-6}	2.00×10^{-5}	1.60×10^{-5}	2.55×10^{-5}
B[a]P	4.87×10^{-6}	1.36×10^{-5}	1.54×10^{-5}	3.25×10^{-5}
Ind	3.29×10^{-6}	9.97×10^{-6}	1.75×10^{-5}	3.44×10^{-5}
D[a,h]A	2.66×10^{-6}	7.04×10^{-6}	9.21×10^{-6}	3.40×10^{-5}
B[g,h,i]P	2.33×10^{-6}	1.35×10^{-4}	1.42×10^{-5}	2.99×10^{-5}

1991). The Superfund Program employs a cancer risk level of 10^{-6} to highlight the point at which risk management decisions should be considered. By applying the same protocol, the possibility of negative cancer-causing impacts in coastal areas of Asian countries falls in the 'acceptable level' range as well as those of world marine sediments (see Table 4.2).

In addition, all cancer and non-cancer risk levels of PAH contents in ACS are generally lower than those of WMS except B[g,h,i]P. According to the 'Role of the Baseline Risk Assessment in Superfund Remedy Selection Decisions' memorandum provided by US EPA (1991), the non-carcinogenic Hazard Index (*HI*) can be applied as an indicator to evaluate the cumulative non-carcinogenic site risk to an individual based on reasonable maximum exposure. In a case where *HI* is less than one, the action basically is not warranted unless there are adverse environmental impacts for both current and future land use. In this study the *HI* of non-cancer risk levels of PAH content in ACS ranged from 1.95×10^{-2} to 3.62×10^{-1} with an average of 1.00×10^{-1}, which is three times less than that of WMS (Table 4.3). It is also crucial to underline that an average *HI* of 1.00×10^{-1} in ACS is ten times lower than the US EPA baseline, highlighting the fact that the degree of *HI* falls in the 'acceptable level' range.

4.4 CONCLUSION

The total concentrations of twelve probably carcinogenic PAHs in ACS ranged from 32 ng g^{-1} to 32 571 ng g^{-1} with an arithmetic mean of 1871 ± 5533 ng g^{-1}. The comparatively high percentage contribution of B[g,h,i]P observed in ACS can be explained by its relatively low photodecomposition rate constants. Diagnostic binary ratios of PAH congeners underline the complexity of emitting source strengths in the coastal areas of Asian countries. The evaluation of soil and sediment toxicity based on the cancer and non-cancer risk principles in oil spill affected Asian coastal areas accentuates the importance of two remaining challenges. Firstly, the lack of ecotoxicity information associated with both spatial and temporal distribution of PAH congeners in oil spilled coastal zones in Asian countries. Secondly, most default values employed in the calculations of both cancer and non-cancer risk assessments were specifically designed for US citizens and thus no effective *Cancer Risk* and *HI* are available for residents in Asian coastal areas. Despite these research limitations, the main findings

Table 4.3 Statistical descriptions of non-cancer risk (Eq. 4.2) by using PAH congeners detected in oil spill affected coastal areas in Asian Coastal Soils & Sediments (ACS) (this study) and World Marine Sediments (WMS).

	ACS		WMS	
	Average	Stdev	Average	Stdev
Phe	9.43×10^{-2}	1.62×10^{-1}	4.28×10^{-1}	1.03×10^{0}
An	1.95×10^{-2}	4.10×10^{-2}	1.15×10^{-1}	2.85×10^{-1}
Fluo	1.07×10^{-1}	2.05×10^{-1}	5.10×10^{-1}	1.45×10^{0}
Pyr	9.04×10^{-2}	1.76×10^{-1}	4.94×10^{-1}	1.33×10^{0}
B[a]A	5.70×10^{-2}	1.11×10^{-1}	3.07×10^{-1}	6.95×10^{-1}
Chry	6.46×10^{-2}	1.29×10^{-1}	2.59×10^{-1}	5.34×10^{-1}
B[b]F	1.44×10^{-1}	5.21×10^{-1}	4.22×10^{-1}	1.22×10^{0}
B[k]F	9.51×10^{-2}	3.11×10^{-1}	2.49×10^{-1}	3.97×10^{-1}
B[a]P	7.58×10^{-2}	2.11×10^{-1}	2.39×10^{-1}	5.05×10^{-1}
Ind	5.13×10^{-2}	1.55×10^{-1}	2.72×10^{-1}	5.35×10^{-1}
D[a,h]A	4.15×10^{-2}	1.10×10^{-1}	1.43×10^{-1}	5.30×10^{-1}
B[g,h,i]P	3.62×10^{-1}	2.10×10^{0}	2.20×10^{-1}	4.65×10^{-1}

of this study underlined that (i) the arithmetic mean contents of PAHs in ACS are much lower than those of WMS, (ii) an average *HI* in ACS falls in the 'acceptable level' range and (iii) the possibility of adverse cancer risk in oil spilled coastal areas of Asian countries lies in between 1×10^{-6} and 1×10^{-4}, indicating that the average *Cancer Risk* can be considered as acceptable based on the Role of the Baseline Risk Assessment in Superfund Remedy Selection Decisions.

ACKNOWLEDGEMENT

This work was performed with the approval of School of Social & Environmental Development (SSED), National Institute of Development Administration (NIDA) and financially supported by Research Centre of National Institute of Development Administration (RC-NIDA), Thailand.

REFERENCES

ACGIH (American Conference of Governmental Industrial Hygienists). (2005). Polycyclic Aromatic Hydrocarbons (PAHs) Biologic Exposure Indices (BEI). American Conference of Governmental Industrial Hygienists, Cincinnati, OH.

Akyüz M. and Çabuk H. (2010). Gas–particle partitioning and seasonal variation of polycyclic aromatic hydrocarbons in the atmosphere of zonguldak, Turkey. *Science of the Total Environment*, **408**(22), 5550–5558, https://doi.org/10.1016/j.scitotenv.2010.07.063

Apiratikul R., Pongpiachan S. and Hashmi M. Z. (2020). Health risk assessment of polycyclic aromatic hydrocarbons in coastal soils of Koh samed island (Thailand) after the oil spill incident in 2013. *Marine Pollution Bulletin*, **150**, 110736, https://doi.org/10.1016/j.marpolbul.2019.110736

Arey J. (1998). Atmospheric reactions of PAHs including formation of nitroarenes. In: PAHS And Related Compounds, A. H. Neilson (ed.), Springer, Berlin, Heidelberg, pp. 347–385.

ATSDR. (2005). Health Consultation. Review of sediment data of Sylvester's (A/K/A) Gilson road site. Nashua, Hillsborough County, New Hampshire. EPA Facility ID: NHD099363541; August 19, 2005.

Bacosa H. P., Erdner D. L. and Liu Z. (2015). Differentiating the roles of photooxidation and biodegradation in the weathering of light Louisiana sweet crude oil in surface water from the deepwater horizon site. *Marine Pollution Bulletin*, **95**(1), 265–272, https://doi.org/10.1016/j.marpolbul.2015.04.005

Bandowe B. A. M., Meusel H., Huang R. J., Ho K., Cao J., Hoffmann T. and Wilcke W. (2014). PM2. 5-bound oxygenated PAHs, nitro-PAHs and parent-PAHs from the atmosphere of a Chinese megacity: seasonal variation, sources and cancer risk assessment. *Science of the Total Environment*, **473**, 77–87, https://doi.org/10.1016/j.scitotenv.2013.11.108

Banjoo D. R. and Nelson P. K. (2005). Improved ultrasonic extraction procedure for the determination of polycyclic aromatic hydrocarbons in sediments. *Journal of Chromatography A*, **1066**(1–2), 9–18, https://doi.org/10.1016/j.chroma.2005.01.033

Bølling A. K., Pagels J., Yttri K. E., Barregard L., Sallsten G., Schwarze P. E. and Boman C. (2009). Health effects of residential wood smoke particles: the importance of combustion conditions and physicochemical particle properties. *Particle and Fibre Toxicology*, **6**(1), 29, https://doi.org/10.1186/1743-8977-6-29

Bonner M. R., Han D., Nie J., Rogerson P., Vena J. E., Muti P., Trevisan M., Edge S. B. and Freudenheim J. L. (2005). Breast cancer risk and exposure in early life to polycyclic aromatic hydrocarbons using total suspended particulates as a proxy measure. *Cancer Epidemiology and Prevention Biomarkers*, **14**(1), 53–60.

Cabrerizo A., Galbán-Malagón C., Del Vento S. and Dachs J. (2014). Sources and fate of polycyclic aromatic hydrocarbons in the antarctic and Southern Ocean atmosphere. *Global Biogeochemical Cycles*, **28**(12), 1424–1436, https://doi.org/10.1002/2014GB004910

Cai X. B., Xue D., Li D., Wang F. and Zhao Z. Y. (2013). Distribution and sources of polycyclic aromatic hydrocarbons in the sediments of Ramsar swamp in inner deep bay of Hong Kong. *Environmental Chemistry*, **32**, 451–458 (in Chinese).

Chen H. Y., Teng Y. G., Wang J. S., Song L. T. and Zuo R. (2013). Source apportionment of sediment PAHs in the pearl river Delta region (China) using nonnegative matrix factorization analysis with effective weighted variance solution. *Science of the Total Environment*, **444**, 401–408, https://doi.org/10.1016/j.scitotenv.2012.11.108

Chen Y. C., Chiang H. C., Hsu C. Y., Yang T. T., Lin T. Y., Chen M. J., Chen N. T. and Wu Y. S. (2016). Ambient PM2.5-bound polycyclic aromatic hydrocarbons (PAHs) in changhua county, central Taiwan: seasonal variation, source apportionment and cancer risk assessment. *Environmental Pollution*, **218**, 372–382, https://doi.org/10.1016/j.envpol.2016.07.016

ChooChuay C., Pongpiachan S., Tipmanee D., Deelaman W., Iadtem N., Suttinun O., Wang Q., Xing L., Li G., Han Y., Hashmi M. Z., Palakun J., Poshyachinda S., Aukkaravittayapun S., Surapipith V. and Cao J. (2020a). Effects of agricultural waste burning on PM2.5-bound polycyclic aromatic hydrocarbons, carbonaceous compositions, and water-soluble ionic species in the ambient air of chiang-Mai, Thailand. *Polycyclic Aromatic Compounds*, **40**, 1–22, https://doi.org/10.1080/10406638.2020.1750436

ChooChuay C., Pongpiachan S., Tipmanee D., Deelaman W., Iadtem N., Suttinun O., Wang Q., Xing L., Li G., Han Y., Hashmi M. Z., Palakun J., Poshyachinda S., Aukkaravittayapun S., Surapipith V. and Cao J. (2020b). Long-range transboundary atmospheric transport of polycyclic aromatic hydrocarbons, carbonaceous compositions, and water-soluble ionic species in southern Thailand. *Aerosol and Air Quality Research*, **20**(7), 1591–1606, https://doi.org/10.4209/aaqr.2020.03.0120

Clay D. R. (1991). Role of the Baseline Risk Assessment in Superfund Remedy Selection Decisions (Memorandum from D. R. Clay, OSWER 9355.0–30, April 1991). Washington, DC, United States.

Dachs J. and Eisenreich S. J. (2000). Adsorption onto aerosol soot carbon dominates gas-particle partitioning of polycyclic aromatic hydrocarbons. *Environmental Science & Technology*, **34**(17), 3690–3697, https://doi.org/10.1021/es991201+

D'adamo R., Pelosi S., Trotta P. and Sansone G. (1997). Bioaccumulation and biomagnification of polycyclic aromatic hydrocarbons in aquatic organisms. *Marine Chemistry*, **56**(1–2), 45–49, https://doi.org/10.1016/S0304-4203(96)00042-4

Deelaman W., Pongpiachan S., Tipmanee D., Choochuay C., Iadtem N. and Suttinun O., *et al.* (2020). Source identification of polycyclic aromatic hydrocarbons in terrestrial soils in Chile. *Journal of South American Earth Sciences*, **99**, 102514, https://doi.org/10.1016/j.jsames.2020.102514

Degrendele C., Audy O., Hofman J., Kučerik J., Kukučka P., Mulder M. D., *et al.* (2016). Diurnal variations of air-soil exchange of semivolatile organic compounds (PAHs, PCBs, OCPs, and PBDEs) in a central European receptor area. *Environmental Science & Technology*, **50**(8), 4278–4288, https://doi.org/10.1021/acs.est.5b05671

de Oliveira Alves N., Brito J., Caumo S., Arana A., de Souza Hacon S., Artaxo P., *et al.* (2015). Biomass burning in the Amazon region: aerosol source apportionment and associated health risk assessment. *Atmospheric Environment*, **120**, 277–285, https://doi.org/10.1016/j.atmosenv.2015.08.059

Fernández P., Grimalt J. O. and Vilanova R. M. (2002). Atmospheric gas-particle partitioning of polycyclic aromatic hydrocarbons in high mountain regions of Europe. *Environmental Science & Technology*, **36**(6), 1162–1168, https://doi.org/10.1021/es010190t

Froehner S., Maceno M. and Machado K. S. (2011). Predicting bioaccumulation of PAHs in the trophic chain in the estuary region of paranagua, Brazil. *Environmental Monitoring and Assessment*, **174**(1–4), 135–145, https://doi.org/10.1007/s10661-010-1444-1

Goswami P., Ohura T., Guruge K. S., Yoshioka M., Yamanaka N., Akiba M. and Munuswamy N. (2016). Spatio-temporal distribution, source, and genotoxic potential of polycyclic aromatic hydrocarbons in estuarine and riverine sediments from southern India. *Ecotoxicology and Environmental Safety*, **130**, 113–123, https://doi.org/10.1016/j.ecoenv.2016.04.016

Gouin T., Mackay D., Jones K. C., Harner T. and Meijer S. N. (2004). Evidence for the 'grasshopper' effect and fractionation during long-range atmospheric transport of organic contaminants. *Environmental Pollution*, **128**(1–2), 139–148, https://doi.org/10.1016/j.envpol.2003.08.025

He X. R., Pang Y., Song X. J., Chen Feng B. L. Z. H. and Ma Y. Q. (2014). Distribution, sources and ecological risk assessment of PAHs in surface sediments from guan river estuary. *Marine Pollution Bulletin*, **80**, 52–58, https://doi.org/10.1016/j.marpolbul.2014.01.051

Ho K. F., Ho S. S. H., Lee S. C., Cheng Y., Chow J. C., Watson J. G., *et al.* (2009). Emissions of gas-and particle-phase polycyclic aromatic hydrocarbons (PAHs) in the shing Mun tunnel, Hong Kong. *Atmospheric Environment*, **43**(40), 6343–6351, https://doi.org/10.1016/j.atmosenv.2009.09.025

Hong W. J., Jia H., Sun Y., Liu X., Wang L. and Li Y. F. (2018). Distribution, source and ecological risk assessment of parent and alkylated PAHs in coastal environment of dalian, China after oil spill. Polycyclic Aromatic Compounds, **38**, 1–15.

Hui Y. M., Zheng M. H., Liu Z. T. and Gao L. R. 2009. Distribution of polycyclic aromatic hydrocarbons in sediments from Yellow River estuary and Yangtze river estuary, China. *Journal of Environmental Sciences*, **21**, 1625–1631, https://doi.org/10.1016/S1001-0742(08)62465-1

International Agency for Research on Cancer. (2010). Some Non-Heterocyclic Polycyclic Aromatic Hydrocarbons and Some Related Exposures. IARC Press, International Agency for Research on Cancer, Lyon, France, Vol. **92**.

Jariyasopit N., McIntosh M., Zimmermann K., Arey J., Atkinson R., Cheong P. H. Y., *et al.* (2014). Novel nitro-PAH formation from heterogeneous reactions of PAHs with NO_2, NO_3/N_2O_5, and OH radicals: prediction, laboratory studies, and mutagenicity. *Environmental Science and Technology*, **48**(1), 412–419, https://doi.org/10.1021/es4043808

Jenkins B. M., Jones A. D., Turn S. Q. and Williams R. B. (1996). Particle concentrations, gas–particle partitioning, and species intercorrelations for polycyclic aromatic hydrocarbons (PAH) emitted during biomass burning. *Atmospheric Environment*, **30**(22), 3825–3835, https://doi.org/10.1016/1352-2310(96)00084-2

Jiao L. X., Meng W., Zheng B. H., Zhang L., Zhao X. R. and Qin Y. W. (2010). Distribution of polycyclic aromatic hydrocarbons (PAHs) in different size fractions of sediments from intertidal zone of Bohai Bay, China. *Chinese Journal of Environmental Science*, **30**, 1241–1248 (in Chinese).

Keshavarzifard M., Zakaria M. P. and Keshavarzifard S. (2019). Evaluation of polycyclic aromatic hydrocarbons contamination in the sediments of the johor strait, peninsular Malaysia. *Polycyclic Aromatic Compounds*, **39**(1), 44–59, https://doi.org/10.1080/10406638.2016.1257997

Kaupp H. and McLachlan M. S. (1999). Gas/particle partitioning of PCDD/Fs, PCBs, PCNs and PAHs. *Chemosphere*, **38**(14), 3411–3421, https://doi.org/10.1016/S0045-6535(98)00554-2

Khairy M. A., Kolb M., Mostafa A. R., Anwar E. F. and Bahadir M. (2009). Risk assessment of polycyclic aromatic hydrocarbons in a Mediterranean semi-enclosed basin affected by human activities (Abu Qir Bay, *Egypt). Journal of Hazardous Materials*, **170**(1), 389–397, https://doi.org/10.1016/j.jhazmat.2009.04.084

Kim E. J., Oh J. E. and Chang Y. S. (2003). Effects of forest fire on the level and distribution of PCDD/Fs and PAHs in soil. *Science of the Total Environment*, **311**(1–3), 177–189, https://doi.org/10.1016/S0048-9697(03)00095-0

Kozak K., Ruman M., Kosek K., Karasiń***ski*** G., Stachnik Ł. and Polkowska Ż. (2017). Impact of volcanic eruptions on the occurrence of PAHs compounds in the aquatic ecosystem of the southern part of west spitsbergen (hornsund fjord, svalbard). *Water*, **9**(1), 42, https://doi.org/10.3390/w9010042

Lallas P. L. (2001). The Stockholm convention on persistent organic pollutants. *The American Journal of International Law*, **95**(3), 692–708, https://doi.org/10.2307/2668517

Lang Y. H., Wang N. N., Gao H. W. and Bai J. (2012). Distribution and risk assessment of polycyclic aromatic hydrocarbons (PAHs) from liaohe estuarine wetland soils. *Environmental Monitoring and Assessment*, **184**, 5545–5552, https://doi.org/10.1007/s10661-011-2360-8

Liao C. M., Chio C. P., Chen W. Y., Ju Y. R., Li W. H., Cheng Y. H., *et al.* (2011). Lung cancer risk in relation to traffic-related nano/ultrafine particle-bound PAHs exposure: a preliminary probabilistic assessment. *Journal of Hazardous Materials*, **190**(1–3), 150–158, https://doi.org/10.1016/j.jhazmat.2011.03.017

Lim L., Wurl O., Karuppiah S. and Obbard J. P. (2007). Atmospheric wet deposition of PAHs to the sea-surface microlayer. *Marine Pollution Bulletin*, **54**(8), 1212–1219, https://doi.org/10.1016/j.marpolbul.2007.03.023

Lohmann R., Northcott G. L. and Jones K. C. (2000). Assessing the contribution of diffuse domestic burning as a source of PCDD/Fs, PCBs, and PAHs to the UK atmosphere. *Environmental Science & Technology*, **34**(14), 2892–2899, https://doi.org/10.1021/es991183w

Ma J., Hung H., Tian C. and Kallenborn R. (2011). Revolatilization of persistent organic pollutants in the Arctic induced by climate change. *Nature Climate Change*, **1**(5), 255–260, https://doi.org/10.1038/nclimate1167

Malloci G., Mulas G. and Benvenuti P. (2003). Testing the 'strong' PAHs hypothesis – I. Profile invariance of electronic transitions of interstellar PAH cations. *Astronomy and Astrophysics*, **410**(2), 623–637, https://doi.org/10.1051/0004-6361:20031211

Marvin C. H., Allan L., McCarry B. E. and Bryant D. W. (1992). A comparison of ultrasonic extraction and soxhlet extraction of polycyclic aromatic hydrocarbons from sediments and air particulate material. *International Journal of Environmental Analytical Chemistry*, **49**(4), 221–230, https://doi.org/10.1080/03067319208027573

Miller J. S. and Olejnik D. (2001). Photolysis of polycyclic aromatic hydrocarbons in water. *Water Research*, **35**(1), 233–243, https://doi.org/10.1016/S0043-1354(00)00230-X

Niu J., Sun P. and Schramm K. W. (2007). Photolysis of polycyclic aromatic hydrocarbons associated with fly ash particles under simulated sunlight irradiation. *Journal of Photochemistry and Photobiology A: Chemistry*, **186**(1), 93–98, https://doi.org/10.1016/j.jphotochem.2006.07.016

Nwaichi E. O., Chuku L. C. and Ighoavwogan E. (2016). Polycyclic aromatic hydrocarbons and selected heavy metals in some oil polluted sites in Delta state Nigeria. *Journal of Environmental Protection*, **7**(10), 1389–1410, https://doi.org/10.4236/jep.2016.710120

Onojake M. C., Osakwe J. O. and Omokheyeke O. (2013). Source distribution of polycyclic aromatic hydrocarbons of an oil spill impacted site in Niger Delta, Nigeria. *European Chemical Bulletin*, **3**(2), 179–182.

Ozaki N., Nitta K. and Fukushima T. (2006). Dispersion and dry and wet deposition of PAHs in an atmospheric environment. *Water Science and Technology*, **53**(2), 215–224, https://doi.org/10.2166/wst.2006.055

Parikh P. V. and Wei Y. (2016). PAHS And PM2. 5 emissions and female breast cancer incidence in metro atlanta and rural Georgia. *International Journal of Environmental Health Research*, **26**(4), 458–466, https://doi.org/10.1080/09603123.2016.1161178

Park J. S., Wade T. L. and Sweet S. T. (2002). Atmospheric deposition of PAHs, PCBs, and organochlorine pesticides to Corpus Christi Bay, Texas. *Atmospheric Environment*, **36**(10), 1707–1720, https://doi.org/10.1016/S1352-2310(01)00586-6

Pongpiachan S. (2011). Analytical Instruments for Environmental Sciences. Athens Publishing, Bangkok, Thailand. ISBN 978-974-492-729-3.

Pongpiachan S. (2015). A preliminary study of using polycyclic aromatic hydrocarbons as chemical tracers for traceability in soybean products. *Food Control*, **47**, 392–400, https://doi.org/10.1016/j.foodcont.2014.07.032

Pongpiachan S. (2016). Assessment of reliability when using diagnostic binary ratios of polycyclic aromatic hydrocarbons in ambient Air PM 10. *Asian Pacific Journal of Cancer Prevention*, **16**(18), 8605–8611, https://doi.org/10.7314/APJCP.2015.16.18.8605

Pongpiachan S., Bualert S., Sompongchaiyakul P. and Kositanont C. (2009). Factors affecting sensitivity and stability of polycyclic aromatic hydrocarbons determined by gas chromatography quadrupole ion trap mass spectrometry. *Analytical Letters*, **42**(13), 2106–2130, https://doi.org/10.1080/00032710903082838

Pongpiachan S., Ho K. F. and Lee S. C. (2010). A study of gas-particle partitioning of PAH according to adsorptive models and season. In: Air Pollution XVIII, C. A. Brebbia and J. W. S. Longhurst (eds), WIT Press, Southampton, pp. 37–48.

Pongpiachan S., Hirunyatrakul P., Kittikoon I. and Khumsup C. (2012). Parameters influencing on sensitivities of polycyclic aromatic hydrocarbons measured by Shimadzu GCMS-QP2010 ultra. Advanced Gas Chromatography–Progress in Agricultural, Biomedical and Industrial Applications, pp. 109–130.

Pongpiachan S., Tipmanee D., Deelaman W., Muprasit J., Feldens P. and Schwarzer K. (2013). Risk assessment of the presence of polycyclic aromatic hydrocarbons (PAHs) in coastal areas of Thailand affected by the 2004 tsunami. *Marine Pollution Bulletin*, **76**(1–2), 370–378, https://doi.org/10.1016/j.marpolbul.2013.07.052

Pongpiachan S., Hattayanone M., Choochuay C., Mekmok R., Wuttijak N. and Ketratanakul A. (2015). Enhanced PM10 bounded PAHs from shipping emissions. *Atmospheric Environment*, **108**, 13–19.

Pongpiachan S., Hattayanone M. and Cao J. (2017). Effect of agricultural waste burning season on $PM_{2.5}$-bound polycyclic aromatic hydrocarbon (PAH) levels in northern Thailand. *Atmospheric Pollution Research*, **8**(6), 1069–1080, https://doi.org/10.1016/j.apr.2017.04.009

Pongpiachan S., Hattayanone M., Tipmanee D., Suttinun O., Khumsup C., Kittikoon I. and Hirunyatrakul P. (2018). Chemical characterization of polycyclic aromatic hydrocarbons (PAHs) in 2013 rayong oil spill-affected coastal areas of Thailand. *Environmental Pollution*, **233**, 992–1002, https://doi.org/10.1016/j.envpol.2017.09.096

Pongpiachan S., Deelaman W., Choochuay C., Iadtem N., Surapipith V., Hashmi M. Z., *et al.* (2019). Data relating to spatial distribution of polycyclic aromatic hydrocarbons in terrestrial soils of Pakistan and king george island, Antarctica. *Data in Brief*, **25**, 104327, https://doi.org/10.1016/j.dib.2019.104327

Sany S. B. T., Hashim R., Salleh A., Safari O., Mehdinia A. and Rezayi M. (2014). Risk assessment of polycyclic aromatic hydrocarbons in the west port semi-enclosed basin (Malaysia). *Environmental Earth Sciences*, **71**(10), 4319–4332, https://doi.org/10.1007/s12665-013-2826-9

Sarigiannis D. A., Karakitsios S. P., Zikopoulos D., Nikolaki S. and Kermenidou M. (2015). Lung cancer risk from PAHs emitted from biomass combustion. *Environmental Research*, **137**, 147–156, https://doi.org/10.1016/j.envres.2014.12.009

Seegar W. S., Yates M. A., Doney G. E., Jenny J. P., Seegar T. C., Perkins C. and Giovanni M. (2015). Migrating tundra peregrine falcons accumulate polycyclic aromatic hydrocarbons along Gulf of Mexico following deepwater horizon oil spill. *Ecotoxicology*, **24**(5), 1102–1111, https://doi.org/10.1007/s10646-015-1450-8

Shankar R., Shim W. J., An J. G. and Yim U. H. (2015). A practical review on photooxidation of crude oil: laboratory lamp setup and factors affecting it. *Water Research*, **68**, 304–315, https://doi.org/10.1016/j.watres.2014.10.012

Simcik M. F., Franz T. P., Zhang H. and Eisenreich S. J. (1998). Gas–particle partitioning of PCBs and PAHs in the Chicago urban and adjacent coastal atmosphere: states of equilibrium. *Environmental Science & Technology*, **32**(2), 251–257, https://doi.org/10.1021/es970557n

Singh A., Kamal R., Mudiam M. K. R., Gupta M. K., Satyanarayana G. N. V., Bihari V., *et al.* (2016). Heat and PAHs emissions in indoor kitchen air and its impact on kidney dysfunctions among kitchen workers in lucknow, north India. *PLoS One*, **11**(2), e0148641.

Smith K. E., Schwab A. P. and Banks M. K. (2008). Dissipation of PAHs in saturated, dredged sediments: a field trial. *Chemosphere*, **72**(10), 1614–1619, https://doi.org/10.1016/j.chemosphere.2008.03.020

Slezakova K., Castro D., Delerue-Matos C., da Conceição Alvim-Ferraz M., Morais S. and do Carmo Pereira M. (2013). Impact of vehicular traffic emissions on particulate-bound PAHs: levels and associated health risks. *Atmospheric Research*, **127**, 141–147, https://doi.org/10.1016/j.atmosres.2012.06.009

Soliman Y. S., Al Ansari E. M. S. and Wade T. L. (2014). Concentration, composition and sources of PAHs in the coastal sediments of the exclusive economic zone (EEZ) of Qatar, Arabian gulf. *Marine Pollution Bulletin*, **85**(2), 542–548, https://doi.org/10.1016/j.marpolbul.2014.04.027

Tao Y., Yu J., Liu X., Xue B. and Wang S. (2018). Factors affecting annual occurrence, bioaccumulation, and biomagnification of polycyclic aromatic hydrocarbons in plankton food webs of subtropical eutrophic lakes. *Water Research*, **132**, 1–11, https://doi.org/10.1016/j.watres.2017.12.053

Tian Y., Liu H. J., Zheng T. L., Kwon K. K., Kim S. J. and Yan C. L. (2008). PAHS Contamination and bacterial communities in mangrove surface sediments of the jiulong river estuary, China. *Marine Pollution Bulletin*, **57**(6–12), 707–715, https://doi.org/10.1016/j.marpolbul.2008.03.011

US EPA. (1991). Role of the baseline risk assessment in superfund remedy selection decisions. Washington, D. C. 20460.

Veiga P., Besteiro C. and Rubal M. (2010). Meiofauna communities in exposed sandy beaches on the galician coast (NW Spain), six months after the prestige oil spill: the role of polycyclic aromatic hydrocarbons (PAHs). *Scientia Marina*, **74**(2), 385–394, https://doi.org/10.3989/scimar.2010.74n2385

Wang X., Yao T., Cong Z., Yan X., Kang S. and Zhang Y. (2007). Concentration level and distribution of polycyclic aromatic hydrocarbons in soil and grass around Mt. Qomolangma, China. *Chinese Science Bulletin*, **52**(10), 1405–1413, https://doi.org/10.1007/s11434-007-0184-2

Wang Z. C., Liu Z. F., Yang Y., Li T. and Liu M. (2012). Distribution of PAHs in tissues of wetland plants and the surrounding sediments in the chongming wetland, shanghai, China. *Chemosphere*, **89**, 221–227, https://doi.org/10.1016/j.chemosphere.2012.04.019

Yang H. H., Lee W. J., Chen S. J. and Lai S. O. (1998). PAH Emission from various industrial stacks. *Journal of Hazardous Materials*, **60**(2), 159–174, https://doi.org/10.1016/S0304-3894(98)00089-2

Yang H. H., Lai S. O., Hsieh L. T., Hsueh H. J. and Chi T. W. (2002). Profiles of PAH emission from steel and iron industries. *Chemosphere*, **48**(10), 1061–1074, https://doi.org/10.1016/S0045-6535(02)00175-3

Yang W., Lang Y. H. and Li G. L. (2014). Cancer risk of polycyclic aromatic hydrocarbons (PAHs) in the soils from jiaozhou Bay wetland. *Chemosphere*, **112**, 289–295, https://doi.org/10.1016/j.chemosphere.2014.04.074

Yim U. H., Oh J. R., Hong S. H., Lee S. H. and Shim W. J. (2002). Identification of PAHs sources in bivalves and sediments 5 years after the Sea prince Oil spill in korea. *Environmental Forensics*, **3**, 357–366, https://doi.org/10.1080/713848380

Zeng X. K., Li F. L., Zhou K., Yang J. D., Hao W. L. and Zan Q. J. (2013). Distribution of polycyclic aromatic hydrocarbons (PAHs) in the sediments and mangrove plants at baguang wetland shenzhen. *Environmental Science & Technology*, **36**, 368–373 (in Chinese).

Zhang Y., Yang B., Meng J., Gao S., Dong X. and Shu J. (2010). Homogeneous and heterogeneous reactions of phenanthrene with ozone. *Atmospheric Environment*, **44**(5), 697–702, https://doi.org/10.1016/j.atmosenv.2009.11.017

Zhang Y., Yang B., Gan J., Liu C., Shu X. and Shu J. (2011). Nitration of particle-associated PAHs and their derivatives (nitro-, oxy-, and hydroxy-PAHs) with NO_3 radicals. *Atmospheric Environment*, **45**(15), 2515–2521, https://doi.org/10.1016/j.atmosenv.2011.02.034

Zhuang W. E., Wang X. X., Yao M. S., Yang L., Song X. K. and Gong Z. B. (2011). Distribution and source of polycyclic aromatic hydrocarbons in the surface sediments of quanzhou Bay China. *Environmental Chemistry*, **30**, 928–934 (in Chinese).

Zuloaga O., Prieto A., Ahmed K., Sarkar S. K., Bhattacharya A., Chatterjee M., *et al.* (2013). Distribution of polycyclic aromatic hydrocarbons in recent sediments of sundarban mangrove wetland of India and Bangladesh: a comparative approach. *Environmental Earth Sciences*, **68**(2), 355–367, https://doi.org/10.1007/s12665-012-1743-7

doi: 10.2166/9781789061123_073

Chapter 5

Microbiological impact of diffuse pollution sources on water quality

Mats Leifels[1], Kwanrawee Sirikanchana[2] and James Ebdon[3]

[1]Singapore Centre for Environmental Life Sciences Engineering, Nanyang Technological University, Singapore
[2]Research Laboratory of Biotechnology, Chulabhorn Research Institute, Bangkok, Thailand
[3]Environment and Public Health Research and Enterprise Group, School of Environment and Technology, University of Brighton, Brighton, United Kingdom

Keywords: Agriculture; catchment; diffuse pollution; faecal indicators; mitigation; pathogens

5.1 INTRODUCTION

This chapter outlines some of the impacts that diffuse pollution inputs can have on water quality and introduces some of the waterborne microbial contaminants and emerging pollutants of human and/or environmental health significance. The chapter also highlights some of the challenges associated with successfully identifying and mitigating diffuse sources of contamination and mentions some of the limitations of traditional faecal indicators (e.g., *E. coli*, intestinal enterococci). The emergence of new and potentially promising indicators is subsequently mentioned including culture-dependent bacteriophage-based approaches (e.g., somatic coliphage, *Bacteroides* phage) and culture-independent molecular-based methods (e.g., Pepper Mild Mottle Virus and CrAssphage). There is considerable focus throughout the chapter on agricultural inputs of diffuse pollution, as they are often a key driver of poor microbiological water quality in many catchments around the world. Specific case studies from different geographical settings are also featured in the chapter and provide the reader with information on the different ways in which diffuse pollution monitoring and modelling is being used to support more effective catchment management.

5.2 WATERBORNE DISEASES

Numerous epidemiological studies have highlighted the association between the presence of excreta-related pathogens within water bodies, and an increased risk in contracting a waterborne disease (e.g., Ford and Hamner, 2018). The causative agents of waterborne disease may be bacterial, viral, protozoal, or fungal (Table 5.1).

Many of these pathogens are stable within aquatic environments and may persist for extended periods (several weeks, months) under temperate and tropical conditions, while some are resistant to industrial treatments and are therefore difficult to remove from waters (FAO, 2006; Kay *et al.*,

Editor(s): Puangrat Kajitvichyanukul and Brian D'Arcy

Table 5.1 Waterborne bacterial, viral, protozoal, and fungal pathogens and the associated diseases (based on Weiskerger and Brandão, 2020; Woodall, 2009).

Organism	Stability in Water	Size (μm)	Associated Disease
Bacteria			
Burkholderia pseudomallei	Can multiply	1.0–4.0	Melioidosis (septicaemia)
Campylobacter jejuni	Moderate	1.5–4.0	Diarrhoea
Escherichia coli	Moderate	0.2–2.0	Diarrhoea, haemolytic uremic syndrome
Salmonella Typhi/Paratyphi A	Moderate	0.2–2.0	Typhoid fever
Shigella spp.	Low	0.2–2.0	Bacillary dysentery
Vibrio cholerae	Low	1.0–3.0	Cholera
Yerisinia enterocolitica	High	0.2–2.0	Diarrhoea
Viruses			
Adenovirus 40/41	High	0.07	Severe childhood gastroenteritis
Enteroviruses	High	0.03	Myalgia, meningitis, paralysis
Hepatitis A virus	High	0.03	Hepatitis
Hepatitis E virus	High	0.03	Hepatitis
Noroviruses	High	0.03	Mild gastroenteritis
Rotaviruses	High	0.06	Severe childhood gastroenteritis
Protozoa			
Cryptosporidium parvum	High	4.0–10.0	Cryptosporidiosis (diarrhoea)
Cyclospora cayetanensis	High	8.0–10.0	Cyclosporiasis (diarrhoea)
Entamoeba histolytica	Moderate	8.0–15.0	Amoebiasis (amoebic dysentery)
Giardia intestinalis	Moderate	7.0–12.0	Giardiasis (diarrhoea)
Toxoplasma gondii	High	2.0–70.0	Toxoplasmosis
Fungi			
Aspergillus spp.	High	900–1600	Aspergillosis (lung infection), onychomycosis (nail infection), allergy
Candida spp.	High	5–10	Candidosis (skin infection)
Dermatophytes	High	6–160	Onychomycosis (nail infection), tineae (skin infection)
Fusarium spp.	High	5–12	Fusariosis (skin lesion and lung infection)

2012; LeClerc *et al.*, 2002). The vast majority of reported cases of water-related disease relate to acute illnesses including self-limiting gastroenteritis, minor respiratory infections and skin and eye infections (Leclerc, 2003), although certain pathogens such as *E. coli* O157:H7 and the hepatitis E virus are capable of causing more serious infections (Belkin & Colwell, 2005). According to Christou (2011), hundreds of thousands of zoonotic (animal to human) infections occur each year due to pathogens from livestock including bacteria such as *Campylobacter* spp., *E. coli* O157:H7, *Salmonella* spp. and *Clostridium botulinum* and parasitic protozoa such as *Giardia lamblia*, *Cryptosporidium parvum* and fungi such as *Microsporidia* spp. Particularly, *Cryptosporidium* spp. in the form of oocysts are shed with faeces of human and animals and have been detected in untreated sewage, surface water and recreational water, groundwater, drinking water, irrigation water, as well as seawater and shellfish (Betancourt, 2019). With high resistance to environmental conditions and disinfectants, *Cryptosporidium* oocysts could not be effectively removed by the traditional water supply treatment processes, including chlorine disinfection. The largest documented drinking water outbreak in the

US was due to *Cryptosporidium* contamination, affecting over 400 000 residents in Milwaukee, Wisconsin, in 1993. Later on, the mandatory monitoring of *Cryptosporidium* was regulated by the US Long Term 2 Enhanced Surface Water Treatment Rule, and UV treatment has been used as a disinfection option. Although regulatory and research actions have been emphasised for drinking water contamination prevention, *Cryptosporidium* was still among the most significant etiological agents causing outbreaks associated with untreated recreational water in the US during 2000–2014 (Graciaa *et al.*, 2018). Furthermore, the understudied waterborne fungi limited available knowledge on their diversity and functions. Aquatic fungi have been reported for a small fraction (3000–4000) of approximately ten million discovered fungal species (Grossart & Rojas-Jimenez, 2016), and spread through water in vegetative (so-called mycelium) and spore forms (Góralska *et al.*, 2020).

Although bacterial and protozoan pathogens are responsible for a significant burden of disease worldwide, viruses are of particular concern due to their remarkably lower minimum infective dose (Leclerc *et al.*, 2002). Haas *et al.* (1993) estimate that the risk of infection when exposed to the same concentration of a virus compared to a bacterium is between 10 and 10^4 times greater. Diagnosis, prediction and monitoring are therefore important requirements for the successful management of aquatic ecosystems and the mitigation of harmful impacts. To understand and prevent the onwards spread of waterborne pathogens associated with point or diffuse pollution sources, it is important to understand their origin(s) and subsequent transmission pathway(s) (Leifels *et al.*, 2019, 2021; Mackowiak *et al.*, 2018).

5.3 TRADITIONAL FAECAL INDICATORS AND THEIR LIMITATIONS

The concentration of enteric microbial pathogens, even in grossly contaminated waters such as those directly impacted by agricultural slurry or by untreated or partially treated municipal wastewater, may be low (or highly variable) making their routine detection problematic and potentially costly. The sheer amount of enteric viruses alone reported within human faeces (over 140 different species of varying public health relevance) increases the difficulty to ascertain risk to health by measuring a single pathogen (Leclerc *et al.*, 2003; Wong *et al.*, 2012). Consequently, faecal indicator organisms (typically bacteria) have been extensively used to establish the sanitary quality of surface and ground waters susceptible to point and/or diffuse pollution and used for drinking water abstraction, recreational pursuits, shellfish harvesting and fish farming. Decades of research and policymaking resulted in the following definition of an 'ideal indicator organism': (i) detectable by simple, rapid and low-cost laboratory techniques, (ii) found in high concentrations in untreated (raw) wastewater, (iii) present in wastewater in higher numbers than pathogens (conservative), (iv) unable to multiply in the environment (or within treatment systems), and (v) show similar survival and removal characteristics to pathogens during water treatment and disinfection processes (e.g., Berg, 1978; NHMRC, 2001; Payment, 1998; WHO, 2011).

Faecal indicator bacteria (FIB) are most widely used for this purpose and include total coliforms, faecal or thermotolerant coliforms, *Escherichia coli* (*E. coli*) and intestinal enterococci. However, with improvements in microbiological test methods and bacterial species identification, there is growing evidence that FIB including *E. coli* and enterococci, do not always provide reliable information regarding the occurrence, removal or inactivation of more resistant pathogens, particularly enteric viruses or parasites (Field & Samadpour, 2007; Jiang *et al.*, 2002; McQuaig *et al.*, 2012; Stoeckel & Harwood, 2007). For example, there is evidence that *E. coli* and enterococci are able to survive, grow, and establish populations in natural environments, such as freshwater lakes and streams, soils, and sediments (Byappanahalli *et al.*, 2006; Liang *et al.*, 2015). Consequently, alternative viral indicators have been proposed such as bacteriophages (like somatic and F-specific coliphages), or those capable of infecting human-specific *Bacteroides* and *Enterococcus* spp. (Chyerochana *et al.*, 2020; Ebdon *et al.*, 2012; IAWPRC, 1991; Payan *et al.*, 2005; Purnell *et al.*, 2018). Recently the plant pathogenic

Pepper Mild Mottle Virus (Hamza *et al.*, 2011; Rosario *et al.*, 2009) and CrAssphage have been proposed as markers of faecal contamination due to their high abundance in faeces and wastewaters (Kongprajug *et al.*, 2019a; Stachler *et al.*, 2017). Further in-depth information about the full spectrum of faecal indicators available for monitoring water quality and treatment efficacy (Momba *et al.*, 2019) is available via the open access *Global Water Pathogen Project* (http://www.waterpathogens.org/book/using-indicators-assess-microbial-treatment-and-disinfection-efficacy).

5.4 AGRICULTURAL SOURCES OF MICROBIOLOGICAL POLLUTION

While crop production is the main agricultural activity contributing to global water pollution through the release of nutrients, pesticides, salts and sediments, livestock is the principal source of pollution by organic matter, pathogens, and emerging pollutants such as hormones and antibiotics (UN FAO, 2018). Domestic animals, such as poultry, cattle, sheep, and pigs, have been estimated to generate 85% of the world's animal faecal waste, proportionally a far greater amount than is contributed by the world's human population. The faecal production rate and contribution to the environment by these animals can be as high as 2.62×10^{13} kg/year (Kay *et al.*, 2012; UN FAO, 2018).

Addressing such agricultural sources of pollution within catchments used for drinking water provision provides substantial benefits for water utilities and resource managers worldwide (Smolders *et al.*, 2015), through the reduction in risk to the safety of drinking water supplies or by decreasing water treatment costs to an acceptable standard (Bryan *et al.*, 2009). Pollution sources can be defined as either point or non-point. Whilst point sources can be relatively easily defined as 'any single identifiable and confided conveyance from where pollutants are or potentially may be discharged from' (USEPA, 2015), non-point or 'diffuse' sources cover a wider range of inputs that cannot be attributed to one singular direct input. Non-point sources include septic tank drainage, surface runoff, re-suspension of sediment or sand, and solid waste and human bather shedding (see Figure 5.1; Fewtrell and Kay, 2015). Other types of land-use activities, such as road construction, mine drainage, rainwater runoff from city streets (collected in storm drains), from agriculture and rural villages can produce water pollution that tends to be dispersed across the landscape (UN FAO, 2018). Faecally polluted recreational water, drinking water, and food also increase public health risk to waterborne diseases (Figure 5.1).

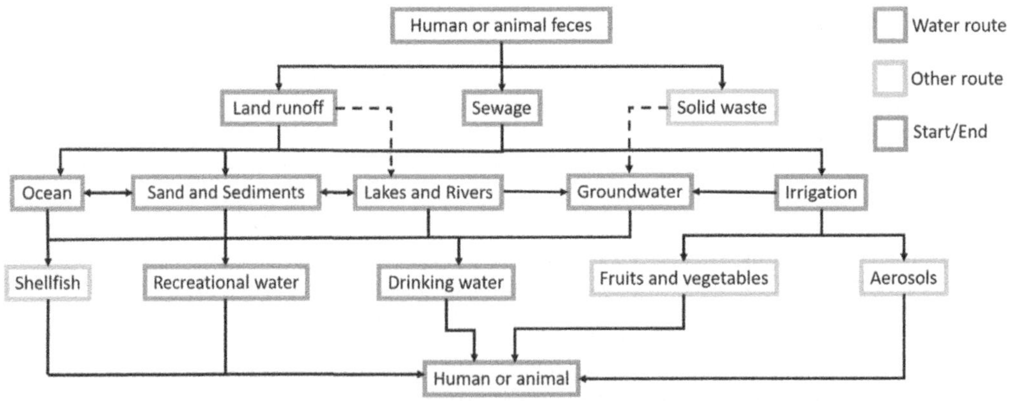

Figure 5.1 Contamination routes for waterborne pathogen of human and animal faecal origin (broken lines indicates scientific dispute in literature; adapted from Rodriguez-Lazaro *et al.* (2012).

5.5 CATCHMENT-SCALE MODELS

The diffuse nature of non-point inputs mean that it is often complex to identify and quantify such pollution sources (e.g., agriculture, wild birds) and establish their relative contribution to the degradation of water quality. Furthermore, establishing risk to human health or the environment in river catchments is also complex as it relies on an understanding of hydrological processes, land management practices, livestock stocking densities, contaminant loading rates, fates of pathogens, and meteorological parameters. This has led to the development and application of catchment-scale models such as the open-source Soil and Water Assessment Tools (SWAT) (Arnold & Fohrer, 2005; Arnold *et al.*, 1998), a river-basin-scale model that has been used to assess water resource and non-point (diffuse) source pollution issues worldwide. The model, already the focus of over 2700 peer-reviewed research articles in over 500 journals (Dile *et al.*, 2016), was developed to predict the impact of land management practices on water, sediment, and agricultural chemical yields (e.g., nitrates, phosphates and pesticides). SWAT can operate at a range of scales, in complex catchments with different soils, land-use and management practices (Arnold *et al.*, 2012). Other screening-tools and models such as Visualizing Pathogen and Environmental Risk (ViPER) approaches have also been developed to identify relative contributions of microbial load from catchments (Palazón *et al.*, 2017) and to predict *E. coli* burdens contributed to agricultural land from grazing livestock in catchments (Oliver *et al.*, 2018). The major components of such models include meteorology, hydrology, soil characteristics, plant growth, nutrients, pesticides, bacteria, bacterial pathogens and land-use, and land management. Further information about catchment-scale water quality and soil erosion models can be found in a recent review by Fu *et al.* (2019).

5.6 MITIGATING DIFFUSE SOURCES OF MICROORGANISMS

The management of diffuse contributions of microbial pollution to receiving waters represents a more difficult challenge for the policy and regulatory community, when compared with point source inputs (Oliver *et al.*, 2018). Changes to on-farm operations, such as the introduction of vegetated riparian buffer strips along riverbanks, have been shown to reduce catchment-derived microbial contamination to aquatic environments and water supply reservoirs (e.g., Burt and Howden, 2013; Kay *et al.*, 2012). Preventing livestock from entering waterways, via the construction of bridges, can also help reduce microbial pollution to streams (Smolders *et al.* 2015). Assessment of the replacement of ford crossings by bridges was predicted to achieve a 0.62 $\log_{10}$ reduction in the daily *E. coli* concentration in New Zealand streams (Davies-Colley *et al.*, 2004; Kay *et al.*, 2012). However, with the exception of a few studies such as case study 1 below (Smolders *et al.*, 2015), there remains a paucity of experimental data to assess the performance of specific contamination mitigation methods arising from livestock movements.

5.7 DIFFUSE POLLUTION CASE STUDIES

According to a recent report from the UK Environment Agency (2018), 86% of river water bodies in the UK had not reached a *good ecological status* under the EU Water Framework Directive and 55% were at less than good status for phosphorus in 2016. The main reasons for this are agriculture and rural land management, the water industry, and urban and transport pressures. However, whilst overall pollutant loads to rivers from water industry discharges have declined in recent years (up to 70% since 1995 in some cases), the number of serious water pollution incidents remained broadly the same. Approximately 60 Category-1 incidents – defined as spills and other pollution events having a major, persistent or significant impact on the environment, people or property – were reported (Environment Agency, 2018).

5.7.1 Case study 1: assessment of stream crossing modification on cattle-derived microbial input to source water catchments (Australia)

Smolders and co-workers (2015) tested the effects of stream bridging on the longitudinal changes in the concentrations of *E. coli*, enterococci and suspended solids (TSS) within stream channels draining into a reservoir, which is the source of 300 000 persons' drinking water across the Sunshine Coast region of Australia. Their study conducted between 2011 and 2013 focused on two streams, which bisected a dairy farm (400 heads of dairy cattle) and were both crossed twice daily by the herd in transit. One stream had a culvert bridge over it; the other (control stream) had a ford through which each animal crossed. Water samples were collected at regular intervals from each stream when cattle were present either in or over (in the case of the control) the stream. Samples were also taken simultaneously upstream and downstream at prescribed intervals during each crossing. Three sub-samples were analysed for the FIB *E. coli*, enterococci as well as TSS. Results showed that whilst stream bridging provided a clear water-quality benefit (significant reductions in *E. coli*, enterococci and TSS in the vicinity of the crossing points), water-quality benefits to source water catchments may be overwhelmed by chronic microbiological and suspended solids pollution from diffuse sources (i.e., faecal inputs) of the same cattle ranging. Therefore, although their findings suggest that reducing direct contact between livestock and drinking water supplies is a simple and cost-effective method to help minimize risks of future waterborne disease outbreaks, chronic pollution arising from diffuse sources by cattle as they range across the surrounding catchment to graze is not necessarily ameliorated by solely addressing point-source inputs. These findings suggest that to achieve a significant reduction in the impact of livestock within such settings, multiple management approaches may be necessary, for example reducing stocking densities, introducing riparian buffers as well as measure to removing transit points across water sources.

5.7.2 Case study 2: prediction of spatial and temporal E. coli burden on pasture from grazing livestock in Scotland (UK)

Oliver and co-workers (2018) evaluated the recently developed Visualising Pathogen & Environmental Risk (ViPER) model, created to predict the *E. coli* burden on agricultural land (Oliver *et al.*, 2012). The model was applied in a spatially distributed manner to two contrasting catchments (the River Ayr and Lunan Water) where livestock numbers were known to represent relatively high and low grazing densities. The outputs from the model allowed exploring where, when, and why *E. coli* burden varied within these two contrasting catchments. Application of the model revealed significant correlations between area of improved grassland and the maximum total *E. coli* per 1 km^2 grid cell (Ayr: $r = 0.57$; $p < 0.001$, Lunan: $r = 0.32$; $p < 0.001$). A significant difference in the predicted maximum *E. coli* burden between seasons in both catchments, with summer and autumn predicted to accrue higher *E. coli* contributions relative to spring and winter ($p < 0.001$), driven largely by livestock presence, could also be identified. The findings of this study suggest that the ViPER model is capable of capturing spatial nuances (at the landscape scale) in the vulnerability of *E. coli* loading to land as driven by stocking density and livestock grazing regimes. The resulting risk maps are therefore of value to catchment managers, as they provide useful evidence to inform spatially targeted decision-making with respect to managing sources of *E. coli* in agricultural environments.

5.7.3 Case study 3: tracking sources of faecal pollution in a mixed-agricultural catchment (Thailand)

Kongprajug and co-workers (2019b) studied the Tha Chin River, which has been ranked one of the most polluted freshwater rivers in Thailand for several decades. Various activities along this 325 km freshwater river include human residential areas, agricultural land use and animal husbandry operations (e.g., swine and cattle farms), especially along the downstream sites prior to where it connects to the Gulf of Thailand. This study utilised validated DNA-based host-specific gut microorganisms, which are known as microbial source tracking markers, to identify sources of faecal pollution that are specifically from human sewage, swine, and cattle as well as non-host-specific (universal) faecal pollution. The results, per universal Bacteroidales 16 s rRNA gene detection, found that this catchment experienced persistent and continuing faecal contamination. Although no significant difference in precipitation was observed between the two wet and two dry samplings during a one-year period, cattle-specific qPCR markers were detected only in the wet season. Conversely, the human sewage- and swine-specific qPCR marker concentrations did not vary among the four sampling events. The different temporal distributions among various faecal sources could imply diverse contamination patterns, which might possibly have dissimilar point and/or diffuse contamination sources. This research was the first to discriminate sources of faecal pollution in freshwater in Thailand. Its results could guide both the pollution surveillance and the water quality management of polluted water bodies towards improvements in water quality.

REFERENCES

Arnold J. G. and Fohrer N. (2005). SWAT2000: current capabilities and research opportunities in applied watershed modelling. *Hydrological Processes*, **19**(3), 563–572, https://doi.org/10.1002/hyp.5611

Arnold J. G., Srinivasan R., Muttiah R. S. and Williams J. R. (1998). Large hydrological modelling and assessment part 1: model development. *Journal of American Water Resources Association*, **24**(1), 73–89, https://doi.org/10.1111/j.1752-1688.1998.tb05961.x

Arnold J. G., Moriasi D. N., Gassman P. W., Abbaspour K. C., White M. J., Srinivasan R., Santhi C., Harmel R. D., van Griensven A., Van Liew M. W., Kannan N. and Jha M. K. (2012). SWAT: model use, calibration, and validation. *Transactions of the ASABE*, **55**(4), 1491–1508, https://doi.org/10.13031/2013.42256

Belkin S. and Colwell R. R. (2005). Ocean and Health: Pathogens in the Marine Environment. Springer, New York.

Berg G. (1978). The indicator system. In: Indicators of Viruses in Water and Food, G. Berg (ed.), Ann Arbor Science Publishers, Ann Arbor, pp. 1–13.

Betancourt W. (2019). Cryptosporidium spp. In: Global Water Pathogen Project, J. B. Rose and B. Jiménez-Cisneros (eds.), http://www.waterpathogens.org (R. Fayer and W. Jakubowski, (eds) Part 3 Protists) http://www.waterpathogens.org/book/cryptosporidium (accessed 14 October 2019). Michigan State University, E. Lansing, UNESCO, MI, USA, pp. 7–9.

Burt T. P. and Howden N. J. K. (2013), North Atlantic Oscillation amplifies orographic precipitation and river flow in upland Britain. *Water Resources Research*, **49**, 3504–3515, https://doi.org/10.1002/wrcr.20297

Byappanahalli M. N., Whitman R. L., Shively D. A., Ting W. T. E., Tseng C. C. and Nevers M. B. (2006). Seasonal persistence and population characteristics of *Escherichia coli* and enterococci in deep backshore sand of two freshwater beaches. *Journal of Water and Health*, **4**, 313–320, https://doi.org/10.2166/wh.2006.018b

Bryan B. A., Kandulu J., Deere D. A., White M., Frizenschaf J. and Crossman N. D. (2009). Adaptive management for mitigating cryptosporidium risk in source water: a case study in an agricultural catchment in South Australia. *Journal of Environmental Management*, **90**, 3122–3134, https://doi.org/10.1016/j.jenvman.2009.05.014

Christou L. (2011). The global burden of bacterial and viral zoonotic infections. *Clinical Microbiology and Infectious Diseases*, **17**, 326–330.

Chyerochana N., Kongprajug A., Somnark P., Kampaengthong P. L., Mongkolsuk S. and Sirikanchana K. (2020). Distributions of enterococci and human-specific bacteriophages of enterococci in a tropical watershed. *International Journal of Hygiene and Environmental Health*, **226**, 1–9, https://doi.org/10.1016/j.ijheh.2020.113482

Davies-Colley R. J., Nagels J. W., Smith R. A., Young R. G. and Phillips C. J. (2004). Water quality impact of a dairy cow herd crossing a stream. *New Zealand Journal of Marine and Freshwater Research*, **38**, 569–576, https://doi.org/10.1080/00288330.2004.9517262

Dile Y. T., Daggupati P., George C., Srinivasan R. and Arnold J. G. (2016). Introducing a new open source GIS user interface for the SWAT model. *Environmental Modelling and Software*, **85**, 129–138, https://doi.org/10.1016/j.envsoft.2016.08.004

Environment Agency. (2018). The State of the Environment: Water Quality. UK Environment Agency Report LIT 10702, Bristol, UK.

Ebdon J. E., Sellwood J., Shore J. and Taylor H. D. (2012). Phages of *bacteroides* (GB-124): a novel tool for viral waterborne disease control? *Environmental Science and Technology*, **46**, 1163–1169, https://doi.org/10.1021/es202874p

FAO. (2006). Livestock's Long Shadow: Environmental Issues and Options. FAO, Rome.
Fewtrell L. and Kay D. (2015). Recreational water and infectious: a review of recent findings. *Current Environmental Health Reports*, **2**(1), 85–94, https://doi.org/10.1007/s40572-014-0036-6
Field K. G. and Samadpour M. (2007). Fecal source tracking, the indicator paradigm, and managing water quality. *Water Research*, **41**, 3517–3538, https://doi.org/10.1016/j.watres.2007.06.056
Ford T. E. and Hamner S. (2018). A perspective on the global pandemic of waterborne disease. *Microbial Ecology*, **76**, 2–8, https://doi.org/10.1007/s00248-015-0629-0
Fu B., Merritt W. S., Croke B. F. W., Weber T. R. and Jakeman A. J. (2019). A review of catchment-scale water quality and erosion models and a synthesis of future prospects. *Environmental Modelling & Software*, **114**, 75–97, https://doi.org/10.1016/j.envsoft.2018.12.008
Góralska K., Błaszkowska J. and Dzikowiec M. (2020). Occurrence of potentially pathogenic filamentous fungi in recreational surface water as a public health risk. *Journal of Water and Health*, **18**, 127–144, https://doi.org/10.2166/wh.2020.096
Graciaa D. S., Cope J. R., Roberts V. A., Cikesh B. L., Kahler A. M. and Vigar M. (2018). Outbreaks associated with untreated recreational water – United States, 2000–2014. *Morbidity and Mortality Weekly Report*, **67**, 2016–2017.
Grossart H. P. and Rojas-Jimenez K. (2016). Aquatic fungi: targeting the forgotten in microbial ecology. *Current Opinion in Microbiology*, **31**, 140–145, https://doi.org/10.1016/j.mib.2016.03.016
Haas C., Rose J., Gerba C. and Stig R. (1993). Risk assessment of viruses in drinking water. *Risk Analysis*, **13**, 545–552, https://doi.org/10.1111/j.1539-6924.1993.tb00013.x
Hamza I. A., Jurzik L., Uberla K. and Wilhelm M. (2011). Methods to detect infectious human enteric viruses in environmental water samples. *International Journal of Hygiene and Environmental Health*, **214**, 424–436, https://doi.org/10.1016/j.ijheh.2011.07.014
IAWPRC (IAWPRC Study Group on Health-Related Water Microbiology). (1991). Bacteriophages as model viruses in water quality control. *Water Research*, **25**, 529–546, https://doi.org/10.1016/0043-1354(91)90126-B
Jiang X., Morgan J. and Doyle M. P. (2002). Fate of *Escherichia coli* O157:H7 in manure-amended soil. *Applied and Environmental Microbiology*, **68**, 2605–2609, https://doi.org/10.1128/AEM.68.5.2605-2609.2002
Kay D., Crowther J., Kay C., McDonald A. T., Ferguson C., Stapleton C. M. and Wyer M. D. (2012). Effectiveness of best management practices for attenuating the transport of livestock-derived pathogens within catchments. In: Animal Waste, Water Quality and Human Health, A. Dufour, J. Bartram, R. Bos, and V. Gannon, (eds.), IWA Publishing, London, pp. 197–225.
Kongprajug A., Mongkolsuk S. and Sirikanchana K. (2019a). CrAssphage as a potential human sewage marker for microbial source tracking in Southeast Asia. *Environmental Science & Technology Letters*, **6**, 159–164, https://doi.org/10.1021/acs.estlett.9b00041
Kongprajug A., Chyerochana N., Somnark P., Kampaengthong P., Mongkolsuk S. and Sirikanchana K. (2019b). Human and animal microbial source tracking in a tropical river with multiple land use activities. *International Journal of Hygiene and Environmental Health*, **222**, 645–654, https://doi.org/10.1016/j.ijheh.2019.01.005
Leclerc H. (2003). Relationships between common water bacteria and pathogens in drinking-water. In: Heterotrophic Plate Counts and Drinking-Water Safety, J. Bartram, *et al.* (eds.), IWA Publishing, London, Chap. 6, pp. 80–118.
LeClerc H., Schwartzbrod L. and Dei-Cas E. (2002). Microbial agents associated with waterborne diseases. *Critical Reviews in Microbiology*, **28**(4), 371–409, https://doi.org/10.1080/1040-840291046768
Leifels M., Shoults D., Wiedemeyer A., Ashbolt N. J., Sozzi E., Hagemeier A. and Jurzik L. (2019). Capsid integrity qPCR–an azo-dye based and culture-independent approach to estimate adenovirus infectivity after disinfection and in the aquatic environment. *Water*, **11**, 1196, https://doi.org/10.3390/w11061196
Leifels M., Dan C., Sozzi E., Shoults D., Wuertz S., Mongkolsuk S. and Sirikanchana K. (2021). Capsid integrity quantitative PCR to determine virus infectivity in environmental and food applications; a systematic review. *Water Research X*, **11**, 100080, https://doi.org/10.1016/j.wroa.2020.100080
Liang L., Goh S. G., Vergara G. G. R. V., Fang H. M., Rezaeinejad S., Chang S. Y., Bayen S., Lee W. A., Sobsey M. D., Rose J. B. and Gin K. Y. H. (2015). Alternative fecal indicators and their empirical relationships with enteric viruses, *Salmonella enterica*, and *Pseudomonas aeruginosa* in surface waters of a tropical urban catchment. *Applied and Environmental Microbiology*, **81**, 850–860, https://doi.org/10.1128/AEM.02670-14
Mackowiak M., Leifels M., Hamza I. A., Jurzik L. and Wingender J. (2018). Distribution of *Escherichia coli*, coliphages and enteric viruses in water, epilithic biofilms and sediments of an urban river in Germany. *Science of the Total Environment*, **626**, 650–659, https://doi.org/10.1016/j.scitotenv.2018.01.114

McQuaig S., Griffith J. and Harwood V. J. (2012). Association of fecal indicator bacteria with human viruses and microbial source tracking markers at coastal beaches impacted by nonpoint source pollution. *Applied and Environmental Microbiology*, **78**, 6423–6432, https://doi.org/10.1128/AEM.00024-12

Momba M., Edbon J., Kamika I. and Verbyla M. (2019). Using indicators to assess microbial treatment and disinfection efficacy. In: Global Water Pathogen Project, J. B. Rose and B. Jiménez-Cisneros (eds.), http://www.waterpathogens.org (A. Farnleitner, and A. Blanch (eds) Part 2 Indicators and Microbial Source Tracking Markers) http://www.waterpathogens.org/book/using-indicators-assess-microbial-treatment-and-disinfection-efficacy (accessed 14 October 2019). Michigan State University, E. Lansing, MI, Unesco, pp. 1–49, https://doi.org/10.14321/waterpathogens.9

NHMRC. (2001). Microbial Indicators of Water Quality: an NHMRC Discussion Paper. National Health and Medical Research Council, Commonwealth of Australia, Canberra, Australia.

Oliver D. M., Page T., Zhang T., Heathwaite A. L., Beven K., Carter H., McShane G., O'Keenan P. and Haygarth P. M. (2012). Determining *E. coli* burden on pasture in a headwater catchment: combined field and modelling approach. *Environment International*, **43**, 6–12, https://doi.org/10.1016/j.envint.2012.02.006

Oliver D., Bartie P. J., Heathwaite A. L., Reaney S. M., Jared A. Q., Parnell J. A. Q. and Quilliam R. S. (2018). A catchment-scale model to predict spatial and temporal burden of *E. coli* on pasture from grazing livestock. *Science of the Total Environment*, **616–617**, 678–687, https://doi.org/10.1016/j.scitotenv.2017.10.263

Palazón A., López I., Aragonés L., Villacampa Y. and Navarro-González F. J. (2017). Modelling of *Escherichia coli* concentrations in bathing water at micro tidal coasts. *Science of the Total Environment*, **593**, 173–181, https://doi.org/10.1016/j.scitotenv.2017.03.161

Payan A., Ebdon J., Taylor H., Gantzer C., Ottoson J., Papageorgiou G. T., Blanch A. R., Lucena F., Jofre J. and Muniesa M. (2005). A method for the isolation of suitable bacteriophage host strains of Bacteroides for tracking sources of faecal pollution in water. *Applied and Environmental Microbiology*, **71**, 5659–5662, https://doi.org/10.1128/AEM.71.9.5659-5662.2005

Payment P. (1998). Water borne viruses and parasites: resistance to treatment and disinfections. Presented at OECD Workshop on Molecular Methods for Safe Drinking Water, 5–8 July 1998, Interlaken, Switzerland, pp. 1–11.

Purnell S. E., Ebdon J. E., Wilkins H. and Taylor H. D. (2018). Human-specific phages infecting *Enterococcus* host strain MW47: are they reliable microbial source tracking markers? *Journal of Applied Microbiology*, **124**, 1274–1282, https://doi.org/10.1111/jam.13700

Rodriguez-Lazaro D., Cook N., Ruggeri F. M., Sellwood J., Nasser A., Nascimento M. S., D'Agostino M., Santos R., Saiz J. C., Rzezutka A., Bosch A., Girones R., Carducci A., Muscillo M., Kovac K., Diez-Valcarce M., Vantarakis A., von Bonsdorff C. H., de Roda Husman A. M., Hernandez M. and van der Poel W. H. (2012). Virus hazards from food, water and other contaminated environments. *FEMS Microbiology Reviews*, **36**, 786–814, https://doi.org/10.1111/j.1574-6976.2011.00306.x

Rosario K., Nilsson C., Lim Y., Wei Ruan Y. and Breitbart M. (2009). Metagenomic analysis of viruses in reclaimed water. *Environmental Microbiology*, **11**, 2806–2820, https://doi.org/10.1111/j.1462-2920.2009.01964.x

Smolders A., Rolls R. J., Ryder D., Watkinson A. and Mackenzie M. (2015). Cattle-derived microbial input to source water catchments: an experimental assessment of stream crossing modification. *Journal of Environmental Management*, **156**, 143–149, https://doi.org/10.1016/j.jenvman.2015.03.052

Stachler E., Kelty C., Sivaganesan M., Li X., Bibby K. and Shanks O. C. (2017). Quantitative CrAssphage PCR assays for human fecal pollution measurement. *Environmental Science & Technology*, **51**, 9146–9154, https://doi.org/10.1021/acs.est.7b02703

Stoeckel D. M. and Harwood V. J. (2007). Performance, design, and analysis in microbial source tracking studies. *Applied and Environmental Microbiology*, **73**(8), 2405–2415, https://doi.org/10.1128/AEM.02473-06

UN FAO. (2018). More People, More Food, Worse Water? A Global Review of Water Pollution From Agriculture. FAO, Rome; and International Water Management Institute on behalf of the Water Land and Ecosystems research program of CGIAR Colombo. ISBN 978-92-5-130729-8.

US Environmental Protection Agency. (2015). Polluted Runoff: Nonpoint Source Pollution. Environmental Protection Agency, Washington, DC, USA, https://www.epa.gov/nps/what-nonpoint-source (accessed 14 October 2019).

Weiskerger C. J. and Brandão J. (2020). Fungal contaminants in water and sand: a new frontier for quantitative microbial risk assessment. *Current Opinion in Environmental Science & Health*, **16**, 73–81, https://doi.org/10.1016/j.coesh.2020.03.001

WHO. (2011). Guidelines for Drinking Water Quality. 4th edn. World Health Organization, Geneva.

Wong K., Fong T., Bibby K. and Molina M. (2012). Application of enteric viruses for faecal pollution source tracking in environment waters. *Environment International*, **45**, 151–164, https://doi.org/10.1016/j.envint.2012.02.009

Woodall C. (2009). Waterborne diseases – what are the primary killers? *Desalination*, **248**(1–3), 616–621, https://doi.org/10.1016/j.desal.2008.05.110

IWA PUBLISHING

doi: 10.2166/9781789061123_083

Chapter 6

Emerging pollutants from various land use types and their impacts on water resources

N. J. Reyes and L. H. Kim

Kongju National University, Civil and Environmental Engineering Department, Cheonan City, Chungcheongnamdo, South Korea

Keywords: Emerging pollutants; land use and land-use changes; nature-based solutions; non-point source; point source

6.1 EMERGING POLLUTANTS (EPS) IN THE ENVIRONMENT

The past decades have been characterized by an extensive production, synthesis, and use of chemical compounds for human consumption and production of goods. The increasing demand for these chemicals consequently resulted to the occurrence of a wide spectrum of substances in the environment. While several degradation processes and dilution may occur after the release in a natural environment, certain compounds remain in stable condition and may circulate in the environment indefinitely. Emerging pollutants (EPs), sometimes referred to as emerging contaminants or contaminants of emerging concern, are special groups of substances that are gaining global attention owing to their relative abundance and persistence in the environment. As defined by the United States Geological Survey (USGS), EPs are synthetic or natural compounds or microorganisms that are not monitored but can have potential adverse ecological or human health effects when introduced into the environment. These substances occur in the environment in minute concentrations, usually ranging from pg/L to ng/L, but their continuous release in a natural setting can pose risks to several organisms due to long-term exposure (Calvo-Flores *et al.*, 2017; Raghav *et al.*, 2013). Some groups of EPs were composed of substances that have just been recently developed and introduced into the environment (e.g. industrial and agricultural chemicals that are used for the production of goods). Various compounds that already exist in the environment for a long time can also be classified as EPs. Environmental concentrations of EPs are very low and thus, conventional sampling and detection techniques may be insufficient to recognize the presence of such compounds. Recent advancements in analytical chemistry and the development of highly sensitive detection apparatuses enabled the identification and quantification of certain EPs in the environment. Since the presence and the potential hazards of some compounds were only recently identified, no guidelines or limited control measures have been implemented across different locations (Geissen *et al.*, 2015; Sardiña *et al.*, 2019; Sharma *et al.*, 2019).

6.1.1 History and progress in EP research

The main limiting factor for the discovery of EPs is the lack of instrumentation and detection techniques. In 1953 Wolfgang Paul developed the concept of the mass filter or spectrometer which paved the

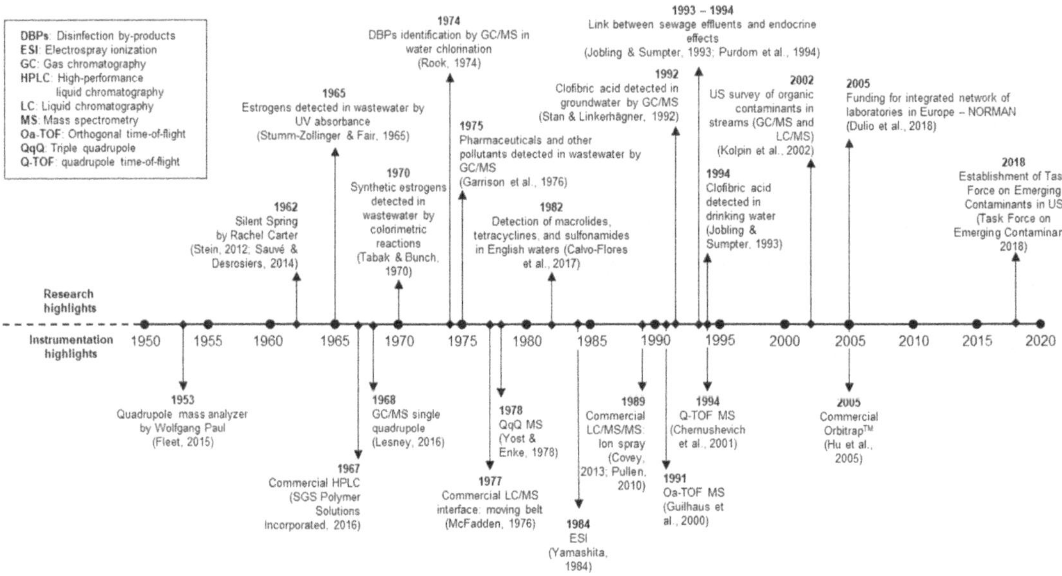

Figure 6.1 Research and instrumentation highlights for EPs (Modified from Noguera-Oviedo & Aga, 2016).

way for advanced chemical analyses (Fleet, 2015). Following the development of instrumentation and analytical methods, studies about the chemicals present in different environmental media started to flourish. The first publication in relation to EPs can be traced back in 1962, wherein Carson emphasized that the use of dichlorodiphenyltrichloroethane (DDT) as a mosquito repellent consequently led to the death or disappearance of many bird species (Sauvé & Desrosiers, 2014; Stein, 1962). The occurrence of steroid hormones in North American wastewater was reported by Stumm-Zollinger and Fair in 1965. They emphasized that steroid hormones, such as ovulation-inhibiting steroids, can reach surface water bodies through wastewater discharge and may gradually be present in drinking water. In response to the commercialization of spectrometry devices capable of detecting low pollutant concentrations, the scientific community showed an increasing interest in the study of EPs. The detection of pharmaceutical, disinfection by-products, and other organic contaminants in water was extensively studied alongside the development of advanced instrumentation. The progress of research on EPs and instrumentation highlights are illustrated in Figure 6.1.

The number of publications concerning EPs increased continuously over the last 25 years. A handful of reports mainly focused on the occurrence of potentially toxic or physiologically active chemical compounds in wastewater (Ahel & Giger, 1985; Pham & Proulx, 1997; Vogelsang *et al.*, 2006). As the primary pathway of EPs in the environment, examination of wastewater components was considered as the first step in understanding the behaviour of different pollutants. The fate, occurrence and distribution of EPs in different environmental compartments were eventually seen as crucial aspects in ecosystem studies. Despite the extensive literature and publications with regard to EPs in varying environmental matrices, research gaps remain with regard to their ecotoxicological effects, removal pathways, and control measures. Understanding the behaviour and properties of EPs can be a daunting challenge for the whole scientific community, but the shifting interest from conventional water pollutants to in-depth analyses of EPs can pave the way for a better understanding of the emerging threats to ecosystems.

6.1.2 Classifications of EPs

EPs can be classified into different groups of chemical compounds or substances. The list includes chemicals from industrial activities and production of goods, disinfection by-products from water treatment, human pharmaceuticals, personal care products, veterinary medicines, and a wide array of substances that can potentially result in detrimental effects to the environment and human health. The NORMAN network, initiated by the European Commission in 2005, was established to enhance the exchange of information among research groups and experts in the field of emerging substances from different countries (NORMAN, n.d.). The NORMAN network holds one of the largest databases for EPs with a total of 1036 identified compounds and their biotransformation products. There is a vast number of EPs that currently persist in various environmental matrices; however, the amounts and types of EPs in the environment may continue to increase due to the introduction of new commercial and industrial compounds and the ever-changing patterns of human consumption (Dey *et al.*, 2019). The characteristics and examples of different EPs are summarized in Table 6.1.

Table 6.1 Classification of and description of major EPs.

Classification	Description	Examples
Perfluorinated Compounds (PFCs)	PFCs are synthetic or manufactured compounds with a 'fully fluorinated hydrophobic linear carbon chain attached to one or more hydrophilic head groups.' This group of chemical compounds is characterized by their ability to repel water and oil, making them ideal chemicals for surface treatment (Petrovic *et al.*, 2013).	Perfluorinated carboxylic acid (PFCA) Perfluorocarbon sulfonic acid (PFSA)
Disinfection By-products (DBPs)	The formation of DBPs occurs when disinfectants (i.e., chlorine, chlorine dioxide, chloramine, and ozone) react with organic matter and inorganic substances present in water (Richardson *et al.* 2007).	Trihalomethane (THM) Haloacetic acid (HAA)
Pharmaceuticals and Personal Care Products (PPCPs)	According to the general definition provided by the US Environmental Protection Agency (US EPA), PPCPs are 'any products used by individuals for personal health or cosmetic reasons or used by agribusiness to enhance growth or health of livestock' (Cizmas *et al.* 2015)	Musk UV Filter Anti-inflammatory drugs Antibiotics
Endocrine Disrupting Compounds (EDCs)	EDCs are explicitly defined by the World Health Organization (WHO) and International Programme on Chemical Safety (IPCS) as 'exogenous substances or mixtures that alter function(s) of the endocrine system and consequently cause adverse health effects in an intact organism, or its progeny, or (sub) populations' (Miyagawa *et al.* 2016)	Bisphenol A (BPA) Estrone (E1) 17β-estradiol (E2) Estriol (E3) 17α-ethinylestradiol (EE2)
Nanomaterials (NMs)	NMs are basically defined as particles having at least one dimension in the nanoscale (usually 1 to 100 nm). NMs are widely used in various applications due to their superb performance attributable to these materials' tuneable physico-chemical characteristics (Gentile *et al.* 2016; Jeevanandam *et al.* 2018)	Fullerenes Carbon nanotubes Quantum dots
Pesticides	US EPA defined pesticides as '(1) any substance intended for preventing, destroying, repelling, or mitigating any pest, (2) any substance or mixture of substances intended for use as a plant regulator, defoliant, or desiccant, and (3) any nitrogen stabilizer' (US EPA, 2013)	Dichlorodiphenyl-ethanes Organophosphorus esters Carbamic acid esters
Industrial chemicals and miscellaneous compounds	Industrial chemicals include a broad spectrum of compounds used as additives or primary components for manufacturing commercial products.	Corrosion inhibitor Flame retardant Plasticizer Artificial sweetener Gasoline additives

6.2 SOURCES, PATHWAYS AND PRESENCE OF EPS IN DIFFERENT LAND-USE TYPES

Land use and land-use changes (LULUCs) are considered as primary factors shaping the present environment. Apart from their economic and socio-cultural implications, LULUCs can directly or indirectly affect environmental quality through the alteration of pollutant deposition and circulation patterns. These effects are associated with the changes in impermeable surface ratio, population density, and anthropogenic activities for each set of land-use type. One concrete example of the effect of LULUCs on environmental quality is the global urbanization trend. The past decades were characterized by drastic conversion of agricultural and forest areas into urban landscapes. Across Europe and Asia, more than half of the urban area expansion was attributed to the conversion of agricultural areas. Approximately 13% of the global loss of forest areas from 1970 to 2010 was also related to urban land-use conversion (Güneralp *et al.*, 2020). This continuous trend of urban expansion resulted in degraded environmental quality as a factor of increased point and non-point sources of pollution alongside the upsurge in population density. Agricultural and livestock regions have relatively lower population densities in comparison to urban areas, but they are also considered as pollution 'hotspots' due to the significant amounts of pollutants generated from cultivation activities and animal husbandry. In the case of industrial sites, highly concentrated chemical discharges from factories can be regarded as major point sources of pollutants in receiving water bodies. Generally, all land-use categories contribute a fair share of pollutants in the environment.

Similarly to conventional water quality pollutants (e.g., particulates, organics, nutrients, and heavy metals), EPs can be introduced into the environment through point and non-point sources (as shown in Figure 6.2). The types, concentrations, and transport mechanisms of EPs may vary for different land uses; however, they exist in the environment as a complex mixture of chemical substances, since water bodies and soil matrices serve as the ultimate sinks of pollutants. It is therefore necessary to determine the dominant types of EPs in each land-use category to understand the sources, environmental occurrence, and deposition pathways of these substances.

6.2.1 Point sources

Wastewater is the main source of EPs in the environment. Industrial processes, agricultural practices, and anthropogenic activities generate wastewater that usually ends up in the environment. A portion of the total generated wastewater can be redirected to treatment facilities but in some cases, wastewater is directly discharged into receiving water bodies. Moreover, wastewater treatment plants (WWTPs) are not designed to eliminate EPs. Some compounds that are resistant to degradation or other physico-chemical wastewater treatment processes may persist in the effluent and therefore, accumulate in the environment.

Urban areas are characterized by vast anthropogenic activities that generate pollutant loads of varying types and concentrations. The majority of EP groups are present in urban wastewaters since the end-products of manufacturing processes and other materials containing EP compounds are mainly utilized by humans. In urban areas, households represent the most important source of EPs (Stefanakis & Becker, 2015). Human excrement contains residual amounts of EPs usually derived from the consumption of PPCPs, artificial sweeteners, and other natural or synthetic EDCs that cannot be completely metabolized by the human body. Typical household activities (e.g., washing the dishes, laundry, and urban gardening) and disposal of items with PFCs, pesticides, and other industrial chemicals also contribute to the EP loads in wastewater. In areas with combined sewer systems, stormwater runoff may also contribute significant EP loads in wastewater due to the washoff of contaminated sediment deposits or chemical residues on impermeable surfaces.

EP loads of industrial origin are mainly by-products of manufacturing processes. The quality and volume of industrial wastewaters are highly variable and dependent on the type of manufactured materials (Cisneros, 2011). Industrial wastewaters may have highly toxic properties that often require advanced treatment mechanisms to substantially reduce harmful pollutant concentrations. Since

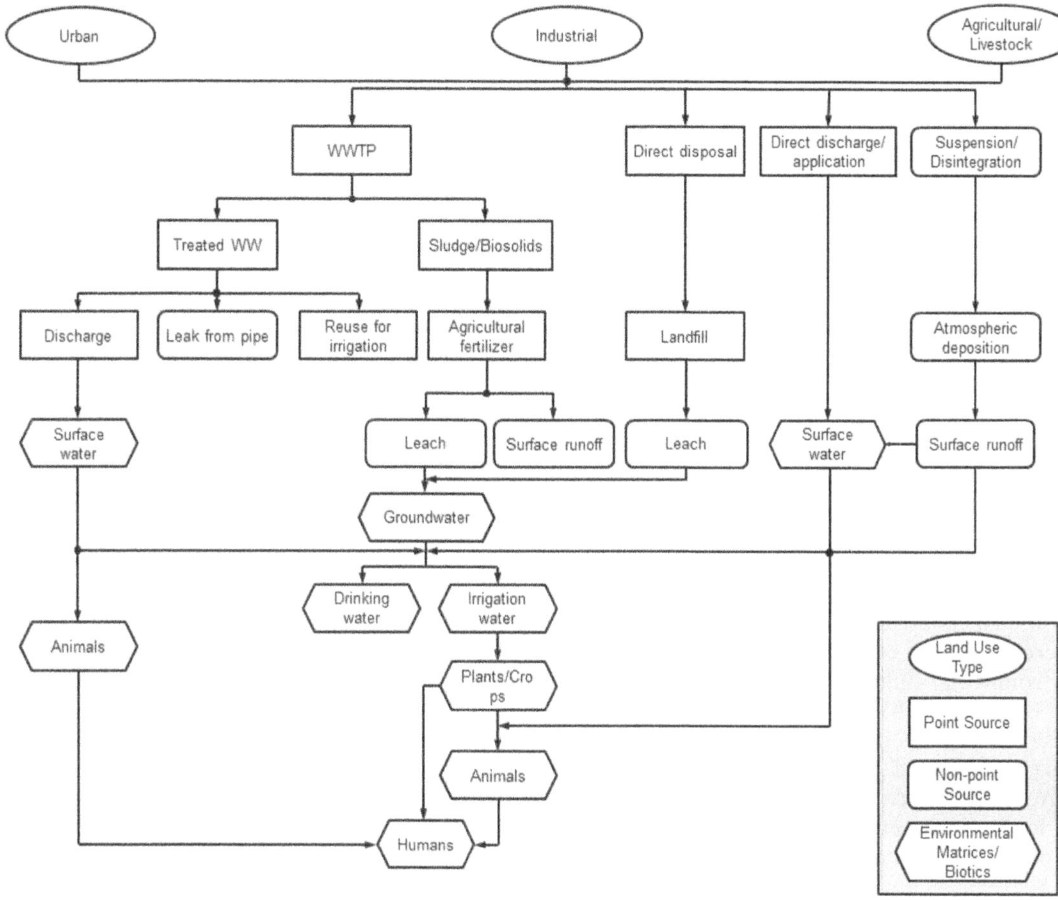

Figure 6.2 Point and non-point sources of EPs.

there are no guidelines or water quality limits for some EPs, their presence in industrial effluents can remain unregulated. Although industrial areas contribute greatly to the presence of EPs in the environment, pollutant concentration data from these point sources are seldom reported or studied as compared with domestic wastewater.

Livestock farms constitute the major point source of EPs in agricultural areas. Veterinary drugs and supplements are used extensively to enhance animal health and well-being. However, animals also cannot completely metabolize these compounds, thereby resulting to the presence of pharmacologically active compounds in the wastewater stream (Khan *et al.*, 2019). Apart from antibiotics and hormones, livestock wastewater may also contain artificial sweeteners. These compounds are added on livestock feeds to increase feed intake and weight gain of animals (Sterk *et al.*, 2008). In general, point sources of EPs in agricultural areas are limited, since wastewater treatment is mainly focused on animal-raising industries.

Point sources serve as major pathways of EPs in the environment. The continuous release of treated or untreated wastewater in the environment ultimately results in the ubiquitous presence and accumulation of EPs in various environmental compartments and organisms (Ebele *et al.*, 2020).

While it is evident that each land-use type contributes a unique set of EPs, it is highly probable that a wide array of compounds remain unexplored. Additionally, further investigations should be conducted on industrial and agricultural point sources to identify the presence of other EP compounds and establish a definitive pattern of EP loadings.

6.2.2 Non-point sources

Non-point, or diffuse, sources remain a critical challenge in managing the occurrence of EPs in the environment. There are two categories of diffuse sources of emerging pollutants: at the point of manufacture (runoff from the factory picking up contamination from roofs, yards and roads there), and from the very many dispersed points of use (D'Arcy *et al.*, 2018). Unlike point sources with definite points of origin, or diffuse sources contaminating runoff from a factory or industrial estate, non-point source EPs associated with the use of the products come from several diffuse sources that usually encompass large geographical areas (Stefanakis & Becker, 2015). This scenario further complicates the complex nature of EPs due to the difficulty in tracing the specific sources of compounds. If EPs are mobilised in similar ways to other contaminants (see Ferrier *et al.*, 2005) overall *loads* will often be significant (Mitchell, 2005). Some studies also suggest that the contribution of non-point sources pose equal or greater importance than actual point sources (Kim *et al.*, 2014; Lamastra *et al.*, 2016). Moreover, non-point pollutants can be directly discharged into receiving water bodies or deposited on various environmental compartments without further treatment.

As illustrated in Figure 6.3, anthropogenic activities in varying land-use types can also influence non-point sources of EPs. Stormwater or surface runoff is the most common medium for the transport of EPs. The contribution of non-point pollutants can be significant, especially during rainfall events wherein deposits of accumulated pollutants on impervious surfaces or soil layers are transported to environmental sinks. EPs can also reach the atmosphere by way of volatilization, aerosol formation, and diffusive exchange processes. Succeeding dry and wet deposition mechanisms can redistribute these compounds in different environmental matrices and thus result in the ubiquitous presence of EPs (Barroso *et al.*, 2019).

Apart from being point-source generators, urban and industrial areas also release significant EP loads from non-point sources. Chemical leaks and spills from degraded conveyance pipes in industrial areas usually result in the contamination of soils or water bodies surrounding the vicinity of

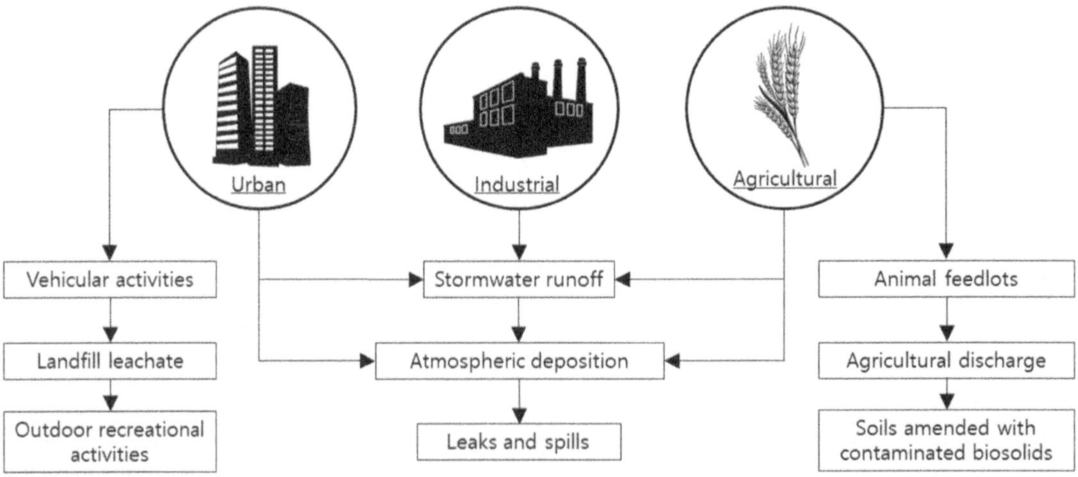

Figure 6.3 Non-point sources of EPs for different land use types.

manufacturing sites. Moreover, weathering and leaching of EP compounds during the manufacturing process leave chemical residues that can be potentially transported in the environment (Barroso *et al.*, 2019). Urban areas are characterized by extensive vehicular activities. Studies found that vehicle exhausts and tyre wear due to abrasion generate nanomaterials and other combustion by-products (Balducci *et al.*, 2012; Sheehan, 2016). Densely populated urban areas produce large amounts of heterogeneous mixtures of solid and liquid wastes that are usually disposed of in landfills. These sites are usually engineered to prevent the unintended release of pollutants in the environment.

At the landfill site, leachates produced by the complex mixture of wastes can still be considered as potential non-point sources of EPs (Masoner *et al.*, 2014). DBP precursors, externally applied personal care products, and PFCs on textiles can also contribute to the presence of EPs in the environment during recreational activities, such as swimming and bathing.

Agricultural areas have greater tendencies to be diffuse sources of EPs in the environment. Trace concentrations of veterinary pharmaceuticals and feed additives are usually found on animal feedlots, especially in areas designated for livestock industries. In addition to animal husbandry, other agricultural practices may also contribute to non-point sources of EPs. Biosolids are commonly used as organic fertilizers on farmlands due to their high nutrient content. Despite being a feasible alternative to chemical fertilizers, biosolids usually contain EP compounds from the wastewater (Mohapatra *et al.*, 2016). Accumulated EP compounds on biosolids can leach and percolate through soil layers or be washed-off along with surface runoff. Recently, water reuse has been a growing practice all over the world. This concept aims to relieve the pressure on water resources in regions experiencing water scarcity. However, some published literature suggests that the use of reclaimed wastewater for irrigating croplands redistributes EPs in the agricultural areas (Estévez *et al.*, 2012; Hyland *et al.*, 2015; Qin *et al.*, 2015).

Non-point sources can greatly contribute to the presence of EPs in the environment. Similarly to point sources, the types and concentration of EP compounds from non-point sources are greatly influenced by land use. EPs from diffuse sources also exhibit complex pathways and patterns of deposition due to environmental factors (e.g., rainfall, temperature, velocity vectors, etc.) affecting the mobility and transport of these compounds. Generally, managing the presence of EP compounds in the environment requires a better understanding of their non-point sources and transport mechanisms.

6.3 IMPACTS OF EPS ON WATER RESOURCES

Water is a vital resource for sustaining the life of all living components of the ecosystem. Water use has been continuously increasing owing to rising demand as a factor of population growth, socio-economic development, and changing patterns of water consumption. This surge in water demand has resulted in water scarcity problems in several parts of the world. In a report published by the United Nations (UN), over 2 billion people experience high water stress in different countries around the world. This situation is aggravated by climate change and water quality degradation (UN, 2019). Altered rainfall patterns and increased occurrence of extreme weather events can significantly affect water resources qualitatively and quantitatively. Long-term droughts limit the available water supply, whereas frequent flooding speeds up water quality deterioration due to the deposition of pollutants in waterways. These constraints pose inevitable consequences to the state of water resources globally.

The discovery of EPs in natural waterways has raised concerns among various water resources sectors and practitioners. The presence of EPs in surface water has been widely studied due to its potentially detrimental effects on humans and aquatic biota. As one of the primary sources of drinking water, rivers, streams, and lakes contaminated by EPs pose high health risks for humans. Some surface water bodies with continuous flow are capable of self-purification, but human impacts and continuous pollutant loading from point and diffuse sources oftentimes exceed this capacity (Houtman, 2010). Aquatic organisms can also be directly affected by these compounds due to their prolonged and constant exposure to these harmful compounds. Scientists and researchers have given

Table 6.2 Concentration of EPs in water sources and treated water.

EP	Compound	Sample	Concentration, ng/L	References
PPCP	Ibuprofen	Surface water	2.4–320	Stackelberg *et al.* (2007); Lin *et al.* (2015); Kot-Wasik *et al.* (2016); Madikizela and Chimuka (2017); Lin *et al.* (2018); Lu *et al.* (2018); Yang *et al.* (2018); Ebele *et al.* (2020); Zhang *et al.* (2020)
		Drinking water	5.7–223.6	
		Groundwater	7–836.7	
	Carbamazepine	Surface water	0.32–9.78	
		Groundwater	<1–50	
		Drinking water	29	
	Diclofenac	Surface water	900–5300	
		Groundwater	<1–42	
	Tonalide	Surface water	2.98–232	
		Drinking water	36	
	Methyl Paraben	Surface water	1.4–3173.9	
	Galaxolide	Surface water	13.2–395	
	Sulfamethoxazole	Surface water	2.57–22.6	
		Groundwater	<1–64	
PFC	PFOS	Surface water	2–5900	Skutlarek *et al.* (2006); Lin *et al.* (2015); Chen *et al.* (2016)
		Groundwater	0.6–76.8	
	PFOA	Surface water	0.1–223.8	
		Groundwater	0.1–40.3	
EDC	E2	Surface water	0.05–0.8	Huang and Sedlak (2001); Kim *et al.* (2007); Lin *et al.* (2015)
	EE2	Surface water	<0.05–0.07	
		Groundwater	1822. 2	
	E1	Groundwater	1–14.7	
		Surface water	1.7–5	
Industrial chemical	BPA	Surface water	107	Achten *et al.* (2002); Kolb and Püttmann (2006); Stackelberg *et al.* (2007); Wolschke *et al.* (2011); Spoelstra *et al.* (2013)
		Drinking water	26	
	Acesulfame	Tap water	50–1590	
	Saccharin	Tap water	180–350	
	MTBE	Surface water	13–2357	
		Drinking water	<10–712	
	Benzotriazole	Surface water	24–304	
NM	C_{60} Fullerene	Surface water	0.008–59	Peters *et al.* (2018); Sanchís *et al.* (2018)
	Ag	Surface water	0.3–2.5	
	CeO_2	Surface water	0.4–5.2	
DBP	THM	Tap water	8.72–86.2	Santa Marina *et al.* (2010); Ioannou *et al.* (2016)
		Drinking Water	3500–22 200	
	HAA	Drinking Water	5800–14 700	
Pesticide	DEET	Drinking Water	78	Stackelberg *et al.* (2007); Hildebrandt *et al.* (2008); Wu *et al.* (2014)
	Heptachlor	Groundwater	4.97–189.2	
	Atrazine	Surface water	10–630	
		Groundwater	10–2460	
	Metolachlor	Surface water	10–270	
		Groundwater	10–2000	

special attention to EDCs due to their negative effects on species biodiversity and aquatic system productivity (Wang & Zhou, 2013).

Groundwater is an indispensable component of water resources that sustain human needs. In the United States alone, 83 billion gallons of water are withdrawn daily to supply domestic, agricultural, and industrial operations. It was also estimated that approximately 50% of the global population relies on groundwater as a potable water source (Kuwayama *et al.*, 2017; UN, 2015). One major issue that threatens groundwater resources is widespread contamination by different compounds. Unlike in the case of surface water bodies, groundwater contamination is nearly irreversible. A handful of studies suggest that EP compounds may reach groundwater by means of leaching, infiltration of contaminated water, and groundwater–surface water exchange (Lapworth *et al.*, 2012, 2015; Sorensen *et al.*, 2015; Stuart *et al.*, 2011). Given the persistent nature of several EPs, these compounds may continue to accumulate on groundwater reserves.

Drinking water undergoes rigorous treatment processes to ensure the safety of consumers. Under strict regulations, drinking water should be free from all types of substances that can possibly endanger public health and well-being; however, the majority of EP compounds are still unregulated. In a national survey conducted in the United States, 148 out of the 247 chemical and microbial analytes were detected at least once in the raw water source. Out of the 148 detected compounds and microorganisms, 121 were still found to exist in treated drinking water, implying that most treatment technologies are inadequate for the removal of EPs (Glassmeyer *et al.*, 2017). Similarly to most WWTPs, conventional water treatment technologies were not designed to eliminate EP compounds. EP concentrations in treated water can be mainly influenced by the water source. A portion of the EP concentrations in the water source can still be traced after treatment, thus resulting in inevitable exposure of humans to EP compounds. A collection of published literature on the concentration of various EPs in water sources and treated water is listed in Table 6.2.

Water can be considered as a renewable resource. The natural water cycle provides an infinite water supply that can be drawn from various sources, but it should also be noted that the global water supply is still subject to limitations. Despite the fact that water is continuously replenished, certain constraints such as pollution, over-extraction, and increasing demand can lead to the depletion of water resources. The widespread occurrence of EPs in different water sources can aggravate water scarcity problems, especially in areas with very limited water supply. In addition to water supply issues, human health and the aquatic ecosystem remain in constant risk due to the incessant exposure to these potentially toxic compounds. Ultimately, this emerging challenge highlights the need for proper land, water resources, and waste management to minimize the negative effects of EPs in the environment.

6.4 CHALLENGES, OPPORTUNITIES AND FUTURE CONCERNS

The environment experiences constant pressure from diverse anthropogenic stressors. EPs set new environmental challenges that can potentially have irreversible effects on biodiversity, water resources, and human health and well-being. These new sets of environmental pollutants are continuously discharged into the environment and can be amplified by the existing LULUCs scenarios. LULUCs accelerate environmental degradation due to the alteration of natural hydrology, increased pollutant generation, and other aspects leading to the loss of ecosystem services. In order to maximize the opportunities and address the future concerns brought about by EPs and LULUCs, the following process can be utilized by the policy-makers and other agencies concerned with environmental preservation.

- *Acknowledge the problem*
  Solving the problem begins with accepting the fact that a problem actually exists. Over the past decades, significant attention has been given to the occurrence and underlying consequences of

EPs in the environment. Research interest in EPs has also grown substantially since it has been acknowledged that these compounds have detrimental impacts on the environment and various living organisms. Progress in addressing environmental issues always begins with recognizing the existence of problems arising from various stressors.

- *Invest in research and development*
  Several technological and societal advancements were considered as products of scientific research and development. Investing in EP research is a crucial part of understanding the nature of these compounds and identifying possible control measures to prevent or mitigate their ubiquitous presence in the environment. Despite the rising number of publications in relation to EPs, it is still recognized that further studies are necessary to fully understand these compounds, especially in terms of their ecotoxicity and human health effects (Farré *et al.*, 2008; Pal *et al.*, 2014; Wei *et al.*, 2015). Development of alternative compounds that are less hazardous should also be encouraged. One concrete example of government initiatives for improving research and development can be reflected in the case of China. In the People's Republic of China, the government supported research and development efforts for devising alternatives to PFCs (OECD, n.d.). Generally, understanding the existing threats and developing alternatives or control measures for EPs can only be accomplished by investing in research.
- *Apply controls and implement regulatory measures*
  The implementation of regulations and control measures remains one of the key strategies to limit the presence of EPs in the environment. A complete set of guidelines should be implemented based on the results of numerous research and monitoring activities. For point sources, advanced wastewater treatment technologies were found to significantly decrease the EPs being discharged in the environment. NBS can also be applied to reduce point and non-point EPs in wastewater. Using nature-based technologies provides a cost-efficient alternative to advanced wastewater treatment schemes, especially for communities with financial constraints. It should also be noted that specific regulatory schemes and control measures for EPs should be carefully planned and evaluated in order to maximize the social, economic, and environmental benefits of the implemented strategies.
- *Evaluate, reconstruct, and adapt*
  It is always necessary to conduct constant monitoring of the adapted management schemes to evaluate their effectiveness. In cases where the adapted schemes are found to be insufficient or ineffective, reconstructing the plans may be deemed necessary to effectively manage the situation. Effective management of EPs involves a continuous process of planning, implementation, evaluation, and revision of regulations until satisfactory results can be obtained.

REFERENCES

Achten C., Kolb A., Püttmann W., Seel P. and Gihr R. (2002). Methyl tert-butyl ether (MTBE) in river and wastewater in Germany. 1. *Environmental Science and Technology*, **36**(17), 3652–3661, https://doi.org/10.1021/es011492y

Ahel M. and Giger W. (1985). Determination of alkylphenols and alkylphenol mono- and diethoxylates in environmental samples by high-performance liquid chromatography. *Analytical Chemistry*, **57**(8), 1577–1583, https://doi.org/10.1021/ac00285a018

Balducci C., Perilli M., Romagnoli P. and Cecinato A. (2012). New developments on emerging organic pollutants in the atmosphere. *Environmental Science and Pollution Research*, **19**, 1875–1884, https://doi.org/10.1007/s11356-012-0815-2

Barroso P. J., Santos J. L., Martín J., Aparicio I. and Alonso E. (2019). Emerging contaminants in the atmosphere: analysis, occurrence and future challenges. *Critical Reviews in Environmental Science and Technology*, **49**(2), 104–171, https://doi.org/10.1080/10643389.2018.1540761

Calvo-Flores, F. G., Isac-García, J. and Dobado, J. A. (eds). (2017). Emerging Pollutants: Origin, Structure and Properties. Wiley-VCH, Weinheim, Germany, https://doi.org/10.1002/9783527691203

Chen S., Jiao X. C., Gai N., Li X. J., Wang X. C., Lu G. H., Piao H. T., Rao Z. and Yang Y. L. (2016). Perfluorinated compounds in soil, surface water, and groundwater from rural areas in eastern China. *Environmental Pollution*, **211**, 124–131, https://doi.org/10.1016/j.envpol.2015.12.024

Chernushevich I. V., Loboda A. V. and Thomson B. A. (2001). An introduction to quadrupole-time-of-flight mass spectrometry. *Journal of Mass Spectrometry*, **36**, 849–865, https://doi.org/10.1002/jms.207

Cisneros B. J. (2011). Safe sanitation in low economic development areas. *Treatise on Water Science*, **4**, 147–200, https://doi.org/10.1016/B978-0-444-53199-5.00082-8

Cizmas L., Sharma V. K., Gray C. M. and McDonald T. J. (2015). Pharmaceuticals and personal care products in waters: occurrence, toxicity, and risk. *Environmental Chemistry Letters*, **13**, 381–394, https://doi.org/10.1007/s10311-015-0524-4

Covey T. R. (2013). LC-MS: a brief history. *The Column*, **9**(16), 11–17. Available online: http://images2.advanstar.com/pixelmags/lctc/pdf/2013-09-06-us.pdf (accessed 14 May 2020)

D'Arcy B. J., Kim L.-H. and Morison P. (2018) Industrial pollution and the water environment: a historical perspective. In: Wealth Creation Without Pollution, B. J. D'Arcy, L.-H. Kim and M. Maniquiz-Redillas (eds.), IWA Publishing, London, chap. 1, pp. 1–24.

Dey S., Bano F. and Malik A. (2019). Pharmaceuticals and personal care product (PPCP) contamination – a global discharge inventory. In: Pharmaceuticals and Personal Care Products: Waste Management and Treatment Technology Emerging Contaminants and Micro Pollutants, M. N. V. Prasad, M. Vithanage and A. Kapley (eds.), Butterworth–Heinemann, Oxford, ch. 1, pp. 1–26, https://doi.org/10.1016/B978-0-12-816189-0.00001-9

Dulio V., van Bavel B., Brorström-Lundén E., Harmsen J., Hollender J., Schlabach M., Slobodnik J., Thomas K. and Koschorreck J. (2018). Emerging pollutants in the EU: 10 years of NORMAN in support of environmental policies and regulations. *Environmental Sciences Europe*, **30**, 5, https://doi.org/10.1186/s12302-018-0135-3

Ebele A. J., Oluseyi T., Drage D. S., Harrad S. and Abou-Elwafa Abdallah M. (2020). Occurrence, seasonal variation and human exposure to pharmaceuticals and personal care products in surface water, groundwater and drinking water in lagos state, *Nigeria. Emerging Contaminants*, **6**, 124–132, https://doi.org/10.1016/j.emcon.2020.02.004

Estévez E., Cabrera M. del C., Molina-Díaz A., Robles-Molina J. and Palacios-Díaz M. del P. (2012). Screening of emerging contaminants and priority substances (2008/105/EC) in reclaimed water for irrigation and groundwater in a volcanic aquifer (gran Canaria, canary islands, Spain). *Science of the Total Environment*, **433**, 538–546, https://doi.org/10.1016/j.scitotenv.2012.06.031

Farré M. la, Pérez S., Kantiani L. and Barceló D. (2008). Fate and toxicity of emerging pollutants, their metabolites and transformation products in the aquatic environment. *TrAC – Trends in Analytical Chemistry*, **27**, 991–1007, https://doi.org/10.1016/j.trac.2008.09.010

Ferrier R., D'Arcy B. J., MacDonald J. and Aitken M. (2005). Diffuse pollution – what is the nature of the problem? *Journal of the Chartered Institution of Water and Environmental Management*, **19**(4), 361–366, https://doi.org/10.1111/j.1747-6593.2005.tb00574.x

Fleet L. (2015). Fly out of the traps. *Nature Methods*, **12**, 9, https://doi.org/10.1038/nmeth.3521

Garrison A. W., Pope J. D. and Allen F. R. (1976). GC/MS analysis of organic compounds in domestic wastewaters. In: Identification and Analysis of Organic Pollutants in Water, C. H. Keith (ed.), Ann Arbor Science, Ann Arbor, MI, USA, pp. 517–556.

Geissen V., Mol H., Klumpp E., Umlauf G., Nadal M., van der Ploeg M., van de Zee S. E. and Ritsema C. J. (2015). Emerging pollutants in the environment: a challenge for water resource management. *International Soil and Water Conservation Research*, **3**(1), 57–65, https://doi.org/10.1016/j.iswcr.2015.03.002

Gentile A., Ruffino F. and Grimaldi M. G. (2016). Complex-morphology metal-based nanostructures: fabrication, characterization, and applications. *Nanomaterials*, **6**(6), 110, https://doi.org/10.3390/nano6060110

Glassmeyer S. T., Furlong E. T., Kolpin D. W., Batt A. L., Benson R., Boone J. S., Conerly O., Donohue M. J., King D. N., Kostich M. S., Mash H. E., Pfaller S. L., Schenck K. M., Simmons J. E., Varughese E. A., Vesper S. J., Villegas E. N. and Wilson V. S. (2017). Nationwide reconnaissance of contaminants of emerging concern in source and treated drinking waters of the United States. *Science of the Total Environment*, **581–582**, 909–922, https://doi.org/10.1016/j.scitotenv.2016.12.004

Guilhaus M., Selby D. and Mlynski V. (2000). Orthogonal acceleration time-of-flight mass spectrometry. *Mass Spectrometry Reviews*, **19**(2), 65–107, https://doi.org/10.1002/(SICI)1098-2787(2000)19:2<65::AID-MAS1>3.0.CO;2-E

Güneralp B., Reba M., Hales B. U., Wentz E. A. and Seto K. C. (2020). Trends in urban land expansion, density, and land transitions from 1970 to 2010: a global synthesis. *Environmental Research Letters*, **15**, 044015, https://doi.org/10.1088/1748-9326/ab6669

Hildebrandt A., Guillamón M., Lacorte S., Tauler R. and Barceló D. (2008). Impact of pesticides used in agriculture and vineyards to surface and groundwater quality (north Spain). *Water Research*, **42**, 3315–3326, https://doi.org/10.1016/j.watres.2008.04.009

Houtman C. J. (2010). Emerging contaminants in surface waters and their relevance for the production of drinking water in Europe. *Journal of Integrative Environmental Sciences*, **7**, 271–295, https://doi.org/10.1080/1943815X.2010.511648

Hu Q., Noll R. J., Li H., Makarov A., Hardman M. and Cooks R. G. (2005). The orbitrap: a new mass spectrometer. *Journal of Mass Spectrometry*, **40**, 430–443, https://doi.org/10.1002/jms.856

Huang C. H. and Sedlak D. L. (2001). Analysis of estrogenic hormones in municipal wastewater effluent and surface water using enzyme-linked immunosorbent assay and gas chromatography/tandem mass spectrometry. *Environmental Toxicology and Chemistry*, **20**, 133–139, https://doi.org/10.1002/and so on.5620200114

Hyland K. C., Blaine A. C., Dickenson E. R. V. and Higgins C. P. (2015). Accumulation of contaminants of emerging concern in food crops – part 1: edible strawberries and lettuce grown in reclaimed water. *Environmental Toxicology and Chemistry*, **34**, 2213–2221, https://doi.org/10.1002/and so on.3066

Ioannou P., Charisiadis P., Andra S. S. and Makris K. C. (2016). Occurrence and variability of iodinated trihalomethanes concentrations within two drinking-water distribution networks. *Science of the Total Environment*, **543A**, 505–513, https://doi.org/10.1016/j.scitotenv.2015.10.031

Jeevanandam J., Barhoum A., Chan Y. S., Dufresne A. and Danquah M. K. (2018). Review on nanoparticles and nanostructured materials: history, sources, toxicity and regulations. *Beilstein Journal of Nanotechnology*, **9**, 1050–1074, https://doi.org/10.3762/bjnano.9.98

Jobling S. and Sumpter J. P. (1993). Detergent components in sewage effluent are weakly oestrogenic to fish: an in vitro study using rainbow trout (*Oncorhynchus mykiss*) hepatocytes. *Aquatic Toxicology*, **27**(3–4), 361–372, https://doi.org/10.1016/0166-445X(93)90064-8

Khan J. A., Sayed M., Khan S., Shah N. S., Dionysiou D. D. and Boczkaj G. (2019). Advanced oxidation processes for the treatment of contaminants of emerging concern. In: Contaminants of Emerging Concern in Water and Wastewater: Advanced Treatment Processes, A. J. Hernández-Maldonado and L. Blaney (eds), Butterworth-Heinemann (Elsevier, Inc.), Oxford, United Kingdom, pp. 299–365. https://doi.org/10.1016/B978-0-12-813561-7.00009-2

Kim S. D., Cho J., Kim I. S., Vanderford B. J. and Snyder S. A. (2007). Occurrence and removal of pharmaceuticals and endocrine disruptors in South Korean surface, drinking, and waste waters. *Water Research*, **41**, 1013–1021, https://doi.org/10.1016/j.watres.2006.06.034

Kim S. K., Li D. H., Shoeib M. and Zoh K. D. (2014). Contribution of diffuse inputs to the aqueous mass load of perfluoroalkyl acids in river and stream catchments in korea. *Science of the Total Environment*, **470–471**, 1430–1440, https://doi.org/10.1016/j.scitotenv.2013.06.040

Kolb A. and Püttmann W. (2006). Methyl *tert*-butyl ether (MTBE) in finished drinking water in Germany. *Environmental Pollution*, **140**, 294–303, https://doi.org/10.1016/j.envpol.2005.07.008

Kolpin D. W., Furlong E. T., Meyer M. T., Thurman E. M., Zaugg S. D., Barber L. B. and Buxton H. T. (2002). Pharmaceuticals, hormones, and other organic wastewater contaminants in U.S. Streams, 1999–2000: a national reconnaissance. *Environmental Science and Technology*, **36**, 1202–1211, https://doi.org/10.1021/es011055j

Kot-Wasik A., Jakimska A. and Śliwka-Kaszyńska M. (2016). Occurrence and seasonal variations of 25 pharmaceutical residues in wastewater and drinking water treatment plants. *Environmental Monitoring and Assessment*, **188**, 661, https://doi.org/10.1007/s10661-016-5637-0

Kuwayama Y., Young R. and Brozović N. (2017). Groundwater scarcity: management approaches and recent innovations. In: Competition for Water Resources: Experiences and Management Approaches in the US and Europe, J. R. Ziolkowska and J. M. Peterson (eds.), Elsevier, Amsterdam, pp. 332–350, https://doi.org/10.1016/B978-0-12-803237-4.00019-7

Lamastra L., Balderacchi M. and Trevisan M. (2016). Inclusion of emerging organic contaminants in groundwater monitoring plans. *MethodsX*, **3**, 459–476, https://doi.org/10.1016/j.mex.2016.05.008

Lapworth D. J., Baran N., Stuart M. E. and Ward R. S. (2012). Emerging organic contaminants in groundwater: a review of sources, fate and occurrence. *Environmental Pollution*, **163**, 287–303, https://doi.org/10.1016/j.envpol.2011.12.034

Lapworth D. J., Baran N., Stuart M. E., Manamsa K. and Talbot J. (2015). Persistent and emerging micro-organic contaminants in chalk groundwater of england and France. *Environmental Pollution*, **203**, 214–225, https://doi.org/10.1016/j.envpol.2015.02.030

Lesney M. (2016). Spectacular spectrometry. In: Enterprise of the Chemical Sciences, J. Ryan (ed.), ACS, Washington, DC, pp. 93–97. Retrieved from https://www.yumpu.com/en/document/read/11832993/spectacular-spectrometry-american-chemical-society-publications (accessed 20 May 2020).

Lin Y. C., Lai W. W. P., Tung H. H. and Lin A. Y. C. (2015). Occurrence of pharmaceuticals, hormones, and perfluorinated compounds in groundwater in Taiwan. *Environmental Monitoring and Assessment*, **187**, 256, https://doi.org/10.1007/s10661-015-4497-3

Lin H., Chen L., Li H., Luo Z., Lu J. and Yang Z. (2018). Pharmaceutically active compounds in the Xiangjiang river, China: distribution pattern, source apportionment, and risk assessment. *Science of the Total Environment*, **636**, 975–984, https://doi.org/10.1016/j.scitotenv.2018.04.267

Lu J., Li H., Luo Z., Lin H. and Yang Z. (2018). Occurrence, distribution, and environmental risk of four categories of personal care products in the Xiangjiang river, China. *Environmental Science and Pollution Research*, **25**, 27524–27534, https://doi.org/10.1007/s11356-018-2686-7

Madikizela L. M. and Chimuka L. (2017). Occurrence of naproxen, ibuprofen, and diclofenac residues in wastewater and river water of KwaZulu–natal province in South Africa. *Environmental Monitoring and Assessment*, **189**, 348, https://doi.org/10.1007/s10661-017-6069-1

Masoner J. R., Kolpin D. W., Furlong E. T., Cozzarelli I. M., Gray J. L. and Schwab E. A. (2014). Contaminants of emerging concern in fresh leachate from landfills in the conterminous United States. *Environmental Sciences: Processes and Impacts*, **16**, 2335–2354, https://doi.org/10.1039/c4em00124a

McFadden W. H., Schwartz H. L. and Evans S. (1976). Direct analysis of liquid chromatographic effluents. *Journal of Chromatography A*, **122**, 389–396, https://doi.org/10.1016/S0021-9673(00)82261-X

Mitchell G. (2005). Mapping hazard from urban non-point pollution: a screening model to support sustainable urban drainage planning, *Journal of Environmental Management*, **74**, 1–9, https://doi.org/10.1016/j.jenvman.2004.08.002

Miyagawa S., Sato T. and Iguchi T. (2016). Chapter 101 – endocrine disruptors. In: Handbook of Hormones, Y. Takei, H. Ando and K. Tsutsui (eds.), Academic, New York, pp. 571–572, https://doi.org/10.1016/B978-0-12-801028-0.00101-X

Mohapatra D. P., Cledón M., Brar S. K. and Surampalli R. Y. (2016). Application of wastewater and biosolids in soil: occurrence and fate of emerging contaminants. *Water, Air, and Soil Pollution*, **227**, 77, https://doi.org/10.1007/s11270-016-2768-4

Noguera-Oviedo K. and Aga D. S. (2016). Lessons learned from more than two decades of research on emerging contaminants in the environment. *Journal of Hazardous Materials*, **316**, 242–251, https://doi.org/10.1016/j.jhazmat.2016.04.058

NORMAN. (n.d.). NORMAN: network of reference laboratories, research centres and related organisations for monitoring of emerging environmental substances. Retrieved 19 May 2020, from https://www.norman-network.net/ (accessed 19 May 2020).

OECD. (n.d.). Portal on Per and Poly Fluorinated Chemicals – People's Republic of China. Retrieved 17 July 2020, from https://www.oecd.org/chemicalsafety/portal-perfluorinated-chemicals/countryinformation/china.htm (accessed July 17, 2020).

Pal A., He Y., Jekel M., Reinhard M. and Gin K. Y. H. (2014). Emerging contaminants of public health significance as water quality indicator compounds in the urban water cycle. *Environment International*, **71**, 46–62, https://doi.org/10.1016/j.envint.2014.05.025

Peters R. J. B., van Bemmel G., Milani N. B. L., den Hertog G. C. T., Undas A. K., van der Lee M. and Bouwmeester H. (2018). Detection of nanoparticles in Dutch surface waters. *Science of the Total Environment*, **621**, 210–218, https://doi.org/10.1016/j.scitotenv.2017.11.238

Petrovic M., Farré M., Eljarrat E., Diaz S. and Barceló D. (2013). Environmental analysis: emerging pollutants. In: Liquid Chromatography, S. Fanali, P. Haddad, C. Poole, P. Schoenmakers and D. Lloyd (eds.), Elsevier, Amserdam, pp. 389–410, https://doi.org/10.1016/C2010-0-68910-3

Pham T. T. and Proulx S. (1997). PCBS And PAHs in the Montreal urban community (quebec, Canada) wastewater treatment plant and in the effluent plume in the St Lawrence river. *Water Research*, **31**, 1887–1896, https://doi.org/10.1016/S0043-1354(97)00025-0

Pullen F. (2010). The fascinating history of the development of LC–MS; a personal perspective. *Chromatography Today*, **3**(1), 4–6.

Qin Q., Chen X. and Zhuang J. (2015). The fate and impact of pharmaceuticals and personal care products in agricultural soils irrigated with reclaimed water. *Critical Reviews in Environmental Science and Technology*, **45**, 1379–1408, https://doi.org/10.1080/10643389.2014.955628

Raghav M., Eden S., Mitchell K. and Witte B. (2013). Contaminants of Emerging Concern in Water. Water Resources Research Center College of Agriculture and Life Sciences, University of Arizona, Tucson, Arizona.

Richardson S. D., Plewa M. J., Wagner E. D., Schoeny R. and DeMarini D. M. (2007). Occurrence, genotoxicity, and carcinogenicity of regulated and emerging disinfection by-products in drinking water: a review and roadmap for research. *Mutation Research – Reviews in Mutation Research*, **636**, 178–242, https://doi.org/10.1016/j.mrrev.2007.09.001

Rook J. J. (1974). Formation of haloforms during chlorination of natural waters. *Water Treatment and Examination*, **23**, 234–243.

Sanchís J., Milačič R., Zuliani T., Vidmar J., Abad E., Farré M. and Barceló D. (2018). Occurrence of C60 and related fullerenes in the sava river under different hydrologic conditions. *Science of the Total Environment*, **643**, 1108–1116, https://doi.org/10.1016/j.scitotenv.2018.06.285

Santa Marina L., Ayerdi M., Lertxundi A., Basterretxea M., Goñi F., Iñaki Alvare J., Arranz L., Blarduni E. and María Ibarluzea J. (2010). Concentración de trihalometanos y de ácidos haloacéticos en el agua de consumo y estimación de su ingesta durante el embarazo en la cohorte INMA–guipúzcoa (españa). *Gaceta Sanitaria*, **24**, 321–328, https://doi.org/10.1016/j.gaceta.2010.03.003

Sardiña P., Leahy P., Metzeling L., Stevenson G. and Hinwood A. (2019). Emerging and legacy contaminants across land-use gradients and the risk to aquatic ecosystems. *Science of the Total Environment*, **695**, 133842, https://doi.org/10.1016/j.scitotenv.2019.133842

Sauvé S. and Desrosiers M. (2014). A review of what is an emerging contaminant. *Chemistry Central Journal*, **8**, 15, https://doi.org/10.1186/1752-153X-8-15

SGS Polymer Solutions Incorporated. (2016). Fabulous at 50: A Brief History of HPLC. Retrieved 14 May 2020, from https://www.polymersolutions.com/blog/a-brief-history-of-hplc/ (accessed 14 May 2020).

Sharma B. M., Bečanová J., Scheringer M., Sharma A., Bharat G. K., Whitehead P. G., *et al.* (2019). Health and ecological risk assessment of emerging contaminants (pharmaceuticals, personal care products, and artificial sweeteners) in surface and groundwater (drinking water) in the Ganges river basin, India. *Science of the Total Environment*, **646**, 1459–1467, https://doi.org/10.1016/j.scitotenv.2018.07.235

Sheehan D. (2016). Nanomaterials as an emerging category of environmental pollutants. In: Transdisciplinary Perspectives on Transitions to Sustainability, E. Byrne, G. Mullally and C. Sage (eds.), Routledge, London, pp. 200–218, https://doi.org/10.4324/9781315550206

Skutlarek D., Exner M. and Färber H. (2006). Perfluorinated surfactants in surface and drinking waters. *Environmental Science and Pollution Research*, **13**, 299–307, https://doi.org/10.1065/espr2006.07.326

Sorensen J. P. R., Lapworth D. J., Nkhuwa D. C. W., Stuart M. E., Gooddy D. C., Bell R. A., *et al.* (2015). Emerging contaminants in urban groundwater sources in Africa. *Water Research*, **72**, 51–63, https://doi.org/10.1016/j.watres.2014.08.002

Spoelstra J., Schiff S. L. and Brown S. J. (2013). Artificial sweeteners in a large Canadian river reflect human consumption in the watershed. *PLoS One*, **8**(12), e82706, https://doi.org/10.1371/journal.pone.0082706

Stackelberg P. E., Gibs J., Furlong E. T., Meyer M. T., Zaugg S. D. and Lippincott R. L. (2007). Efficiency of conventional drinking-water-treatment processes in removal of pharmaceuticals and other organic compounds. *Science of the Total Environment*, **377**, 255–272, https://doi.org/10.1016/j.scitotenv.2007.01.095

Stan H. J. and Linkerhägner M. (1992). Identification of 2-(4-chlorophenoxy)-2-methyl-propionic acid in groundwater using capillary-gas chromatography with atomic emission detection and mass spectrometry. *Vom Wasser*, **79**, 75–88.

Stefanakis A. I. and Becker J. A. (2015). A review of emerging contaminants in water: classification, sources, and potential risks. In: Impact of Water Pollution on Human Health and Environmental Sustainability, A. McKeown and G. Bugyi (eds.), Hershey, PA, USA, pp. 55–80, https://doi.org/10.4018/978-1-4666-9559-7.ch003

Stein K. F. (1962). Silent spring. In: Rachel Carson, K. F. Stein (ed.), SensePublishers, Rotterdam, pp. 61–106, https://doi.org/10.1007/978-94-6209-068-2_5

Sterk A., Schlegel P., Mul J., Ubbink-Blanksma M. and Bruininx E. M. A. M. (2008). Effects of sweeteners on individual feed intake characteristics and performance in group-housed weanling pigs. *Journal of Animal Science*, **86**, 2990–2997, https://doi.org/10.2527/jas.2007-0591

Stuart M. E., Manamsa K., Talbot J. C. and Crane E. J. (2011). Emerging contaminants in groundwater. British Geological Survey, Open Report OR/11/013.

Stumm-Zollinger E. and Fair G. M. (1965). Biodegradation of steroid hormones. *Journal of the Water Pollution Control Federation*, **37**, 1506–1510.

Tabak H. H. and Bunch R. L. (1970). Steroid hormones as water pollutants I. Metabolism of natural and synthetic ovulation-inhibiting hormones by microorganisms of activated sludge and primary settled sewage. *Developments in Industrial Microbiology*, **11**, 367–376.

Task Force on Emerging Contaminants. (2018). Plan for Addressing Critical Research Gaps Related to Emerging Contaminants in Drinking Water. Retrieved from https://www.whitehouse.gov/wp-content/uploads/2018/11/Plan-for-Addressing-Critical-Research-Gaps-Related-to-Emerging-Contaminants-in-Drinking-Water.pdf (accessed 20 May 2020).

United Nations (UN). (2015). The United Nations World Water Development Report 2015: Water for A Sustainable World, Facts and Figures. Unesco, Paris.

United Nations (UN). (2019). World Water Development Report 2019: Leaving No One Behind. Unesco, Paris. Retrieved from United Nations website: https://unesdoc.unesco.org/ark:/48223/pf0000367306/PDF/367306eng.pdf.multi (accessed 9 July 2020).

US EPA. (2013). Chapter 6 – Insecticides and Environmental Pesticide Control Subchapter II – Environmental Pesticide Control. US EPA, Washington, DC.

Vogelsang C., Grung M., Jantsch T. G., Tollefsen K. E. and Liltved H. (2006). Occurrence and removal of selected organic micropollutants at mechanical, chemical and advanced wastewater treatment plants in Norway. *Water Research*, **40**, 3559–3570, https://doi.org/10.1016/j.watres.2006.07.022

Wang Y. and Zhou J. (2013). Endocrine disrupting chemicals in aquatic environments: a potential reason for organism extinction? *Aquatic Ecosystem Health and Management*, **16**, 88–93, https://doi.org/10.1080/14634988.2013.759073

Wei G. L., Li D. Q., Zhuo M. N., Liao Y. S., Xie Z. Y., Guo T. L., Li J. J., Zhang S. Y. and Liang Z. Q. (2015). Organophosphorus flame retardants and plasticizers: sources, occurrence, toxicity and human exposure. *Environmental Pollution*, **196**, 29–46, https://doi.org/10.1016/j.envpol.2014.09.012

Wolschke H., Xie Z., Möller A., Sturm R. and Ebinghaus R. (2011). Occurrence, distribution and fluxes of benzotriazoles along the German large river basins into the north Sea. *Water Research*, **45**, 6259–6266, https://doi.org/10.1016/j.watres.2011.09.028

Wu C., Luo Y., Gui T. and Huang Y. (2014). Concentrations and potential health hazards of organochlorine pesticides in shallow groundwater of taihu lake region, China. *Science of the Total Environment*, **470–471**, 1047–1055, https://doi.org/10.1016/j.scitotenv.2013.10.056

Yamashita M. and Fenn J. B. (1984). Electrospray ion source. Another variation on the free-jet theme. *Journal of Physical Chemistry*, **88**, 4451–4455, https://doi.org/10.1021/j150664a002

Yang Y. Y., Zhao J. L., Liu Y. S., Liu W. R., Zhang Q. Q., Yao L., Hu L. X., Zhang J. N., Jiang Y. X. and Ying G. G. (2018). Pharmaceuticals and personal care products (PPCPs) and artificial sweeteners (ASs) in surface and ground waters and their application as indication of wastewater contamination. *Science of the Total Environment*, **616–617**, 816–823, https://doi.org/10.1016/j.scitotenv.2017.10.241

Yost R. A. and Enke C. G. (1978). Selected ion fragmentation with a tandem quadrupole mass spectrometer. *Journal of the American Chemical Society*, **100**, 2274–2275, https://doi.org/10.1021/ja00475a072

Zhang H., Bu Q., Wu D. and Yu G. (2020). Polycyclic musks in surface water and sediments from an urban catchment in the megacity Beijing, China. *Environmental Pollution*, **263A**, 114548, https://doi.org/10.1016/j.envpol.2020.114548

doi: 10.2166/9781789061123_099

Chapter 7

Marine debris: the Australian perspective

Daniel Burns, Wally Smith and Heidi Tait
Tangaroa Blue Foundation, PO Box 1235, Dunsborough, WA 6281, Australia

Keywords: Litter; plastic waste; marine waste; waste management

7.1 INTRODUCTION

Marine debris represents one of the major pollutants of the aquatic environment, with associated impacts that are of global environmental and economic concern (Agamuthu *et al.*, 2019; Beaumont *et al.*, 2019; Forrest *et al.*, 2019). The vast scale of the marine debris problem makes it logistically difficult, time-consuming and expensive to collect rigorous scientific data about the issue at the state, national and international scales. However, citizen science data can significantly increase the sample size, temporal scale and spatial coverage of datasets to help solve the problem. This is particularly true for marine debris data, which is relatively easy to collect without requiring extensive training by volunteers, is able to be collected by a wide range of coastal users, and provides useful information about litter distribution and composition at varying scales (Bergmann *et al.*, 2017; Hidalgo-Ruz & Thiel, 2015; Hidalgo-Ruz *et al.*, 2018; Nelms *et al.*, 2017; van der Velde *et al.*, 2017).

7.2 AUSTRALIAN MARINE DEBRIS INITIATIVE

One example of the utility of citizen science data can be seen through the growth of the Tangaroa Blue Foundation (TBF) in Australia. After observing increasing amounts of litter on beaches in the south west of Australia in 2004, TBF was founded with the premise 'if all we do is clean-up, that's all we'll ever do'. TBF focused beyond just the removal of marine debris and towards litter source identification as a means of preventing debris from entering the environment in the first place. To assist with this task, TBF created the Australian Marine Debris Initiative (AMDI) as an on-ground network of communities, schools, indigenous ranger groups, government agencies and organisations that contribute data from litter collected during beach and river clean-up events to the AMDI Database. Stakeholders and partners then collaborate on solutions to stop the flow of litter at the source. From its beginnings in Western Australia in 2004, the AMDI and its associated Database has grown to a national level, with an increasing number of contributors submitting data from clean-up events around Australia (Figure 7.1). Of the 3202 sites from which data have been submitted to date,

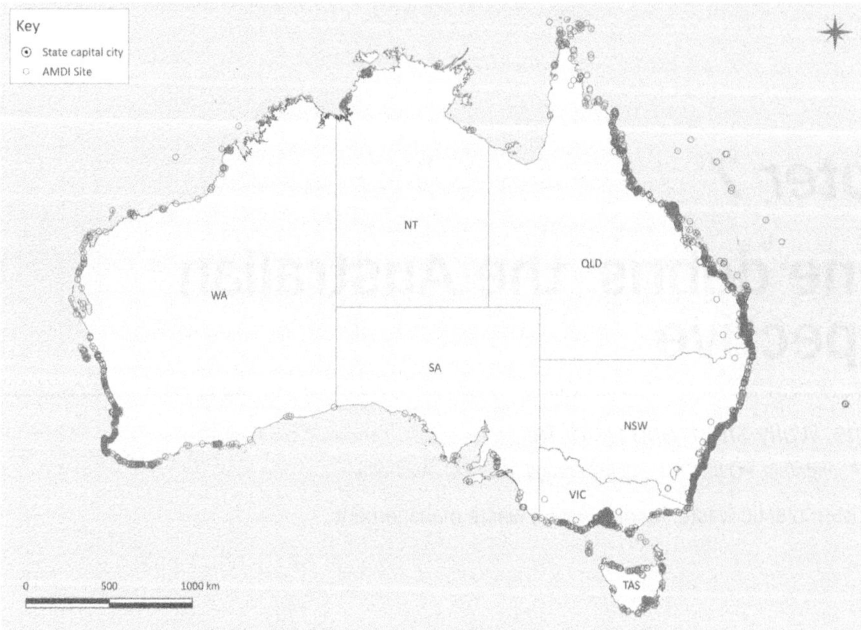

Figure 7.1 Clean-up locations ($n = 3202$) from which data have been submitted to the AMDI database from 2004–2020.

some sites have been sampled at least once per month since 2004, while other are represented by a single clean-up event.

AMDI participants remove debris (generally >5 mm in size) and sort it into 140 categories grouped primarily by material (e.g., plastics, glass and ceramics, metal, rubber, etc.). Standardised data sheets are used to record the total weight of debris removed, item counts for each category, and effort (number of participants, length of beach cleaned, duration of clean-up event). Specific characteristics that assist in source identification, such as barcodes, labels and serial numbers, can also be recorded, enabling, for example, the tracking of strapping bands, tags, cable ties or brands of bottled water. Training materials in the form of marine debris identification guides and videos are also made available to assist participants with data collection. AMDI participants then submit data for inclusion in the database via an online portal, a purpose-built app or hard-copy data sheets.

Since 2004, the AMDI Database has grown substantially and represents the largest source of information about the distribution and composition of marine debris around Australia (Figure 7.2). Having surpassed 15.5 million items removed and entered into the database by April 2020, this constitutes the largest marine debris database in the Southern Hemisphere. Such large sample sizes and temporal coverage help to provide a meta-view of debris around Australia. They also provide data for comparison with shorter-term rigorous, scientific surveys that may produce biased results due to their 'snapshot' nature (Imhof *et al.*, 2017), or to 'ground truth' results of modelling studies.

The benefits of engaging a broad section of the community to collect and record marine debris data goes beyond the removal of items of litter and debris from the natural environment. Other direct benefits of these efforts include improving the aesthetics and health of the environment, reducing risks to humans and fauna from interactions with harmful debris, reducing clean-up costs for management authorities, increasing practical stewardship opportunities for community members and increasing

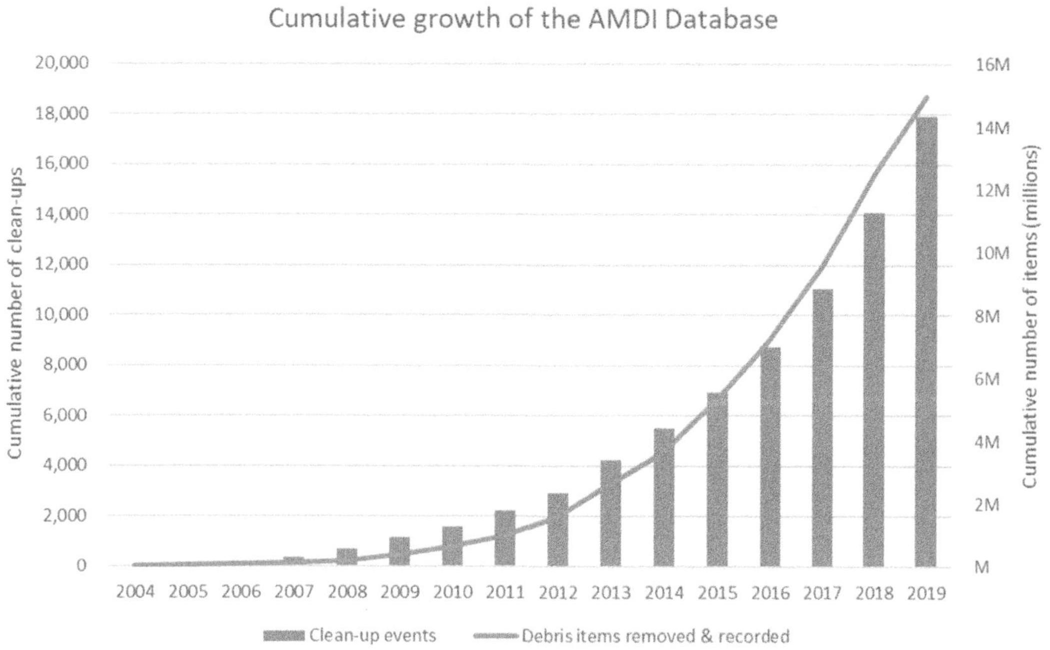

Figure 7.2 Cumulative growth in clean-up events and marine debris items removed and entered into the AMDI database from 2004–2019.

public awareness of environmental issues. By recording data at clean-up events, participants also provide valuable information that can be utilised at various temporal and spatial scales to identify sources of debris, facilitate source reduction plans and measure the effectiveness of implemented actions.

Analysis of AMDI data at a national scale shows that, of the total of 5 378 758 items removed and entered into the AMDI Database in 2018–2019, 'remnant plastic bits and pieces, hard and solid' was the most common item found during clean-up events around Australia during that period (Figure 7.3).

As might be expected, AMDI data also show that the composition of marine debris varies according to location. The 'marine debris signature' of a location is likely to be affected by various factors, including the type of waterway (e.g., coastal beaches versus inland rivers and creeks), prevailing currents, wind, weather and geography of the location, presence of shipping, fishing and industry in the area, type and mobility of debris materials and proximity to built-up areas and access points. Figure 7.4 shows the top five items found based on percentage of total items found at each location type for 2018–2019.

It is notable that nearly three-quarters (74%) of all items found during clean-up events around Australia in 2018–2019 were made of plastic. With plastic waste entering the ocean from all sources now estimated to be more than 15 million tons per year (Forrest *et al.*, 2019), and the problem exacerbated by the persistence of plastic once it enters the marine environment (Andrady, 2015), the scale of the plastic problem is already vast. Additionally, the degradation of larger plastic items into abundant smaller pieces is likely to result in an increasing problem of secondary microplastics over time (Andrady, 2015; Galloway *et al.*, 2017; Jang *et al.*, 2014; Kane *et al.*, 2020; Lebreton *et al.*, 2018; Song *et al.*, 2017; Thompson, 2015).

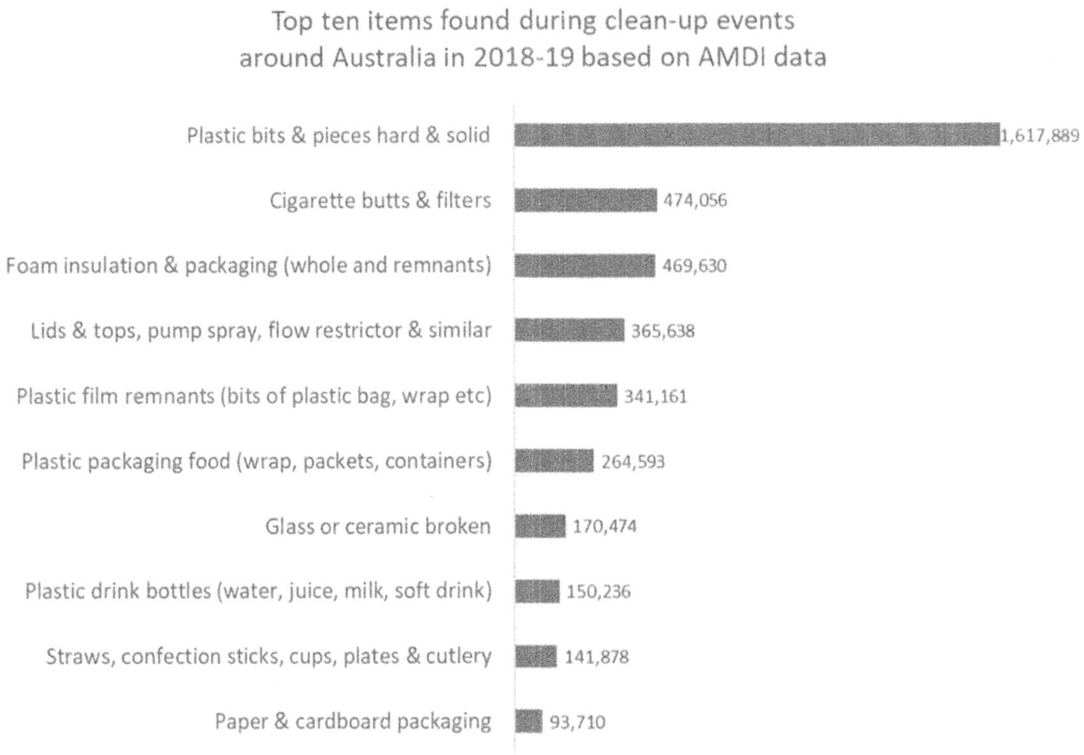

Figure 7.3 Top ten items found during clean-up events entered into the AMDI database for all of Australia in 2018–2019.

Although some targeted surveys have been conducted, the AMDI Database contains only limited information about microplastics. Instead, it primarily provides data about macroplastics, namely debris greater than 5 mm in size. Further analyses of the AMDI Database provide metrics on the mean density of plastic found at different location types. Figure 7.5 shows the mean number of plastic items found per 1000 m of shoreline for different location types for 2018–2019.

It should be noted that the data for islands in this analysis were partially skewed by substantial debris loads at the Cocos (Keeling), Christmas and Whitsunday Islands. However, these metrics are still able to provide indicative measures. While they don't account for seasonal variation, the number of participants or the frequency and duration of clean-up events in the calculations, and therefore should be interpreted with caution, the data can still help to identify possible stakeholders and to inform source reduction efforts at a broader scale.

Focusing on the plastic component of marine debris, more than one-third of all items removed from Australian beaches in 2018–2019 were categorised as plastic remnants (e.g., hard and solid bits and pieces of plastic, bits of plastic bag or wrap, etc.). Expanding the dataset to analyse a four-year period from 2016–2019, the mean density of plastic remnants by location type in 2016–2019 is shown in Figure 7.6.

As per the plastic-debris-by-location measure, these data should be interpreted with caution. However, they support the assertion that failing to address plastic pollution at or close to the source results in the accumulation of large amounts of plastic in remote coastal and island locations. These

Figure 7.4 Top five items found during clean-up events in 2018–2019 based on percentage of total items found at different location types around Australia: (a) clean-up events at all locations combined; (b) dry land; (c) inland waterways; (d) coastal beaches in built areas; (e) coastal beaches away from built areas; (f) islands; and (g) underwater or at sea. Data sourced from the AMDI database.

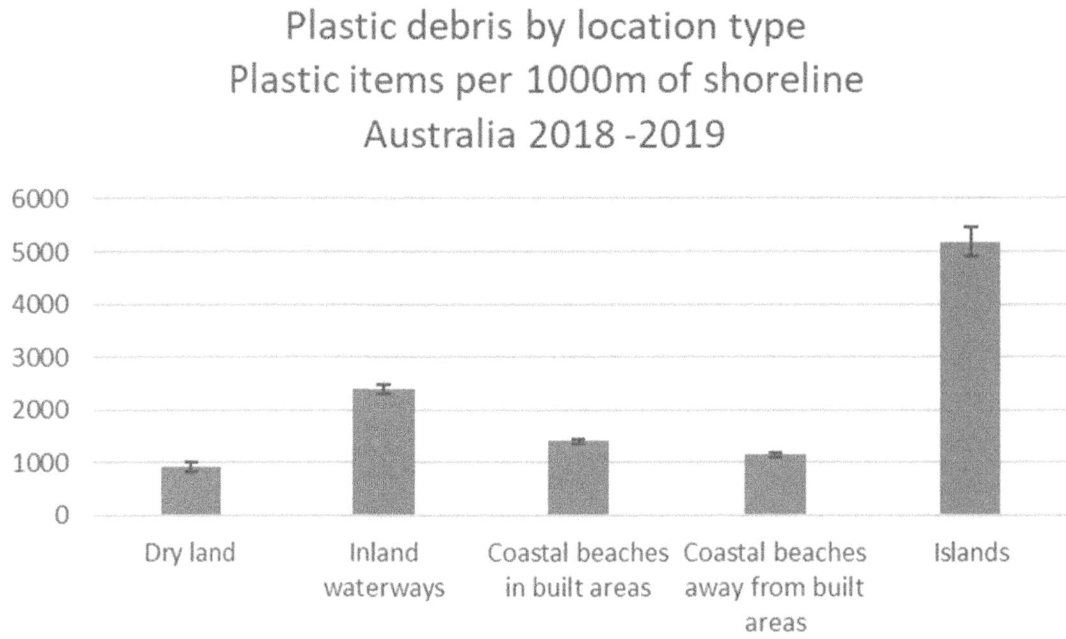

Figure 7.5 Mean number (and standard deviation) of plastic items found per 1000 m of beach cleaned at different location types in Australia, 2018–2019.

locations are more logistically difficult to clean-up, and they are often of high ecological value, highlighting the economic cost and environmental impact of weather-driven movement of plastic through the environment (Lavers & Bond, 2017).

7.3 LAND SEA SOURCE INDEX AND THE MARINE DEBRIS SOURCE INDEX

To better identify the sources of debris, and therefore intervene as close to the source as possible, Tangaroa Blue Foundation has developed indices to estimate the sources of debris at a local level. These indices, the Land Sea Source Index and the Marine Debris Source Index, are based on counts of items, the nature and mobility of debris materials and the location of the clean-up event:

1. *The Land Sea Source Index (LSSI)* is used to identify the onshore and offshore proportions of sources of debris in a sample of beach clean-up data. This statistic is relevant to coastal and island sites, with all inland waterway sites assumed to have 100% land-sourced debris. The LSSI is most accurate with respect to the given beach where the data were collected. The LSSI from an isolated, seldom-visited beach is shown in Figure 7.7 and that from a metropolitan beach in Figure 7.8.

By estimating the proportions of onshore and offshore sources of debris, this provides the first broad indication of where achievable source reduction efforts might be directed. With the given LSSI at Hillview Beach (Figure 7.7), efforts would be directed to targeting offshore sources of debris, requiring a focus on, for example, legislation such as MARPOL regulations. Conversely, the LSSI data at Cottesloe Beach (Figure 7.8) suggest efforts would be better directed at a local level to target

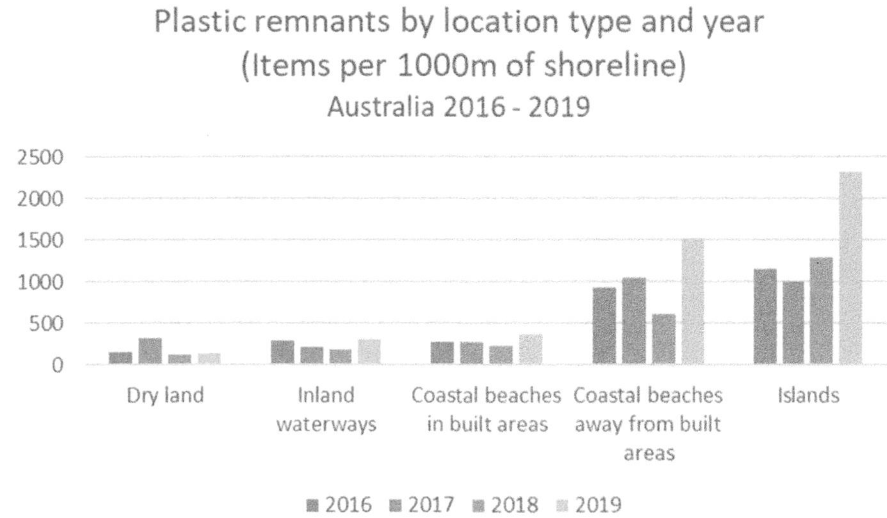

Figure 7.6 Mean number of plastic remnant items found per 1000 m of beach cleaned at different location types in Australia, 2016–2019.

locally generated waste and litter. Further details of the LSSI concept, which is also used to develop the Marine Debris Source Index, are provided in Figure 7.9.

2. *The Marine Debris Source Index (MDSI)* provides an estimate of the proportions of items coming from six general sources of marine debris (Figure 7.10). This statistic is relevant to data from coastal and island clean-up sites but not from inland waterways. The shipping trace category is most accurate for data collected away from heavily populated areas.

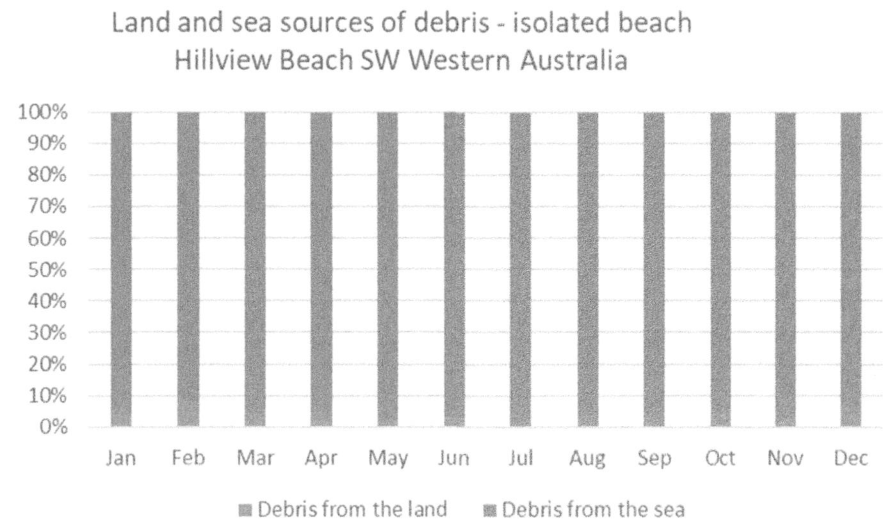

Figure 7.7 Land Sea source Index for Hillview Beach, south-west WA.

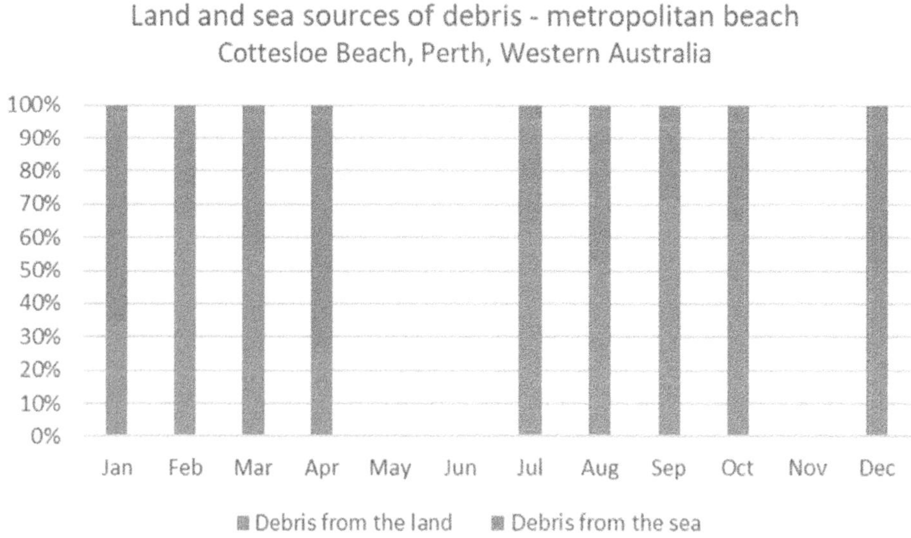

Figure 7.8 Land Sea Source Index for Cottesloe Beach, Perth, WA.

The LSSI and MDSI are used to assess which sources are readily addressed locally or regionally and which require joint action, such as regional data collection, to support state or national actions. Marine Debris Source Index estimates by coastal type are shown for Australian data from 2018–2019 in Figure 7.11. Similar estimates for inland waterways are shown in Figure 7.12. These data suggest that debris from shipping, commercial fishing and garbage washed ashore are major contributors to the debris loads on remote beaches and islands in Australia. Such sources cannot be addressed locally

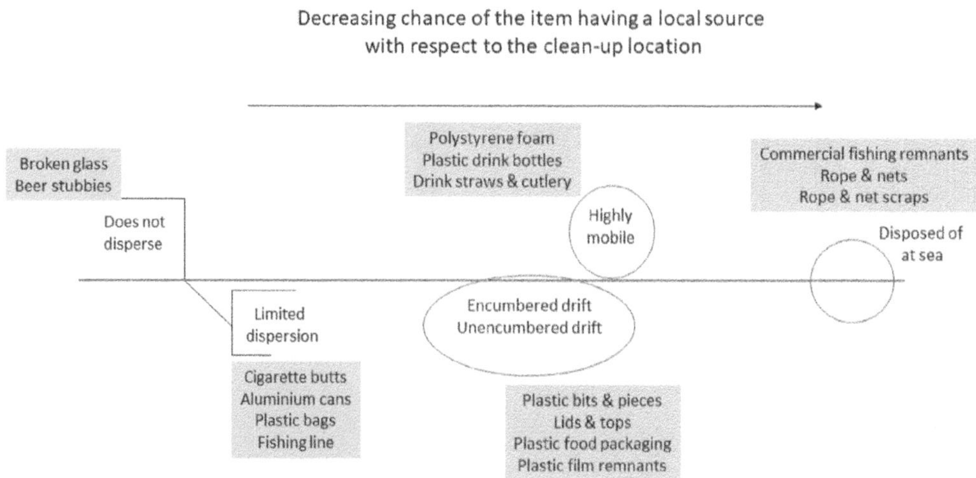

Figure 7.9 AMDI's Land Sea Source Index concept.

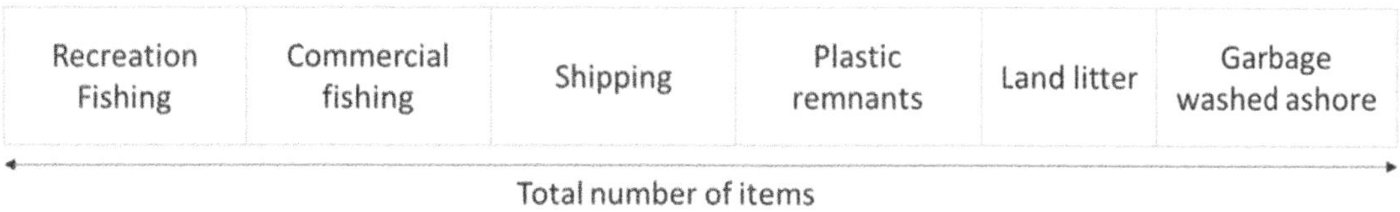

Figure 7.10 AMDI's Marine Debris Source Index categories.

at the given locations and require broader action and collaboration than, for example, plastic pollution that can be mitigated or prevented more easily at dry land, inland waterway and coastal locations in built up areas.

It should be noted that basing analyses on item counts potentially underestimates the impact of some items and overestimates others. For example, abandoned fishing and aquaculture gear represents only a small proportion of the number of items found at inland waterway locations, but the volume and weight of these items is substantial. Source reduction efforts are therefore better directed by considering more than one metric.

Targeted data collection and analyses can also be used to inform source reduction at a finer scale. For example, initiatives to collect and analyse data at the catchment level can result in successful source reduction measures by monitoring and addressing problems at multiple points in the litter-movement pathway (Figure 7.13). Source reduction might then include measures ranging from physical barriers, education campaigns, signage, industry codes of practice, or legislation and policy initiatives.

Several projects are underway in Australia which address different approaches to monitoring and source reduction in the upstream segment of the litter movement pathway. One example is the establishment of *Operation Clean Sweep® Australia* in the Port Phillip Bay region of Victoria. This project combines monitoring of plastic resin pellets extending from pellet loss at the on-ground point source, through the drainage and waterway systems, and into Port Phillip Bay. This is coupled with source reduction activities that involve engagement of plastic industry representatives in a program

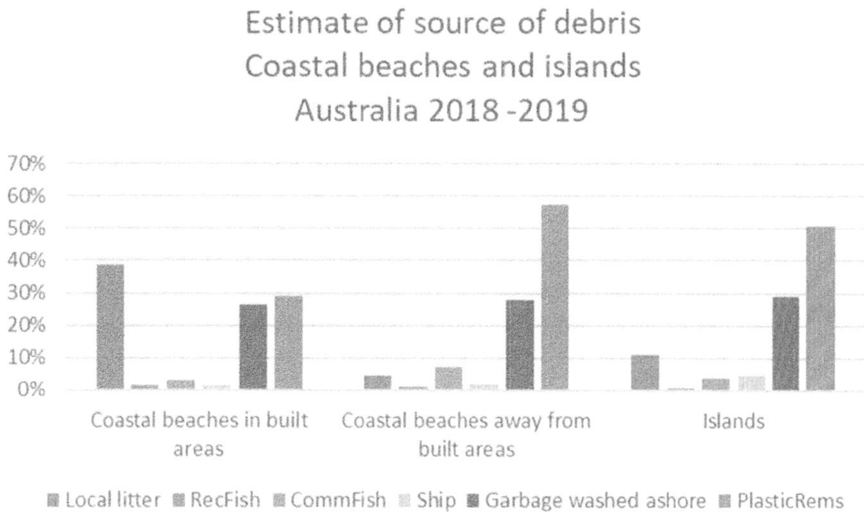

Figure 7.11 Australian marine debris source estimates by coastal type for 2018–2019.

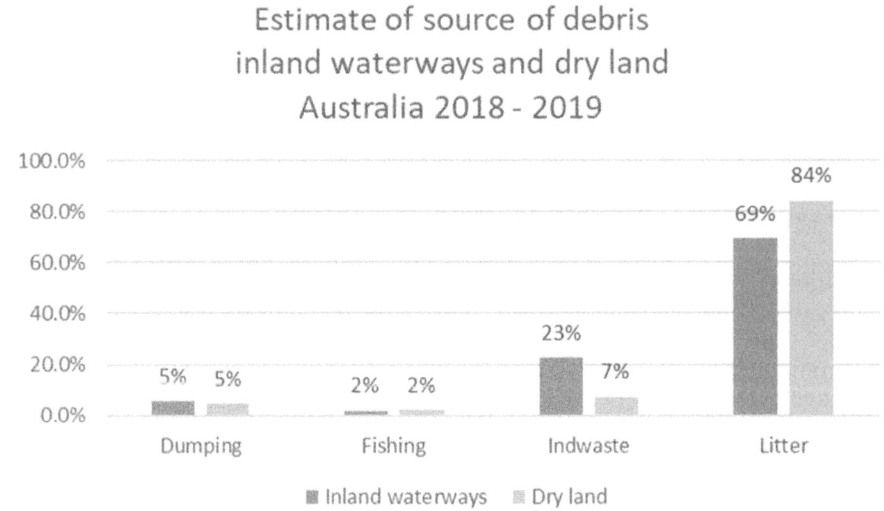

Figure 7.12 Estimates of sources of debris from clean-ups at inland waterways and dry land sites in 2018–2019.

aimed at stopping the loss of pellets into the environment. Another example is the *Clean4Shore* project, which has been documenting the removal of litter and waste from two major estuarine systems north of Sydney: the Hawkesbury River and Brisbane Waters, NSW. The accumulated data show significant levels of waste, litter, dumping and abandoned aquaculture gear and watercraft. This project provides a graphic demonstration of failing to capture waste and litter prior to its entry into sensitive, ecologically important waterbodies.

At a range of levels from local to national, one of the primary uses of AMDI data is to inform and facilitate actions in the form of Source Reduction Plans. Through this process, AMDI participants provide data to identify sources, plan and implement actions and measure the success or failure of those actions on reducing debris loads over time. By bringing together stakeholders to target specific items or locations, more than 200 source reduction plans have been developed and implemented around Australia to date. These range from local initiatives to target specific items (e.g. cigarette butts or single-use plastics), to industry-wide plans (e.g. eliminating packing tape used in commercial fishing operations), to support for state or national policies (e.g. container deposit schemes and plastic

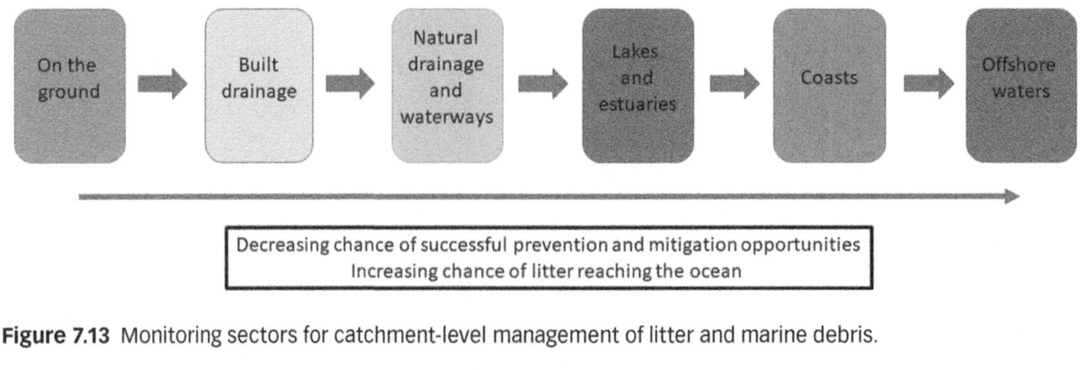

Figure 7.13 Monitoring sectors for catchment-level management of litter and marine debris.

bag bans). This highlights the need to address the marine debris problem through actions at multiple scales, although it is likely that high-level initiatives to develop a circular plastics economy will provide the greatest broad-scale solution to the problem (Agamuthu *et al.*, 2019; Forrest *et al.*, 2019).

ACKNOWLEDGEMENTS

We thank all of the AMDI partners and participants who have contributed data and worked on source reduction plans around Australia.

REFERENCES

Agamuthu P., Mehran S. B., Norkhairah A. and Norkhairiyah A. (2019). Marine debris: a review of impacts and global initiatives. *Waste Management and Research*, **37**(10), 987–1002, https://doi.org/10.1177/0734242X19845041

Andrady A. L. (2015). Persistence of plastic litter in the oceans. In: Marine Anthropogenic Litter, M. Bergmann, L. Gutow and M. Klages (eds.), Springer, Berlin, pp. 57–72.

Beaumont N. J., Aanesen M., Austen M. C., Börger T., Clark J. R., Cole M., Hooper T., Lindeque P. K., Pascoe C. and Wyles K. J. (2019). Global ecological, social and economic impacts of marine plastic. *Marine Pollution Bulletin*, **142**, 189–195, https://doi.org/10.1016/j.marpolbul.2019.03.022

Bergmann M., Lutz B., Tekman M. B. and Gutow L. (2017). Citizen scientists reveal: marine litter pollutes Arctic beaches and affects wild life. *Marine Pollution Bulletin*, **125**, 535–540, https://doi.org/10.1016/j.marpolbul.2017.09.055

Forrest A., Giacovazzi L., Dunlop S., Reisser J., Tickler D., Jamieson A. and Meeuwig J. J. (2019). Eliminating plastic pollution: how a voluntary contribution from industry will drive the circular plastics economy. *Frontiers in Marine Science*, **6**, 627. https://doi.org/10.3389/fmars.2019.00627

Galloway T. S., Cole M., Lewis C., Atkinson A. and Allen J. I. (2017) Interactions of microplastic debris throughout the marine ecosystem. *Nature Ecology & Evolution*, **1**, 0116. https://doi.org/10.1038/s41559-017-0116

Hidalgo-Ruz V. and Thiel M. (2015). The contribution of citizen scientists to the monitoring of marine litter. 3. In: Marine Anthropogenic Litter, M. Bergmann, L. Gutow and M. Klages (eds.), Springer, Berlin, pp. 433–451.

Hidalgo-Ruz V., Honorato-Zimmer D., Gatta-Rosemary M., Nuñez P., Hinojosa I. A. and Thiel M. (2018). Spatio-temporal variation of anthropogenic marine debris on Chilean beaches. *Marine Pollution Bulletin*, **126**, 1516–1524. https://doi.org/10.1016/j.marpolbul.2017.11.014

Imhof H. K., Sigl R., Brauer E., Feyl S., Giesemann P., Klink S., Leupolz K., Löder M. G., Löschel L. A., Missun J. and Muszynski S. (2017). Spatial and temporal variation of macro-, meso- and microplastic abundance on a remote coral island of the Maldives, Indian ocean. *Marine Pollution Bulletin*, **116**(1–2), 340–347, https://doi.org/10.1016/j.marpolbul.2017.01.010

Jang Y. C., Lee J., Hong S., Lee J. S., Shim W. J. and Song Y. K. (2014). Sources of plastic marine debris on beaches of Korea: more from the ocean than the land. *Ocean Science Journal*, **49**, 151–162, https://doi.org/10.1007/s12601-014-0015-8

Kane I. A., Clare M. A., Miramontes E., Wogelius R., Rothwell J. J., Garreau P. and Pohl F. (2020). Seafloor microplastic hotspots controlled by deep-sea circulation. *Science*, **368**(6495), 1140–1145, https://doi.org/10.1126/science.aba5899

Lavers J. L. and Bond A. L. (2017). Exceptional and rapid accumulation of anthropogenic debris on one of the world's most remote and pristine islands. *Proceedings of the National Academy of Sciences*, **114**, 6052–6055, https://doi.org/10.1073/pnas.1619818114

Lebreton L., Slat B., Ferrari F., Sainte-Rose B., Aitken J., Marthouse R., Hajbane S., Cunsolo S., Schwarz A., Levivier A. and Noble K. (2018). Evidence that the Great Pacific Garbage Patch is rapidly accumulating plastic. *Scientific Reports*, **8**(1), 1–15, 4666, https://doi.org/10.1038/s41598-018-22939-w

Nelms S. E., Coombes C., Foster L. C., Galloway T. S., Godley B. J., Lindeque P. K. and Witt M. J. (2017). Marine anthropogenic litter on British beaches: a 10-year nationwide assessment using citizen science data. *Science of the Total Environment*, **579**, 1399–1409, https://doi.org/10.1016/j.scitotenv.2016.11.137

Song Y. K., Hong S. H., Jang M., Han G. M., Jung S. W. and Shim W. J. (2017). Combined effects of UV exposure duration and mechanical abrasion on microplastic fragmentation by polymer type. *Environmental Science & Technology*, **51**, 4368–4376, https://doi.org/10.1021/acs.est.6b06155

Thompson R. C. (2015). Microplastics in the marine environment: sources, consequences and solutions. In: Marine Anthropogenic Litter, M. Bergmann, L. Gutow and M. Klages (eds.), Springer, Berlin, pp. 185–200.

van der Velde T., Milton D. A., Lawson T. J., Wilcox C., Lansdell M., Davis G., Perkins G. and Hardesty B. D. (2017). Comparison of marine debris data collected by researchers and citizen scientists: Is citizen science data worth the effort? *Biological Conservation*, **208**, 127–138, https://doi.org/10.1016/j.biocon.2016.05.025

doi: 10.2166/9781789061123_111

Chapter 8

Diffuse water pollution in the United States: trends, regulations and outlooks

Mi-Hyun Park[1], Eakalak Khan[2], Xavier Swamikannu[3,4], Michael K. Stenstrom[1] and Joo-Hyon Kang[5]

[1]*Department of Civil and Environmental Engineering, University of California, Los Angeles, USA*
[2]*Department of Civil and Environmental Engineering, University of Nevada, Las Vegas, USA*
[3]*Institute of the Environment & Sustainability, University of California, Los Angeles, USA*
[4]*California Water Quality Control Board, Los Angeles, CA, USA*
[5]*Department of Civil and Environmental Engineering, Dongguk University, Seoul, Korea*

Keywords: Water pollution; nonpoint source; diffuse pollution; contaminants; water quality

8.1 INTRODUCTION

In the United States (US) nonpoint source pollution (often referred to as diffuse pollution in Europe) is defined as any pollution that is not from point source under the Clean Water Act (Graham *et al.*, 2011). It is 'caused when rainfall or snowmelt, moving over and through the ground, picks up and carries natural and human-made pollutants, depositing them into lakes, rivers, wetlands, coastal waters and groundwaters' (US EPA, 2003). It includes stormwater pollution flows from urban and agricultural lands and atmospheric deposition on land. For regulatory purposes however, urban stormwater pollution that is discharged into and from storm drainage conveyance systems has been considered a point source, after the 1987 amendments to the Clean Water Act. Stormwater runoff remains one of the largest contributors of surface water pollution and impairment of beneficial uses in the US and continues to grow in importance because of population growth, watershed changes, and land-use modifications without adequate mitigation (WEF, 2015).

In the US, surface water quality nationwide has significantly improved over the past several decades. However, many surface waters remain impaired. For example, 53% of rivers and streams, 71% of lakes, ponds and reservoirs, and 79% of bays and estuaries are identified as impaired (Figure 8.1) (US EPA, 2019a). Many of these waters, that is 46% of rivers and streams, 21% of lakes, ponds and reservoirs, and 18% of coastal waters, are impaired by nonpoint source pollution (US EPA, 2017). Nonpoint sources were identified as being among the top five (except unknown) source groups of rivers/streams, lakes/reservoirs/ponds, and wetlands; and two out of top three (except unknown) source groups for the Great Lakes and bays/estuaries (Table 8.1).

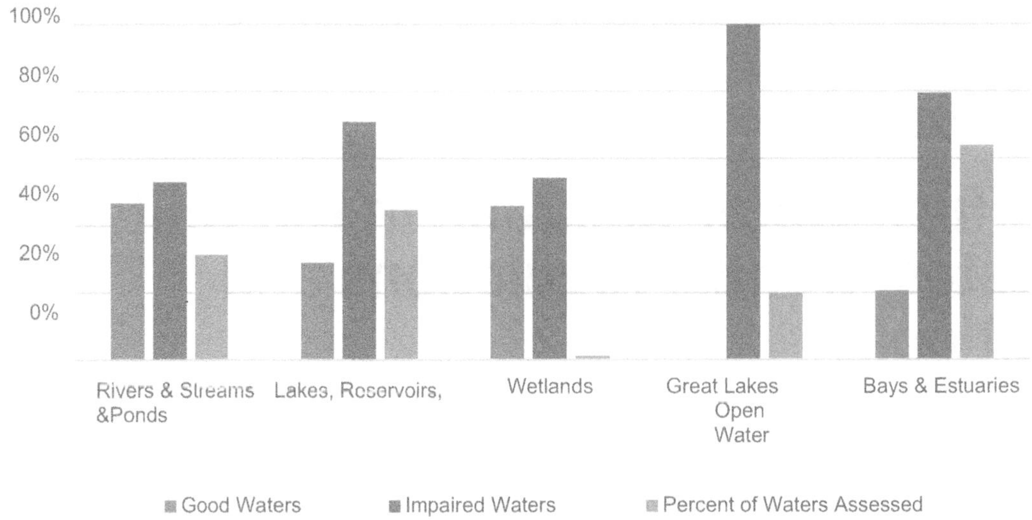

Figure 8.1 Natural water quality situation in the US (adapted from US EPA, 2019a).

8.2 NONPOINT SOURCE TYPES

According to the US Environmental Protection Agency (US EPA), types of nonpoint sources include urban areas, roads/highways/bridges, agriculture, forestry, hydromodification and habitat alteration, marinas and boating, resource extraction (abandoned mine drainage), and wetland/riparian areas (US EPA, 2019b). Atmospheric deposition is also regarded as nonpoint source pollution (Dressing *et al.*, 2016) and is the primary pollution source of lakes/reservoirs/ponds, and great lakes at 30% and 100%, respectively (Table 8.1).

Urbanization has an adverse impact on the nation's water quality. Impervious surfaces in urban areas increase the volume and velocity of stormwater runoff and pollutant loading (US EPA, 2019c). Urban watersheds generate more frequent high-flow pulses and elevated peak discharges than other land uses (Dressing *et al.*, 2016; Schoonover *et al.*, 2006). Typical pollutants from urban runoff include sediments; oil and grease and toxic chemicals from motor vehicles; pesticides and nutrients from lawns and gardens; pathogens and nutrients from pet waste and failing septic systems; road salts; trace metals from rooftops, motor vehicles and other sources; and thermal pollution from streets and rooftops (US EPA, 2019c). Urban runoff is the second-leading source of impaired bays/estuaries and coastal shorelines by 38% and 9%, respectively, and pollution sources of rivers/streams, lakes/reservoirs/ponds by 8% and 6%, respectively (Table 8.1). Table 8.2 presents a summary of pollutant concentrations from urban stormwater runoff in the National Stormwater Quality Database (NSQD) Version 4.02, compiled from about 600 sampling locations across the nation (Pitt, 2015, 2018). Each urban land use generates different pollutant concentrations. For example, residential areas generate more pathogen indicators and phosphorus; freeways generate more metals and nitrogen; and open space generates more suspended solids, cadmium and mercury (Table 8.3).

Nonpoint source pollution also originates from roads, highways, bridges and construction sites (US EPA, 2019c). Urban road runoff contains trace metals, for example, copper, lead and zinc, and organic pollutants, for example, polycyclic aromatic hydrocarbons (PAHs), petroleum hydrocarbons, and herbicides from either vehicles or road management practices (Hwang *et al.*, 2016, 2018). Pollutants from vehicles are deposited on the road surfaces during dry periods and conveyed into a storm sewer system or receiving waters during storm events (Hwang *et al.*, 2016). Large traffic volumes on

Table 8.1 Top 10 probable source groups of impaired waters (% of impaired waters).

Rivers and Streams		Lakes, Reservoirs, and Ponds		Wetlands		Great Lakes Open Water		Bays and Estuaries		Coastal Shoreline	
Unknown	25%	Atmospheric deposition	32%	Unknown	58%	Atmospheric deposition	100%	Legacy/ historical pollutants	49%	Municipal Discharges/ Sewage	12%
Agriculture	23%	Unknown	29%	Natural/wildlife	43%	Unknown	24%	Urban-related runoff/ stormwater	38%	Urban-related runoff/ stormwater	9%
Hydromodification	15%	Agriculture	8%	Agriculture	30%	Agriculture	11%	Unknown	37%	Unknown	5%
Atmospheric deposition	15%	Natural/ wildlife	8%	Atmospheric deposition	30%	Municipal discharges/ sewage	0.3%	Atmospheric deposition	31%	Recreational Boating and Marinas	4%
Habitat alterations	11%	Unspecified nonpoint source	8%	Resource extraction	5%	Unspecified nonpoint source	0.3%	Municipal discharges/ sewage	13%	Hydromodi-fication	4%
Unspecified Nonpoint Source	10%	Other	6%	Hydromodi-fication	1%			Unspecified nonpoint source	9%	Industrial	3%
Municipal discharges/sewage	10%	Urban-related runoff/ Stormwater	6%	Unspecified nonpoint source	0.3%			Other	9%	Unspecified nonpoint source	2%
Natural/wildlife	9%	Legacy/ historical Pollutants	6%	Other	0.1%			Natural/ wildlife	8%	Agriculture	2%
Urban-related runoff/Stormwater	8%	Municipal discharges/ sewage	5%	Land application/ waste sites/ tanks	0.1%			Agriculture	8%	Legacy/ historical pollutants	1%
Silviculture (forestry)	7%	Hydromodi-fication	4%	Groundwater loadings/ Withdrawals	0.1%			Industrial	8%	Land application, waste sites, tanks	1%

(adapted from US EPA, 2019a).

Table 8.2 Statistical summary of event mean concentrations from urban runoff in the US.

Constituents	Unit	Number of EMCs	Percent of NDs	25th Percentile	Median	75th Percentile	Mean	Standard Deviation
E. coli	MPN/100 mL	125	2.40	600	1815	4193	6186	12 119
Fecal Streptococcus	MPN/100 mL	1285	2.65	5000	18 400	50 000	74 462	244 847
Fecal Coliform	MPN/100 mL	2057	3.40	1000	5000	26 000	62 557	289 041
Total Coliform	MPN/100 mL	282	2.84	3950	24 000	170 000	255 168	629 391
Nitrate-N	mg/L	1028	7.88	0.39	0.62	1.07	0.97	1.35
Nitrite-N	mg/L	714	35.85	0.05	0.1	0.16	0.17	0.37
TKN	mg/L	6723	2.31	0.84	1.39	2.3	2.03	3.2
TP	mg/L	7692	2.29	0.14	0.25	0.45	0.4	0.69
TDS	mg/L	3850	0.44	52	80	135	147.67	462.72
TSS	mg/L	7353	0.56	29	66	144	144.8	315.38
Total Cadmium	μg/L	3754	60.18	0.5	1	2.65	3.63	12.18
Total Copper	μg/L	5549	10.81	8.7	17	33	33.7	121.05
Lead	μg/L	5051	24.31	6.13	17	49	48.41	94.69
Total Zinc	μg/L	6275	4.25	50	100	199	200.36	617.06
Total Mercury	μg/L	1597	92.67	0.1	0.2	0.32	2.85	16.95

TKN, Total Kjeldahl Nitrogen; TP, Total Phosphorus; TDS, Total Dissolved Solids; TSS, Total Suspended Solids (adapted from Pitt, 2015).

Table 8.3 Mean values of event mean concentrations from runoff by urban land use.

Constituents	Unit	Residential	Commercial	Industrial	Institutional	Freeway	Open Space
E. coli	MPN/100 mL	7582	5904	2429		6002	
Fecal Streptococcus	MPN/100 mL	97 834	49 775	72 812	23 000	38 148	69 317
Fecal Coliform	MPN/100 mL	89 215	29 563	49 410	3100	23 870	56 314
Total Coliform	MPN/100 mL	426 776	333 893	201 814		203 500	109 213
Nitrate-N	mg/L	1.17	0.62	0.86	0.57	1.32	1.13
Nitrite-N	mg/L	0.17	0.21	0.13	0.14	0.21	0.05
TKN	mg/L	2.05	1.94	2.3	1.54	2.57	1.39
TP	mg/L	0.44	0.39	0.42	0.2	0.41	0.4
TDS	mg/L	140.13	146.94	157.28	139.26	109.69	230.87
TSS	mg/L	141.45	87.16	156.4	144.96	126.27	256.71
Total Cadmium	μg/L	3.27	3.58	4.13	0.91	3.52	13.13
Total Copper	μg/L	29.75	38.92	39.59	10.93	66.51	15.82
Lead	μg/L	50.54	53.97	54.35	9.19	55.9	25.67
Total Zinc	μg/L	137.8	217.49	421.54	88.83	241.23	97.48
Total Mercury	μg/L	2.42	0.44	0.62	ND	9.35	32.32

ND, non detection (adapted from Pitt, 2015).

highway contribute to higher concentrations of trace metals in the runoff (Hwang *et al.*, 2016). In the Lake Tahoe basin, California, 10–20% of fine solids less than 20 μm in highway stormwater runoff came from the wearing off of asphalt pavement (Hwang *et al.*, 2018). In Nebraska, concentrations of metals and nutrients in bridge-deck runoff were similar to those in road runoff (Swadener *et al.*, 2014). In Massachusetts, concentrations of suspended sediments and nutrients (total nitrogen and total phosphorus) in bridge-deck runoff were higher than those in highway runoff (Smith *et al.*, 2018).

Stormwater runoff from construction activities contain pollutants including sediment, nutrients, chemicals (e.g., paints, acids for cleaning masonry surfaces, cleaning solvents, asphalt products, soil additives used for stabilization, compounds in wash water from concrete mixers, and concrete-curing compounds), petroleum products (e.g., fuels and lubricants for vehicles, power tools, and general equipment maintenance), oils from asphalt paving, fertilizers, pesticides, and pH-modifying contaminants, for example bulk cement (Dressing *et al.*, 2016; WA DOE, 2014), resulting in aquatic habitat impairment, stream bank erosion, and clogging waterways. Soil loss rates from construction sites were significantly higher than the rates from natural soil erosion by up to 1000 times (Keener *et al.*, 2007). High stream sediment concentrations were reported at least 100 m below the discharge basin on construction sites (Ehrhart *et al.*, 2002). Construction site runoff, even with stormwater control measures, contains high concentrations of suspended solids, especially fine particles (Dressing *et al.*, 2016; Przepiora, *et al.*, 1998). These pollutant particles come from products such as fertilizers, pesticides, fuels, and so on. (Keener *et al.*, 2007) that can be easily soluble in water, which is often more reactive and bioavailable to aquatic organisms (Dressing *et al.*, 2016; Faucette *et al.*, 2009). Construction site runoff impairs rivers/streams, lakes/reservoirs/ponds, and bays/estuaries by 4%, 2% and 2%, respectively (US EPA, 2019a).

Agricultural nonpoint sources include animal feed-lot operations and manure, grazing, and pesticide and fertilizer (US EPA, 2019c). The main pollutants from agricultural runoff are nutrients from fertilizer, manure, crop residues, irrigation water and atmospheric deposition; sediment from erosion; organic matter, and pathogens from animal wastes; salts and selenium from irrigation water; and pesticides and herbicides (US EPA, 2003). Agricultural runoff can transport particulate-bound and soluble nutrients while groundwater infiltration from agricultural area can contain soluble nutrients (US EPA, 2003). Animal wastes/manure generate runoff with excess nutrients, organismic materials, and pathogens. In North Carolina, a total of 40 million litres of animal waste spill occurred in 1995, which caused high turbidity, nutrients, fecal coliform and pathogens, leading to severe damage to water quality and algal blooms (Mallin *et al.*, 1997). Similar incidents recurred after Hurricane Floyd in 1999, Hurricane Matthew in 2016 and Hurricane Florence in 2018. Irrigation runoff transports salts and selenium that are toxic at excessive concentrations (US EPA, 2003). Pesticide degradation products were more frequently found than the parent compounds. For example, alachlor ethanesulfonic acid was detected almost 10 times more frequently than alachlor in 153 wells (US EPA, 2003). Agriculture runoff is a significant cause of impaired rivers/streams, lakes/reservoirs/ponds, wetlands, and great lakes at 23%, 8%, 30% and 11%, respectively (Table 8.1).

Forestry activities cause nonpoint source pollution during vegetation removal, timber harvesting, mechanical preparation for the planting of trees, road construction and traffic activity (US EPA, 2019c). Main pollutants from forestry operations include sediment from road construction, yarding areas and log dragging; organic debris from harvesting operations; nutrients from fertilization and application of biosolids and animal wastes; and silvicultural chemicals (Binkley *et al.*, 1999; Dressing *et al.*, 2016; Ryan & Grant, 1991). Road construction and erosion from roads are a major source, contributing up to 90% of the total sediment from forestry activities despite its small area (US EPA, 2005a). Timber harvesting, especially clear-cut harvesting, can lead to landslides, stream water temperature increase, and elevated mercury concentration in runoff (US EPA, 2005a). Site preparation can discharge sediments and chemicals for example, fertilizers, herbicides, and pesticides, when applied, and forest operation can discharge petroleum compounds and other machinery-related chemicals (US EPA, 2005a). Scheduled burning for site regeneration can affect water quality by increasing soil erosion and discharging nutrients. However, wildfires have a more severe effect than scheduled burning due to its intensity (US EPA, 2005a). Forestry is listed in the top ten among pollution sources of impaired rivers and streams by 7% (Table 8.1).

Hydrologic modification is defined as the alteration of the hydrologic characteristics of waterways, and grouped into three categories: channelization and channel modification, dams/reservoirs, and streambank and shoreline erosion (US EPA, 2007). Hydrologic modification has an adverse impact on aquatic habitat and hydraulic regime (US EPA, 2019c). Channelization increases the flow and transport of nonpoint source pollutants from upstream with uniform cross section and steeper stream gradients (US EPA, 2007). Channelization and channel modification increase sediments eroded from stream banks, causing increased turbidity, stream temperature, and ultimately reduced dissolved oxygen (DO) in the water column (US EPA, 2007). Eroding sediments can introduce various chemicals with adsorbed nutrients, metals, organic materials and these chemicals can become bioavailable with temperature change in the water (US EPA, 2007). Dams and reservoirs can alter both water quality and quantity, such as sedimentation, and lead to change in temperature, alkalinity, pH, nutrients, metals, and organic matter loads (US EPA, 2007). Streambanks and shorelines can affect sediment concentrations because of increased erosion (US EPA, 2007). Hydrologic modification is one of the main causes of the impairment of rivers and streams and is listed in the top ten for lakes/reservoirs/ponds, wetlands, and coastal shorelines by 4%, 1% and 4%, respectively (Table 8.1).

Mining activity was identified as a significant source of environmental pollution prior to the enactment of the Surface Mining Control and Reclamation Act (SMCRA) of 1977 (Stewart & Skousen, 2003). For example, legacy silver and gold mining activity in the western US generated large amounts of mercury that were used to recover silver and gold. This has resulted in the implementation of the Total Maximum Daily Loads (TMDLs) for mercury in the local waters to reverse the impairment (Alpers *et al.*, 2005). Cyanide was used in gold recovery operations and the spills of cyanide in South Carolina in 1990 caused fish kills as far as 50 miles downstream of the mining site (US EPA, 2000a). Despite the improved mining practices since the enactment of the SMCRA, the adverse impact of abandoned mines on water quality continues. Acid mine drainage, the most prevalent abandoned mine discharge to surface waters or groundwater, is formed by a chemical reaction between water and rocks containing iron sulfide, releasing pollutants including sulfates, acidity, and toxic metals such as copper, lead, zinc, silver, manganese, cadmium, iron, arsenic and mercury (Bigham & Cravotta, 2016; US EPA, 2000a). In Pennsylvania, acid mine drainage from abandoned anthracite mines contaminated Swatara Creek, causing high concentrations of sulfate, dissolved iron, and dissolved aluminum, and pH as low as 3.5 from 1959 to 1986 and no change in sulfate or dissolved iron concentrations immediately downstream from the mined area from 1997 to 2006 (Cravotta *et al.*, 2010). In Tennessee, acid mining drainage discharged suspended solids attached by large amounts of trace metals into Ocoee Lake, resulting in elevated levels of copper, zinc, lead, cadmium, nickel and cobalt in the sediment (Lee *et al.*, 2008). Acid mine drainage can disrupt aquatic life and corrode infrastructure (Lee *et al.*, 2008). Mining (resource extraction) is the fifth largest source of pollutants to wetlands, amounting to 5% (Table 8.1) and the eleventh largest pollution source for rivers/streams, lakes/reservoirs/ponds with 6% and 3%, respectively (US EPA, 2019a).

Marina waters can be polluted from spillage, boat cleaning, fueling operations and marine head discharge, and stormwater runoff from parking lots and hull maintenance and repair areas (US EPA, 2019c). Nonpoint source pollution into marina waters and sediments can contain organic matters, trace metals, pathogens, petroleum hydrocarbons and solvents (McCoy & Johnson, 2010; US EPA, 2001). Organic matters derived from sewage discharged from boats, pet waste in storm water runoff, and fish waste disposed into surface waters can lower DO (US EPA, 2001). In North Carolina, DO concentrations in the marinas were significantly lower than those in the adjacent waterbodies (NCDEM, 1990). Metals come from antifoulants and boat operations: arsenic in paint pigments, pesticides, and wood preservatives; copper in antifouling paints; zinc from anti-corrosion treatment of metal hulls and engine parts; mercury in float switches and thermostats; nickel in brake linings and pavement material; and aluminum, iron, and chromium in the construction of marinas and boats (US EPA, 2001). Petroleum hydrocarbons at marinas come from fuel, oils, lubricants, finishes, and cleansers discharged from fueling stations, boat engines, and stormwater runoff from the marina property (US EPA, 2001). Solvents including methylene chloride, tetrachloroethane, and trichloroethene (trichlorethylene) are derived from degreasing agents, varnishes, paint removers, and lacquers for engine and vessel maintenance, repair and cleaning (McCoy & Johnson, 2010). Pathogens originate from pet waste and overboard sewage discharge (US EPA, 2001) and fecal coliform can be discharged from boats in areas with high boat densities (NCDEM, 1990). Recreation and boating are the fourth largest pollution source for coastal shoreline amounting to 4% (Table 8.1) and the cause of impaired bays and estuaries responsible for 1% (US EPA, 2019a).

Atmospheric deposition is one of the most important nonpoint sources of pollutants especially for urban stormwater pollution. Atmospheric deposition can be either wet or dry deposition depending on whether it occurs during storm events or during dry weather and the characteristics of wet and dry deposition can be distinct (Zannetti, 1990). In urban catchments in Los Angeles, atmospheric deposition contributed to trace metal loads in stormwater runoff by upwards of 57–100%, with dry deposition being the dominant source (Sabin *et al.*, 2005). In Florida, wet deposition of mercury was found to be dominant (Dvonch *et al.*, 2005) and atmospheric deposition was a significant source of nutrients and peaked with increase in agricultural activity (Anderson & Downing, 2006). In urban watersheds in Nebraska, atmospheric deposition was found to be a primary source of nitrate nitrogen in stormwater for small storms whereas its contribution was smaller for large storms (Baral *et al.*, 2018).

8.3 NONPOINT SOURCE POLLUTANTS

Nonpoint sources generate pollutants similar to point sources such as sediment, nutrients, pathogens, pesticides, salts, toxic substances, and petroleum products. The main pollutants associated with the impairment based on a national survey include nutrients, mercury, polychlorinated biphenyls (PCBs), and pathogens (Table 8.4). These pollutants are from various sources in both urban and agricultural runoff (Table 8.5).

Organic debris decomposed by microorganisms depletes DO in water bodies (Dressing *et al.*, 2016), which is the leading cause for the impairment of wetlands attributed to 70% and impairments of rivers/streams, lakes/reservoirs/ponds, bays/estuaries and coastal shoreline by 11–17% (Table 8.4). In the Nationwide Urban Runoff Program (NURP) study (US EPA, 1983), urban runoff was found to transport oxygen-depleting substances at concentrations similar to those in treated wastewater discharges (US EPA, 2005b).

Sediment is one of the most common pollutants in rivers and streams and accounts for 24% (Table 8.4). Potential sources of sediment pollution are landslides, erosion, deforestation, overgrazing, urban runoff, construction and mining activities, and atmospheric deposition, including direct sources such as a waterbody's shoreline, bank, or bottom erosion (Dressing *et al.*, 2016). A study in the Issaquah Creek watershed, an urbanizing watershed near Seattle in Washington, showed urban development increased about 50% of annual sediment yield (Nelson & Booth, 2002). Fine sediment particles are the primary carriers of other attached pollutants, for example, metals, ammonium ions, phosphates, and toxic organic compounds (US EPA, 2005b). The annual financial loss caused by sediment is reported to be approximately $16 billion in North America (Osterkamp *et al.*, 1998; US EPA, 2005b). The damage can be categorized as physical damage to water conveyance, treatment, storage facilities, and loss of recreational and navigational use; chemical damages such as deposition and storage of other pollutants; and biological damages to aquatic habitats (Dressing *et al.*, 2016).

Nutrients such as nitrogen and phosphorus are prevalent in most impaired waters and continue to be of great concerns at present and in the future. Nutrients are the second leading cause of impairments in lakes/reservoirs/ponds, and bays/estuaries by 30% and 41%, respectively and the third leading cause in

Table 8.4 Top ten pollutants in impaired waters (% of impaired waters).

Rivers and Streams		Lakes, Reservoirs, and Ponds		Wetlands		Great Lakes Open Water		Bays and Estuaries		Coastal Shoreline	
Pathogens	32%	Mercury	64%	Organic Enrichment/DO Depletion	70%	PCBs	100%	PCBs	63%	Mercury	71%
Sediment	24%	Nutrients	30%	Mercury	47%	Mercury	100%	Nutrients	41%	Pathogens	24%
Nutrients	20%	PCBs	24%	Metals (excl. Hg)	14%	Dioxins	99%	Mercury	38%	Turbidity	15%
Organic Enrichment/ DO Depletion	17%	Turbidity	11%	Salinity/TDS/ Chlorides/Sulfates	12%	Pesticides	76%	Turbidity	34%	Organic Enrichment/ DO Depletion	12%
Temperature	16%	Organic Enrichment/ DO Depletion	11%	Pathogens	11%	Toxic Organics	57%	Dioxins	23%	pH/Acidity/ Caustic conditions	6%
Metals (excl. Hg)	16%	Metals (excl. Hg)	9%	Nutrients	10%	Nutrients	1%	Toxic Organics	21%	Nutrients	4%
PCBs	14%	pH/Acidity/ Caustic Conditions	8%	Toxic Inorganics	4%			Metals (excl. Hg)	19%	Oil & Grease	3%
Mercury	12%	Salinity/TDS/ Chlorides/ Sulfates	7%	Temperature	2%			Pesticides	17%	Temperature	3%
Habitat Alterations	11%	Algal Growth	5%	pH/Acidity/ Caustic Conditions	1%			Pathogens	13%	Unknown-Impaired Biota	3%
Turbidity	8%	Nuisance Exotic Species	5%	Turbidity	1%			Organic Enrichment/ DO Depletion	12%	Algal Growth	3%

Hg, Mercury; PCBs, Polychlorinated Biphenyls; TDS, Total Dissolved Solids; DO, Dissolved Oxygen. (Adapted from US EPA, 2019a).

Table 8.5 Sources of pollutants in stormwater runoff.

Pollutants	Urban Sources	Agricultural Sources
Sediments	Streets, lawns, driveways, roads, construction activities, atmospheric deposition, drainage channel erosion	Washoff fields
Nutrients	Lawn fertilizers, atmospheric deposition, automobile exhaust, soil erosion, animal waste, detergents	Fertilizers from farms, animal manure, animal feeding operations, irrigation
Pathogens	Lawns, roads, leaky sanitary sewer lines, sanitary sewer cross-connections, animal waste, septic systems	Animal feeding operations, irrigation
Metals	Automobiles, bridges, atmospheric deposition, industrial areas, soil erosion, corroding metal surfaces, combustion processes	Irrigation
Hydrocarbons/Oil & Grease	Roads, driveways, parking lots, vehicle maintenance areas, gas stations, illicit dumping to storm drains	
Organic Materials	Residential lawns and gardens, commercial landscaping, animal wastes	Animal feeding operations
Pesticides & Herbicides	Residential lawns and gardens, roadsides, utility right-of-ways, commercial and industrial landscaped areas, soil wash-off	Agricultural use
Salts	Deicing agents from roads	Irrigation

(adopted from US EPA, 1999, 2005c).

impaired rivers/streams by 20% (Table 8.4). While point sources such as wastewater discharges contributed significantly to nutrient loading in the past, recent increases in phosphorus loads have arisen from nonpoint sources (US EPA, 2019e). The main nutrient sources are fertilizers from cropland and lawn application, manure from animal husbandry and feed-lot facilities, leaking septic systems, wildlife feces, and atmospheric deposition from industrial and automobile emissions (US EPA, 2003). In the Johnson Creek Watershed, Oregon, nitrogen concentrations were higher in agricultural runoff during the wet season, while total phosphorus was higher from urban areas during the dry season (Sonoda *et al.*, 2001). Vegetation, such as lawns and trees lining streets, were found to be significant sources of phosphorus in urban watersheds in Madison, Wisconsin and in Minneapolis–St. Paul, Minnesota (Janke *et al.*, 2017; Waschbusch, 1999). Excessive nutrients cause eutrophication that leads to hypoxia and algal blooms, resulting in fish kills and taste and odor problems. For example, in Lake Champlain, bordered with Vermont, New York and Quebec, has been impaired by phosphorous loads, experiencing severe algal blooms. A sediment core assessment study indicated that the eutrophication driver for Lake Champlain Basins shifted from point source inputs from domestic sewage to nonpoint sources largely from agriculture (Levine *et al.*, 2018). Therefore, 90% of phosphorus is contributed by nonpoint sources and among them 46% of the load is from urban land uses and 38% (352 tons) is from agricultural areas (LCBP, 2018, 2019; Moore *et al.*, 2015). The Great Lakes basin, which generates approximately $15 billion annually in livestock, dairy, grain and corn products, is also impaired because of nonpoint source nutrient loads, especially of bioavailable phosphorus through surface runoff, soil erosion, near-surface interflow and tile drainage systems (Kerr *et al.*, 2016). During the water year 2007, the Maumee and Sandusky Rivers, Ohio, draining to Lake Erie, discharged the largest loads of total phosphorous, dissolved reactive phosphorous, total Kjeldahl nitrogen, and chloride due primarily to the heavy rainfall runoff from agricultural activity (Richards *et al.*, 2010). Dissolved reactive phosphorous, which is bioavailable to algal growth has increased dramatically in Lake Erie tributaries from agricultural sources (Kerr *et al.*, 2016). The water quality continues to decline despite the investment of funds in enhanced implementation of management practices and conservation programs (Kerr *et al.*, 2016). In the Gulf of Mexico, more than 70% of the nutrient loads are from agricultural nonpoint sources, doubling the nutrient load and tripling the nitrate load in the second half of the 20th century (Faust *et al.*, 2018). Modeling results indicated that more than 70% of nutrients from the Mississippi River Basin to the Gulf were transported from agricultural sources: nitrogen contributed by corn and soybean cultivation (52%), atmospheric deposition (16%), other cropland (14%), and manure on pasture and rangeland (5%); and phosphorus from animal manure on pasture and rangeland (37–39%), corn and soybeans (21–25%), other cropland (18%), and forest (8%) (Alexander *et al.*, 2008). The average total nitrogen: total phosphorus ratios in the Lower Mississippi River rose from less than 10:1 in the late 1970s, to around 40:1 in the 1980s, and then declined to 20:1 by the 2000s (Bargu *et al.*, 2019; Turner *et al.*, 2007). In Chesapeake Bay, total nitrogen from the Susquehanna River increased 2.5 times between 1945 and 1990, and approximately 152 million kg of nitrogen per year has entered the Bay since 1990, leading to algal blooms and hypoxia, which has dramatically increased in both duration and frequency (Bilkovic *et al.*, 2019).

Pathogens are the primary cause of the impairment of rivers and streams by 32% and wetlands and bays/estuaries and coastal shoreline by 11–24% (Table 8.4). Pathogens in urban stormwater runoff include bacteria, viruses, and protozoa, particularly because they adversely impact public health and lead to beach closures and health advisories, and the contamination of drinking water sources (Dorfman, 2006; Parker *et al.*, 2010). In 2005, 75% of beach advisories were in response to elevated fecal indicator bacteria (FIB) levels (Dorfman, 2006; Parker *et al.*, 2010). Epidemiological studies suggested strong relationships between gastrointestinal symptoms and FIB counts, and between urban runoff and illnesses (Parker *et al.*, 2010; Pruss, 1998). Some indicator organisms have been used to assess fecal contamination. The US EPA recommended the use of fecal coliforms as indicator for human pathogens until 1986 and then changed the recommendation to *Escherichia coli* (*E. coli*) for fresh recreational waters and *Enterococci* for fresh and marine recreational waters (US EPA, 1986, 2005b). The USGS National Water Quality Assessment program identified land use to have the greatest effect on FIBs in 136 streams (Francy *et al.*, 2000). In Southern California, the FIB levels at all beach sampling locations at the river mouth exceeded California recreational water quality limits because of the presence of indicator bacteria in urban runoff (Jiang *et al.*, 2001). FIBs including *E. coli* levels increased after storm events in Lincoln Beach, New Orleans; Hoosic River, north-eastern US; and southern Lake Michigan (Jeng *et al.*, 2002; Traister & Anisfeld, 2006; Whitman *et al.*, 2006). Two major protozoans found in stormwater were *Giardia* and *Cryptosporidium* (Cizek *et al.*, 2008; Selvakumar & Borst, 2006) and the leading sources of pathogens in urban runoff were human waste, pet waste, and wildlife waste (US EPA, 2005b). Human waste from illicit connections to sanitary sewers, and improperly disposed pet waste can contaminate urban runoff and be discharged from storm drain systems (US EPA, 2005b).

Trace metals are found in sediment and the water column in impaired rivers/streams, lakes/reservoirs/ponds, wetlands and bays/estuaries and account for 9–21% of the cause of impairment (Table 8.4). Trace metals have toxic effects on aquatic life and the potential for groundwater contamination. The most prevalent metals found in urban runoff are copper, lead and zinc (>5000 observations, Pitt, 2018), often originating from motor vehicles. Copper in stormwater runoff mainly originates from brake-pad wear (Hwang *et al.*, 2016). In the US, the copper content in brake pads increased in the mid-1980s as metallic disk brakes replaced asbestos brake pads (Lifset *et al.*, 2012). The increase of copper release from brake pads was estimated from <5 tons to 2550 tons between 1975 and 2000 while that from vehicle fluid leaks increased from 10 tons to 20 tons during the same period (Lifset *et al.*, 2012). In San Francisco Bay, California, an estimated annual load of copper in stormwater runoff was approximately 56 tons and the contribution from brake pads was estimated to be from 10–35% of total load of copper (Donigian *et al.*, 2009). In Oregon, higher copper concentrations were found from urban/high-traffic site runoff and in first-flush samples (Nason *et al.*, 2012). In Puget Sound, Washington, copper was primarily from brake pads and pesticide use on urban lawns and gardens, while zinc and cadmium were mainly from roofing materials (Roberts *et al.*, 2011). The major sources of zinc from urban watersheds are tire wear and galvanized surface (Hwang *et al.*, 2016). The estimate of zinc released from tire wear was approximately 8000 tons in 2012 (Hwang *et al.*, 2016). Zinc concentrations in the roof runoff from a traditional galvanized metal panel were 5–30 mg/L for the first two years, which was high compared to NSQD values of 60–300 μg/L (Clark *et al.*, 2008; NRC, 2009). The major source of lead is from motor vehicles because lead remained a gasoline additive from 1920 until the mid-1990s in the US (Hwang *et al.*, 2016). Lead in urban stormwater runoff peaked in the mid-1970s with three to four orders of magnitude higher than background levels (Hwang *et al.*, 2016). The reported lead concentrations in stormwater runoff in 1970s were 100–12 600 μg/L and 3600–8500 μg/L in North Carolina and Oklahoma, respectively and the mean concentrations after 1990 were reduced to less than 100 μg/L (Hwang *et al.*, 2016). Although lead concentrations have declined over the last couple of decades since the phase out of leaded gasoline, lead concentrations on roadside particles and in some urban watershed sediments are still high due to lead wheel-balancing weights, which contain 95% lead (Hwang *et al.*, 2016). In the Puget Sound Basin, Washington, the lead released from the loss of tire wheel-balancing weights contributed to about 13% of the total load of lead (Roberts *et al.*, 2011).

Mercury is the most prevalent pollutant for all assessed water bodies: the leading cause of pollution for lakes/reservoirs/ponds, Great Lakes, coastal shoreline, and ocean and near coastal waters by 64–100% and impaired rivers/streams, wetlands, and bays/estuaries and by 12% 47%, and 38%, respectively (Table 8.4, US EPA, 2019a). More than 60% of the impairments showed high mercury levels in fish, in which mercury accumulates as methylmercury that is a potent neurotoxin posing a high risk to human health (Copeland, 2012; Mason *et al.*, 1995). As a consequence, over 7000 mercury TMDLs have been developed since 1996 (Copeland, 2012). Mercury comes mainly from aerial deposition from local, regional, and international sources such as coal-fired plants, waste combustors and incinerators and wildfires; legacy practices such as gold mining; and geologic deposits (Burke *et al.*, 2010; Copeland, 2012; Gabriel *et al.*, 2002; Pirrone *et al.*, 2010; Schroeder & Munthe, 1998). In the Pacific Northwest, which receives large atmospheric mercury loads, filtered mercury concentrations and loads in the runoff from timber harvested catchments were higher by 28% and 80%, respectively, than those in the runoff from forest area catchments despite there being no significant difference in particulate-bound mercury concentrations or loads (Eckley *et al.*, 2018).

Toxic organic compounds are known to occur in urban runoff. In the NURP study (US EPA, 1983), at least 10% of urban runoff samples contained toxic compounds. Most common toxic organics found in urban stormwater were chloroethylvinylether, ethylbenzene, toluene, chloroform, acrolein, and methylene chloride (>450 observations, Pitt, 2018). PCBs are the leading cause of impairments for wetlands and bays/estuaries by 100% and 63%, respectively, and impaired rivers/streams and lakes/reservoirs/ponds by 14% and 24%, respectively (Table 8.4). In the Chesapeake Bay, toxic organics as well as metals have been reported to impair approximately 70% of the tidal waters (Bilkovic *et al.*, 2019). Petroleum hydrocarbons such as PAHs, oil and grease in urban runoff come from tire wear, asphalt pavement degradation, deposition from automobile exhaust, leaked engine oils and vehicle fluids (Hwang *et al.*, 2018; US EPA, 2000b, 2005b). PAH concentrations in the receiving waters during storm events were found to be two to three orders of magnitude higher (Hwang *et al.*, 2018). Petroleum hydrocarbons that persist in sediments for long periods are toxic to aquatic life at low concentrations (US EPA, 2000b, 2005b). In Puget Sound, Washington, the total annual releases of petroleum and PAHs were 9200 tons and 310 tons, respectively (Roberts *et al.*, 2011). Oil and grease impaired coastal shoreline by 3% (Table 8.4).

Pesticides and their degradation products cause impairment of surface and groundwater (US EPA, 2003). Pesticides in urban stormwater runoff are of increasing concern as they can contribute to stream toxicity (Wolfand *et al.*, 2019). Pesticides were detected in surface waters, groundwaters, sediments and fishes in urban watersheds (Burton & Pitt, 2001). Pesticides were detected in most of urban stream samples (97%) nationwide and organochlorine compounds were detected in most of fish tissue samples (94%) in urban waters (NRC, 2009). In the Calleguas Creek Watershed, Southern California, organophosphorus insecticide concentrations varied in agricultural runoff as when compared with residential runoff (Pedersen *et al.*, 2006). Residential stormwater runoff in California contains a greater number of pesticides than in dry weather runoff, such that a single storm event in 3 hours can discharge as much as the amount discharged from dry weather runoff over a 6 month period (Weston *et al.*, 2009). Some pesticide degradation by-products can be more problematic because of their persistence. In the Midwest, alachlor ethanesulfonic acid was found more frequently than the parent compound, alachlor, by approximately a factor of ten in wells (US EPA, 2003). Pyrethroid insecticides are frequently found in sediments of urban waters across the nation and reported to cause water column toxicity at low concentrations (Kuivila *et al.*, 2012; Weston & Lydy, 2012).

Micropollutants including endocrine disrupting compounds (EDCs), pharmaceuticals and personal care products (PPCPs) are of great concern. Major sources of endocrine disrupting compounds in the aquatic environment are traditional point source discharges such as municipal and industrial wastewater treatment plants, but nonpoint sources including stormwater runoff can be potential sources of significant input to receiving waters. In New Orleans stormwater canals and Bayou St. John, Louisiana, PPCPs and EDCs such as naproxen, bisphenol A, ibuprofen and triclosan were detected and their concentrations increased with cumulative rainfall, indicating nonpoint source contamination (Boyd *et al.*, 2004). In eleven lakes in Minnesota, where wastewater treatment plant discharges were absent, EDCs such as bisphenol A, 17β-estradiol, estrone, and 4-nonylphenol, were detected in 90% of the lakes, indicating the impact of nonpoint sources (Writer *et al.*, 2010). The concentrations found were similar to those downstream of point source discharges, which have been reported to cause endocrine disruption in fishes in the lakes (Writer *et al.*, 2010). Poly- and perfluoroalkyl substances (PFASs), which contaminate drinking water sources throughout the US, were detected in runoff from a firefighter training area (Dauchy *et al.*, 2019). Long- and short-chain PFASs were found in 33 urban runoff samples from the San Francisco Bay, CA (Houtz & Sedlak, 2012).

Salts are carried in irrigation return flow and road runoff with deicing agent application (Dressing *et al.*, 2016). Salts and deicing agents (salinity/TDS/chlorides/sulfates) impaired lakes/reservoirs/ponds and wetlands by 7% and 12%, respectively (US EPA, 2019a). Road salt has become a problem especially in the Northeast and Midwest, for example, New York, Ohio and Illinois (Church, 1996). In the US, roughly $10 million was spent annually to resolve road salt contamination (US EPA, 2005b). In 2015, the use of deicing salts in the US was 22.7 million tons and increased in the following few years, directly related to winter weather conditions (Bolen, 2017). In New York, road salts were found to have moved up to 172 m in the Adirondack Mountain Region and wetlands in the Hudson River Valley had elevated levels of conductivity (250–1500 μS) likely due most probably to the increased use of road salts (Karraker *et al.*, 2008).

Trash and debris can be discharged to local water bodies by direct disposal, wind carriage and urban runoff (LASAN, 2002; Liang *et al.*, 2019). Floatable and settleable trash can suffocate aquatic habitats and contaminate sediment, respectively (LASAN, 2002). In a highway study in Los Angeles, captured gross pollutants (above 0.5 cm in size) comprised 10% of trash and 90% of vegetation (Kim *et al.*, 2006). Trash on beaches in Orange County, Southern California, in the summer of 1998 (approximately 12 tons in weight) was dominated by plastic, wood, metal and glass (Moore *et al.*, 2001) while trash from highway storm-drain outlets in Los Angeles County, Southern California in wet seasons between 1998 and 2000 was dominated by plastic, wood, metal, cardboard/chipboard and cigarette butts (Caltrans, 2000). Plastic was the largest component in both cases (LASAN, 2002). Plastics and microplastics have been a growing concern in recent years, impacting aquatic life. Many plastics enter the waters through stormwater runoff and illegal dumping. In the US, the of plastics in marine debris emanate from land-based sources, ocean-based sources, and unidentified sources at 49%, 18% and 33%, respectively (Sheavly, 2007). The presence of microplastics was reported in surface waters including ocean, lakes and rivers (Bilkovic *et al.*, 2019). Microplastic concentrations in lakes and rivers were reported to be higher than concentrations in oceans (Baldwin *et al.*, 2016).

For example, approximately 10 million kg/year of plastic debris entered the Great Lakes (Gawlowicz, 2016). In the Great Lakes, the first microplastic study reported a median surface abundance of 5350 particles/km^2, with a maximum of 466 000 particles/km^2 (Eriksen *et al.*, 2013). In San Francisco Bay, California, the average microplastic concentration was 700 000 particles/km^2, which was higher than levels in other urban waterbodies in North America (Sutton *et al.*, 2016).

Several nonpoint source pollutants are associated with wildfires. Temperature increase, DO reduction, pH changes, ash loading, and increased concentrations of nutrients and cyanide in surface water were linked to wildfire (Crouch *et al.*, 2006). Common wildfire retardants are ammonium sulfate and ammonium phosphate. In the past, some wildfire retardants contained sodium ferrocyanide as a corrosion inhibitor (USDA, 2011). In the Western region of the US, including Oregon, Idaho and California, the application of ammonium-based retardants showed initial concentration increases in streams followed by sharp decreases with time after the application and distance downstream, with no significant damage on fish or insects (Norris & Webb, 1989). The 2003 wildfires in Southern California impacted 160 km downstream of fire with increased sediment loads (Meixner, 2004). Data collected from the Cerro Grande Fire in northern New Mexico, the Silver Creek Fire in the Salmon River Valley in British Columbia, Canada, the Viveash Fire near Pecos, New Mexico, and the Rodeo-Chediski Fire in Central Arizona, showed that the concentrations of ammonia, phosphorus, and total cyanide were similar whether or not retardants were applied, indicating minimal effects of wildfire retardants on nearby surface water quality (Crouch *et al.*, 2006). The 2010 Fourmile Canyon Fire near Boulder, Colorado increased stream discharge and nitrate concentration during snowmelt runoff, and turbidity, dissolved organic carbon, nitrate and metals by one to four orders of magnitude during and after high-intensity storm events within and downstream of the burned area (Writer & Murphy, 2012).

8.4 IMPACTS OF STORMWATER POLLUTION

Stormwater adversely impacts both surface water and groundwater. The short-term impacts include temporary increases in pollutant concentrations or toxicant levels whereas long-term impacts can be caused by cumulative contamination and bioaccumulation (US EPA, 1999). The impact on surface waters can be threefold: physical, chemical and biological impacts. These impacts can be significant on aquatic life and public health. Through porous pavements and bioswales, stormwater runoff can reach and contaminate groundwater.

8.4.1 Impact on surface water

8.4.1.1 Physical impacts

Increased frequency and volume of runoff can result in physical impacts such as stream channel erosion and channel bank scour, which introduce more sediments and cause aesthetic degradation (US EPA, 1999). This problem can be exacerbated with more frequent storms with larger volumes because of increased urbanization. Physical changes in streams such as lower depth, sediment deposition, loss of vegetation, and higher temperature, can result in habitat changes (US EPA, 1999). Sedimentation can cause harmful impacts on aquatic life because of decreased light penetration, clogging of gills of fish, and toxic pollutants attached to sediment particles (Schueler, 1987). In Wisconsin, population growth and urbanization led to stream channel erosion and sediment (Krug & Goddard, 1986).

Urban runoff can cause increased temperature in receiving waters. Urban impervious surfaces can increase stream temperature by up to 5 to 10°C during storms (Galli & Dubose, 1990; US EPA, 2005b). In the US, water temperatures have increased across the nation by 0.009–0.077°C per year due to climate change and urbanization (Kaushal *et al.*, 2010). Elevated temperatures disrupt aquatic life especially for cold water fish, such as salmon and trout, and aquatic insects by reducing DO in the water column, as well as changing the composition of algal species (Galli, 1991; Herb *et al.*, 2008; US EPA, 2005b). Sudden change in temperature will stress aquatic life and an increase in water temperature by 5–10°C will cause the dominance of temperature-tolerant fish over temperature-sensitive fish (US

EPA, 1999). Stormwater thermal pollution is more severe with higher air/dew point temperatures than stream temperature for short and intense rainfall events from highly impervious watersheds (Herb *et al.*, 2008).

8.4.1.2 Chemical impacts

Pollutants in stormwater discharges can impact chemistry of aquatic system. One example is acid precipitation and runoff/snowmelt with low pH levels. Urban runoff tends to be more acidic (US EPA, 1999). Changes in pH in receiving waters can have a detrimental impact on aquatic organisms such as fish kills and change in species (US EPA, 1999). In Virginia, episodic acidification of streams was contributed to stormflow (Eshleman *et al.*, 1995).

Large volumes of stormwater runoff can reduce the DO level in the water. In Montana, continuous monitoring over 24 hours at an urban stormwater drain showed reduced levels of DO (Gammons *et al.*, 2005). Nutrient loads from watersheds have resulted in eutrophication and hypoxia in the waters. The Gulf of Mexico has experienced midsummer hypoxia since 1970s and as a result of excessive nutrients from the Mississippi/Atchafalaya Rivers is the largest hypoxic zone in the US (US EPA, 2003). The hypoxic zone ranged between 40 and 22 730 km^2 between 1985 and 2017 (Turner & Rabalais, 2019). Hypoxia in the gulf causes stress and death to aquatic life.

Dioxins and furans are among the most toxic chemicals in surface waters because they are known as carcinogens and endocrine disruptors (Fisher *et al.*, 1999). A one-year survey in Santa Monica Bay, Southern California showed the presence of dioxins and furans in urban runoff and their peak concentrations during storm events (Fisher *et al.*, 1999). In Texas, dioxins, as carcinogens and endocrine disruptors, in runoff demonstrated higher toxicity in the suspended phase than the dissolved, and suspended concentrations were similar to those in dry air deposition (Suarez *et al.*, 2006). PCBs are established toxic compounds and endocrine disruptors, and are resistant to breakdown (Hornbuckle *et al.*, 2006). PCBs contaminated the Great Lakes and accumulated in sediments and fish tissue (Hornbuckle *et al.*, 2006). PCB contamination was initially from industrial discharges such as paper mills until the use of PCBs was banned in the late 1970s because of their impact on public health and ecosystems (Guo *et al.*, 2017). Since then, PCB concentrations in the Great Lakes have declined but the concentrations are still of concern because of their resuspension from the sediment, evaporation to the air and air deposition, and stormwater runoff from the watershed (Khairy *et al.*, 2015).

Salts in highway runoff can pollute soil, roadside vegetation, surface and groundwater, and are toxic to plants and aquatic life (Table 8.6) (Bolen, 2017). In New York, road salts elevated chloride concentrations in streams, lakes and groundwaters (Albright, 2005) and the application of road salts increased sodium and chloride concentrations of surface waters of the Mohawk River Basin by 130% and 243% between the 1950s and the 1990s (Godwin *et al.*, 2005). In Colorado, road salts damaged roadside conifers, causing needle tissue defect, depression of photosynthesis rates, and possibly sapling mortality (Trahan & Peterson, 2008). In Lake Tahoe between California and Nevada, deicing agents damaged 15–55% of roadside trees each year with more damage on pine trees (Munck *et al.*, 2010). The damage was increased with the inverse distance from roads, downhill slope steepness and tree density (Munck *et al.*, 2010).

8.4.1.3 Biological impacts and toxicity

Aquatic life can be adversely affected by stormwater runoff contaminants. Changes in temperature and pollutants in stormwater runoff can have a severe effect on sensitive species, resulting in community composition change. Stormwater runoff reduces benthic species diversity in streams (Hwang *et al.*, 2018). A study in the Adirondack region, New York showed that high concentrations of road salt reduced survival in embryos and larvae of two common vernal-pond-breeding amphibian species, the spotted salamander (*Ambystoma maculatum*) and the wood frog (*Rana sylvatica*), at conductivities of 500 μS (145 mg/L chloride) (Karraker *et al.*, 2008). In Erie, Pennsylvania, road salts from highway runoff to constructed wetlands significantly lower the abundance of chironomid in winter and spring

Table 8.6 Environmental impact of deicing agents.

Environmental Impact	Road Salt (NaCl)	Calcium Chloride ($CaCl_2$)	Magnesium Chloride ($MgCl_2$)	Ca-Mg Acetate (CMA)
Soils	Na^+ accumulation breaks down soil structure and decreases permeability and soil stability.Potential for Cl^- complexes with metals to mobilize.	Potential for Cl^- complexes with metals to mobilize. Ca^{2+} improves soil structure; increases permeability.	Potential for Cl^- complexes with metals to mobilize. Mg^{2+} improves soil structure; increases permeability.	Ca^{2+} and Mg^{2+} improve soil structure; increase permeability
Groundwater	Potential elevated levels of Cl^- during periods of low flow or spring thaws.	Similar to NaCl, potential metal contamination from cation exchange action of Ca^{2+}	Similar to NaCl and $CaCl_2$	Potential trace metal release from soil to groundwater
Surface Water	Potential excessive Cl^- loading in small water bodies. Anoxia in bottom waters due to saline stratification in small water bodies	Similar to NaCl	Similar to NaCl	Decrease DO in small water bodies due to BOD associated with acetate degradation
Vegetation	Negatively effects on vegetation up to 120 m downwind from heavily traveled roadways. May influence spread of salt-tolerant or non-native species.	Similar to NaCl. Ca^{2+} is an important macronutrient for plant growth.	Similar to NaCl. Mg^{2+} is an important element in plant physiology.	Little or no adverse effects, osmotic stress can occur at very high levels
Automobiles and Highway Structures	Initiates and accelerates corrosion of exposed metal and concrete reinforcement bars. Exacerbates scaling.	Similar to NaCl, surfaces stay wet longer, potential increasing corrosion rate	Similar to NaCl. Risk of cement paste deterioration due to Mg^{2+} reactions.	Accelerate metal corrosion due to increased conductivity, less corrosive than chloride-based deicers.
Human Health	Skin and eye irritant. Na^+ >20 mg/L can lead to hypertension	Skin and eye irritant. Ca^{2+} can increase water hardness	Skin and eye irritant. Mg^{2+} can increase water hardness	Skin and eye irritant. Ca^{2+} and Mg^{2+} can increase water hardness

(adapted from Fay and Shi, 2012; MPCA, 2017).

(Silver *et al.*, 2009). In the St. Joseph River, Indiana, the composition of macroinvertebrates (species level) changed due to sedimentation (Kohlhepp & Hellenthal, 1992). In Delaware, the percentage of pollution sensitive macroinvertebrate population in urban watershed was lower than that in forested watersheds (Maxted & Shaver, 1997). In the Provo River, Utah, macroinvertebrates decreased in abundance and species diversity with urbanization with tolerant species such as snails and leeches being dominated in urban reaches of the river (Gray, 2004).

In California, urban stormwater runoff was toxic to *Ceriodaphnia dubia* due to residential use of organophosphate pesticides (Jones-Lee & Lee, 2000). Pyrethroids in sediments near residential watersheds in California showed acute toxicity to *Hyalella azteca* with 30–91% mortality (Weston *et al.*, 2009). In Hawaii, coral cover significantly declined by about 33% between 1992 and 2002 throughout Honolua Bay that received storm runoff from pineapple fields due to periodic sedimentation events whereas the other two open coastal sites without human effects showed increased coral cover (Dollar & Grigg, 2004). In the Anacostia River, Washington DC and the Tuckahoe River, Maryland, elevated tumor prevalence was observed in brown bullhead due to PAHs and other carcinogenic compounds in stormwater runoff (Pinkney *et al.*, 2004). Dissolved copper in stormwater runoff has neurotoxic effects on larval zebrafish and juvenile coho salmon (*Oncorhynchus kisutch*) (Linbo *et al.*, 2006; Sandahl *et al.*, 2007). PAH in stormwater runoff and sediments showed cardiac toxicity to zebrafish (Incardona *et al.*, 2011).

Nutrients in nonpoint source in runoff and aerial deposition can lead to harmful algal blooms (HABs) that eventually cause fish kills and toxic to shellfish, threatening both aquatic habitat and public health. HABs also raise costs for drinking water treatment and impact local economy. The magnitude, frequency and duration of HABs have increased in both inland and coastal waters nationwide (Figure 8.2). In 1997, toxic *Pfiesteria piscicida* caused 50 000 fish kills in the Chesapeake Bay and 1.2 million fish kills in Albemarle-Pamlico estuary, North Carolina, which is the second

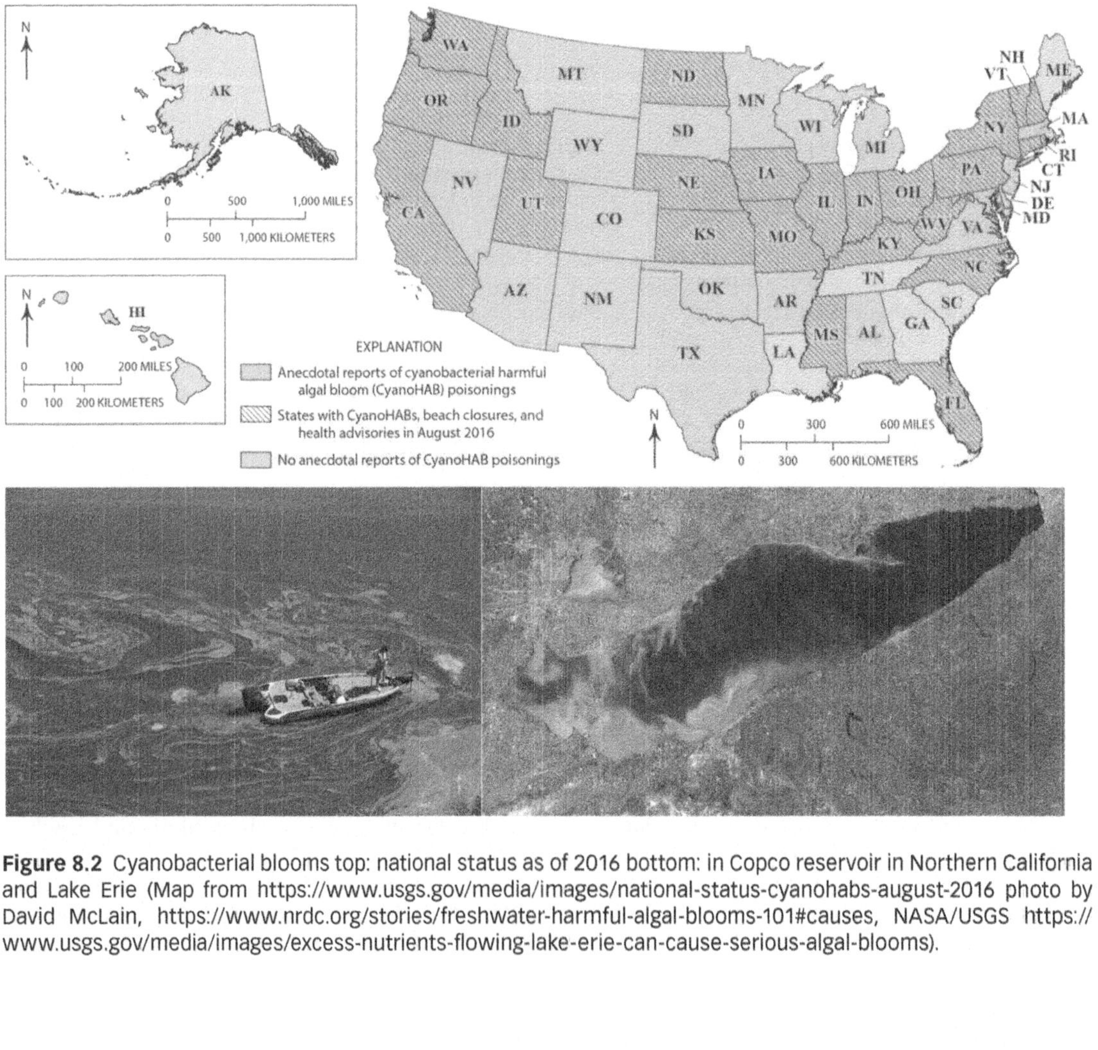

Figure 8.2 Cyanobacterial blooms top: national status as of 2016 bottom: in Copco reservoir in Northern California and Lake Erie (Map from https://www.usgs.gov/media/images/national-status-cyanohabs-august-2016 photo by David McLain, https://www.nrdc.org/stories/freshwater-harmful-algal-blooms-101#causes, NASA/USGS https://www.usgs.gov/media/images/excess-nutrients-flowing-lake-erie-can-cause-serious-algal-blooms).

largest estuary in the US mainland (Burkholder & Glasgow, 2001). In South Carolina, stormwater retention/detention ponds along the coastal area had over 200 HABs with 23 species in 4 years, which might contribute to estuarine HABs (Lewitus *et al.*, 2008).

Lake Erie in the Great Lakes experiences recurring blooms of toxic algae caused by runoff pollution, resulting in the issuing of a 'Do Not Drink Tap Water' advisory for over 500 000 residents of Toledo, Ohio, in summer 2014 (Fitzsimmons, 2014). The City of Tulsa, Oklahoma, spent an additional $ 100 000 per year to deal with taste and odor from algal blooms in the city's drinking water source (US EPA, 2003). Capital expenditure to address Lake Erie algal blooms has to date reached $85 million (Key-log Economics, 2019). Phosphorus reduction in the Great Lakes by 40% is estimated to: bring economic benefits of $117–437 million to recreational anglers and up to $43 million to beach-related recreation; avoid property value loss up to $1 billion; and reduce water treatment costs of $3 million per year in both the US and Canada (Key-log Economics, 2019).

8.4.2 Public health impacts

Stormwater exposure can cause potential public human health problems by contaminating recreational areas, drinking water supplies, and fish and shellfish harvests (Burton & Pitt, 2001).

The impacts on human health of PCBs in Lake Michigan contributed by stormwater runoff include deficits in birth weight and gestational age of children born to mothers who consumed contaminated fish (Hornbuckle *et al.*, 2006). More than half of the reported cases of waterborne disease outbreaks occurred after extreme rainfalls (Curriero *et al.*, 2001). For example, the Milwaukee *Cryptosporidium* outbreak followed spring storm events and snowmelt (Barwick *et al.*, 2000). Stormwater runoff has also been associated with elevated concentrations of bacteria, *Giardia*, *Cryptosporidium*, and other microorganisms (Gaffield *et al.*, 2003). A study in Wisconsin showed the increase of acute gastrointestinal illness in children after stormwater events (Drayna *et al.*, 2010). In the US, there are 90 million recreational waterborne illnesses such as gastrointestinal and respiratory diseases from exposure to polluted water annually, which cost $2.2–3.7 billion annually (DeFlorio-Barker *et al.*, 2018). Antibiotic resistance genes (ARGs), threatening public health, are discharged from agricultural runoff (Chee-Sanford *et al.*, 2009; Fahrenfeld *et al.*, 2014; Joy *et al.*, 2013), and urban stormwater (Garner *et al.*, 2017; McLellan *et al.*, 2007). In the Hudson River Estuary, New York, antibiotic resistant bacteria were positively correlated with sewage indicators, implying the effect of combined sewer system (Young *et al.*, 2013).

The most common impacts in recreation areas are beach closures near stormwater discharge outfalls in many locations because of the risk of pathogen contamination. The usual effects on human health of pathogens in stormwater runoff are ear and eye discharges, skin rashes, and gastrointestinal problems (Gaffield *et al.*, 2003). Skin infections are possibly caused by *Pseudomonas aeruginosa* (Burton & Pitt, 2001). The occurrence of *Pseudomonas aeruginosa* in stormwater runoff is frequent at densities greater than 10 organisms/100 mL although the concentrations and types of pathogens vary at different sites and times (Burton & Pitt, 2001).

8.4.3 Groundwater impact

Stormwater may enter groundwater through stormwater control measures such as porous pavements, bioswales, and infiltration trenches (Burton & Pitt, 2001). Although the infiltration through top soils would not cause major groundwater contamination, subsurface infiltration can increase the risk of groundwater contamination (Pitt *et al.*, 1996). Stormwater with high nitrate concentrations can potentially contaminate groundwater with nitrate (Burton & Pitt, 2001). It was reported that high nitrogen input from both agricultural and urban areas causes high risk of nitrate contamination of shallow groundwater in the Midwest and parts of the Western and Northeastern areas (Nolan *et al.*, 1997). In Big Spring Basin, Iowa, the nitrogen concentration was proportional to the land farmed for corn because of the nitrogen input from fertilizers (Rowden *et al.*, 1995).

8.5 REGULATIONS FOR STORMWATER POLLUTION

The development of US federal regulations for stormwater control is outlined in Figure 8.3. While traditional point source pollution such as discharges from industrial and municipal wastewater treatment plants has been regulated since the Clean Water Act (CWA) in 1972 (33 USC § 1251), nonpoint sources were excluded from the permit scheme and made voluntary. In 1987, the US Congress amended CWA Section 319 to provide funds to develop management programs for nonpoint source pollution, and Section 402 to reclassify stormwater runoff from urbanized areas and areas of industrial activity as point sources (Graham *et al.*, 2011; US EPA, 2019d). Although Section 208 of the CWA in 1972 aimed for a water quality management program to control both point and nonpoint source pollutants, its success was limited because the importance of nonpoint source pollution was underestimated (Ribaudo, 2009).

The CWA Section 402 was amended in 1987, specifically to regulate municipal stormwater pollution and stormwater discharges associated with industrial activity under the National Pollutant Discharge

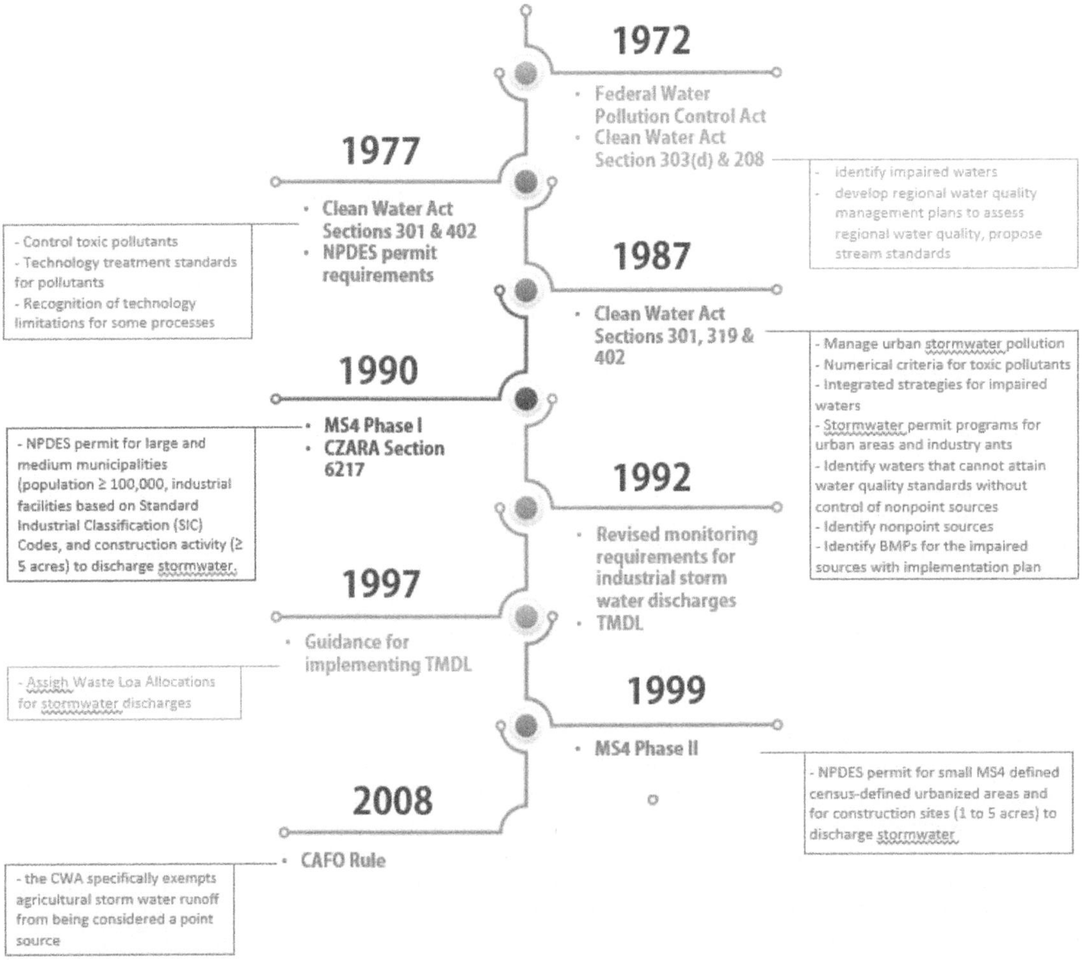

Figure 8.3 Federal regulations for stormwater control in the US.

Elimination System (NPDES; 40 CFR Pt. 122.3) permit program (3333 USC§ 1342), in addition to traditional point source pollution (NASEM, 2019). The regulatory definition of point source was broadened to include stormwater runoff discharged from conveyances (US EPA, 2003). The US EPA is the federal agency granted the statutory authority to implement the NPDES permit program but may delegate the NPDES authority to a State that requests delegation after the US EPA certifies regulatory adequacy and executes an agreement (Craig & Roberts, 2015). Stormwater from municipal separate storm sewer systems (MS4) were phased into the NPDES permit program: Phase I for MS4s serving populations of 100 000 or more and Phase II for MS4s serving smaller communities in 1990 and 1999, based on population sizes under the 1990 and 2000 census respectively (Shimabuku *et al.*, 2018). The NPDES permits were to require structural and non-structural controls to reduce the discharge of pollutants (US EPA, 2019d).

Under Section 303(d) of the CWA, states are required to identify impaired water bodies and develop TMDLs for the impaired water bodies for selected pollutants, addressing both point and nonpoint sources as a total load amount (US EPA, 2019d). The TMDL is defined as the maximum amount of a pollutant that can enter a water body without violating water quality standards (Ribaudo, 2009). The impetus to develop TMDLs arose in 1992 as a result of US Court decisions, with US EPA guidance for implementing TMDLs issued in 1997 and the final rule for developing and implementing TMDL published in 2000 (Copeland, 2012).

In addition to the CWA, the Coastal Zone Act Reauthorization Amendments (CZARA) of 1990 were implemented for federal nonpoint source pollution control programs by adding nonpoint source pollution requirements (Ribaudo, 2009). CZARA requires each coastal state to develop a Coastal Nonpoint Pollution Control Program (CNPCP) in order to receive federal CZARA Section 306 and CWA Section 319 funds (Scott, 2017).

According to the CWA, the US Department of Agriculture (USDA) directs implementation of practices to reduce nonpoint source pollution from agricultural areas (Ribaudo, 2009). One of the USDA programs is the Environmental Quality Incentive Program (EQIP) for installing management practices that protect water resources from farm runoff. EQIP funds from 1997 to 2004 were spent on water quality and water conservation-related practices (37%) and managing nutrients from manure that is a major source of water pollution from agriculture (Ribaudo, 2009).

The water quality regulations in the US are source-based rather than water quality-based (Craig & Roberts, 2015). Nonpoint source pollution regulatory requirements for implementation of controls or the management of pollutants focus on sources (Craig & Roberts, 2015).

8.6 EMERGING ISSUES AND FUTURE OUTLOOK

Nonpoint source pollution was not recognized as an important source of water contaminants until the late 1980s/early 1990s. Even now the general public is much more concerned by point source pollution and/or tends to perceive that nonpoint source pollution is not a major threat to the environment. This compromising view of nonpoint source pollution will remain until effective education and outreach are put in place and implemented for one to two decades. Lack of understanding that humans are major contributors of nonpoint source pollution is a basic impediment for the public to recognize its importance. Another hurdle for public acceptance is that contributions are mostly collective and indirect through human activities such as roadway runoff. This is aggravated by the fact that the polluter pays principle tends to work better for point source pollution problems than for nonpoint source issues. Therefore, it is likely nonpoint source pollution programs will be underfunded for at least the next decade or more. On a positive note, the recognition and resource allocation for nonpoint source pollution have much improved in the last 15 to 20 years. It is projected that awareness of the importance of nonpoint activities will have to rely heavily on outreach and education. For example, minimizing the uses of chemicals such as antibiotics and fertilizers associated with agricultural activities will occur if farmers understand the need and are cooperative. If people are aware of sources and effects of microplastics, it is more likely that they will dispose of plastic trash properly. Every bit helps when it comes to nonpoint source pollution prevention, and outreach and education intensification is a key factor for success. There is a paradigm shift in the point-source pollution field that wastewater should be viewed as a resource for energy and clean water rather a problem. Wastewater treatment plants now are known as water resources reclamation facilities. A similar concept can be used for stormwater runoff that it is not a problem but is a water source by default as it ends up in rivers and lakes. Therefore, the cleaner it is, the better it is.

8.7 CONCLUSIONS

This article covers historical and recent developments on key issues related to nonpoint source water pollution in the US including types/sources, contaminants, impacts and regulation. It also offers a future outlook on different aspects of nonpoint source pollution such as emerging contaminants and growing and promising research areas. Although the quality of natural waters in the US has improved substantially in the past few decades, more than half of rivers, streams, lakes, ponds, reservoirs, bays and estuaries in the country are impaired and nonpoint source pollution contributes significantly to the impairments. The types of nonpoint sources have not changed and include mainly urban, agricultural, construction, forestry, hydrological modifications, mining, marinas, and atmospheric deposition. However, new types of pollutants in stormwater runoff have been identified in recent years such as contaminants of emerging concern, ARGs and microplastics. Although these contaminants in stormwater runoff are not regulated, they can cause significant adverse effects on aquatic ecology. There has been an increasing interest in potential human health impacts from these new contaminants. Uptake of microplastics by humans through the food chain and prevalence of antibiotic resistant pathogens due to the ubiquity of ARGs are among the concerns. While these new contaminants and concerns have emerged, traditional ones such as nutrients and eutrophication are still issues to be reckoned with.

The Clean Water Act was passed in 1972 but nonpoint sources were not included until 1987. Since then, nonpoint source pollution has been regulated along with point source pollution under the same NPDES permit program. The TMDL, which has been in place since 2000, does not distinguish between point and nonpoint sources and lumps them together for a total load. Its significant impact on water quality remains to be seen. Overall, the nonpoint source water pollution situation in the US has improved but progress has been slow. More commitments on funding and resources particularly for outreach and education with a focus on prevention or source reduction are needed. A major change in perception that sees stormwater runoff as a valuable resource contributing to water sources for consumption instead of problematic polluted water, is required to affect the stature of nonpoint source water pollution. This drastic change along with effective outreach and education will be the key to success in nonpoint source water pollution management.

REFERENCES

Albright M. (2005). Changes in water quality in an urban stream following the use of organically derived deicing products. *Lake and Reservoir Management*, **21**(1), 119–124, https://doi.org/10.1080/07438140509354419

Alexander R. B., Smith R. A., Schwarz G. E., Boyer E. W., Nolan J. V. and Brakebill J. W. (2008). Differences in phosphorus and nitrogen delivery to the Gulf of Mexico from the Mississippi river basin. *Environmental Science and Technology*, **42**, 822–830, https://doi.org/10.1021/es0716103

Alpers C. N., Hunerlach M. P., May J. T. and Hothem R. L. (2005). Mercury contamination from historical gold mining in California. US Geological Survey Fact Sheet, 2004–3014, https://pubs.usgs.gov/fs/2005/3014/ (accessed 5 July 2019).

Anderson K. A. and Downing J. A. (2006). Dry and wet atmospheric deposition of nitrogen, phosphorus and silicon in an agricultural region. *Water, Air, and Soil Pollution*, **176**(1–4), 351– 374, https://doi.org/10.1007/s11270-006-9172-4

Baldwin A. K., Corsi S. R. and Mason S. A. (2016). Plastic debris in 29 Great Lakes tributaries: relations to watershed attributes and hydrology, *Environmental Science and Technology*, **50**, 10377–10385, https://doi.org/10.1021/acs.est.6b02917

Baral D., Fisher J. R., Florek M. J. and Dvorak B. I. (2018). Atmospheric contributions of nitrate to stormwater runoff from two urban watersheds. *Journal of Environmental Engineering*, **144**(2), 05017009, https://doi.org/10.1061/(ASCE)EE.1943-7870.0001323

Bargu S., Justic D., White J. R., Lane R., Day J., Paerl H. and Raynie R. (2019). Mississippi River diversions and phytoplankton dynamics in deltaic Gulf of Mexico estuaries: a review. *Estuarine, Coastal and Shelf Science*, **221**, 39–52, https://doi.org/10.1016/j.ecss.2019.02.020

Barwick R. S., Levy D. A., Craun G. F., Beach M. J. and Calderon R. L. (2000). Surveillance for waterborne-disease outbreaks – United States, 1997–1998. *Morbidity and Mortality Weekly Report, CDC Surveillance Summaries*, **49**(SS-4), 1–35.

Bigham J. M. and Cravotta C. A. (2016). Acid Mine Drainage, In: Encyclopedia of Soil Science, R. Lal (ed.), 3rd edn., CRC Press, Chap. 3, pp. 6–10. Boca Raton, FL.

Bilkovic D. M., Mitchell M. M., Havens K. J. and Hershner C. H. (2019). Chesapeake bay. In: World Sea: An Environmental Evaluation, Volume I: Europe, the Americas and West Africa, C. Sheppard (ed.), 2nd edn, Academic Press, pp. 445–464. London.

Binkley D., Burnham H. and Allen H. L. (1999). Water quality impacts of forest fertilization with nitrogen and phosphorus. *Forest Ecology and Management*, **121**, 191–213, https://doi.org/10.1016/S0378-1127(98)00549-0

Bolen W. P. (2017). 2015 Minerals Yearbook: Salt. US Department of the Interior, Washington, DC, and US Geological Survey, https://s3-us-west-2.amazonaws.com/prd-wret/assets/palladium/production/mineral-pubs/salt/myb1-2015-salt.pdf (accessed 19 July 2019).

Boyd G. R., Palmeri J. M., Zhang S. and Grimm D. A. (2004). Pharmaceuticals and personal care products (PPCPs) and endocrine disrupting chemicals (EDCs) in stormwater canals and Bayou St. John in New Orleans, Louisiana, USA. *Science of the Total Environment*, **333**(1–3), 137–148, https://doi.org/10.1016/j.scitotenv.2004.03.018

Burke M. P., Hogue T. S., Ferreira M., Mendez C. B., Navarro B., Lopez S. and Jay J. A. (2010). The effect of wildfire on soil mercury concentrations in southern California watersheds. *Water, Air, and Soil Pollution*, **212**, 369–385, https://doi.org/10.1007/s11270-010-0351-y

Burkholder J. M. and Glasgow H. B. (2001). History of toxic *Pfiesteria* in north Carolina Estuaries from 1991 to the present: many toxic *Pfiesteria* outbreaks have plagued the Albemarle–Pamlico Estuarine System, including events both before and after the 1997 outbreaks in Chesapeake Bay. *BioScience*, **51**(10), 827–841, https://doi.org/10.1641/0006-3568(2001)051[0827:HOTPIN]2.0.CO;2

Burton G. A. and Pitt R. E. (2001). Stormwater Effects Handbook, A Toolbox for Watershed Managers, Scientist, and Engineers. Lewis Publishers, CRC Press, Boca Raton.

California Department of Transportation (Caltrans) (2000). California Department of Transportation District 7 litter management pilot study, Caltrans Document No. CT-SW-RT- 00-013, Sacramento, CA.

Chee-Sanford J. C., Mackie R. I., Koike S., Krapac I. G., Lin Y. F., Yannarell A. C., Maxwell S. and Aminov R. I. (2009). Fate and transport of antibiotic residues and antibiotic resistance genes following land application of manure waste. *Journal of Environmental Quality*, **38**, 1086–1108, https://doi.org/10.2134/jeq2008.0128

Church P. E. (1996). Effectiveness of highway-drainage systems in preventing road-salt contamination of ground water, Southeastern Massachusetts, FS 115-96, US Department of the Interior, Washington, DC, and US Geological Survey, Marlborough, MA.

City of Los Angeles, Department of Public Works, Bureau of Sanitation Watershed Protection Division (LASAN) (2002). *High Trash-Generation Areas and Control Measures*. http://www.lastormwater.org/wp-content/files_mf/trash_gen_study.pdf (accessed 10 July 2019).

Cizek A. R., Characklis G. W., Krometis L. A., Hayes J. A., Simmons O. D., Di Lonardo S., Alderisio K. A. and Sobsey M. D. (2008). Comparing the partitioning behavior of *Giardia* and *Cryptosporidium* with that of indicator organisms in stormwater runoff. *Water Research*, **42**, 4421–4438, https://doi.org/10.1016/j.watres.2008.06.020

Clark S. E., Steeke K. A., Spicher J., Siu C. Y. S., Lalor M. M., Pitt R. and Kirby J. (2008). Roofing materials' contributions to stormwater runoff pollution. *Journal of Irrigation and Drainage Engineering*, **134**(5), 638–645, https://doi.org/10.1061/(ASCE)0733-9437(2008)134:5(638)

Copeland C. (2012). Clean Water Act and pollutant total maximum daily loads (TMDLs), 7-5700, R42752, Congressional Research Service, https://fas.org/sgp/crs/misc/R42752.pdf (accessed 5 July 2019).

Craig R. K. and Roberts A. M. (2015). When will governments regulate nonpoint source pollution? A comparative perspective, *Boston College Environmental Affairs Law Review*, **42**(1), https://pdfs.semanticscholar.org/28a5/a763f8c7743788b7e2c5f521cc0f7ef67314.pdf (accessed 20 July 2019).

Cravotta C. A., Brightbill R. A. and Langland M. J. (2010). Abandoned mine drainage in the Swatara Creek Basin, Southern Anthracite coalfield, Pennsylvania, USA: 1. Stream water quality trends coinciding with the return of fish. *Mine Water and the Environment*, **29**(3), 176–199, https://doi.org/10.1007/s10230-010-0112-6

Crouch R. L., Timmenga H. J., Barber T. R. and Fuchsman P. C. (2006). Post-fire surface water quality: comparison of fire retardant versus wildfire-related effects. *Chemosphere*, **62**(6) 874–889, https://doi.org/10.1016/j.chemosphere.2005.05.031

Curriero F. C., Patz J. A., Rose J. B. and Lele S. D. (2001). The association between extreme precipitation and waterborne disease outbreaks in the United States, 1948–1994. *American Journal of Public Health*, **91**, 1194–1199, https://doi.org/10.2105/AJPH.91.8.1194

Dauchy X., Boiteux V., Colin A., Bach C., Rosin C. and Munoz J.-F. (2019). Poly- and perfluoroalkyl substances in runoff water and wastewater sampled at a firefighter training area. *Archives of Environmental Contamination and Toxicology*, **76**(2), 206–215, https://doi.org/10.1007/s00244-018-0585-z

DeFlorio-Barker S., Wing C., Jones R. M. and Dorevitch S. (2018). Estimate of incidence and cost of recreational waterborne illness on United States surface waters. *Environmental Health*, **17**(3), 3, https://doi.org/10.1186/s12940-017-0347-9

Dollar S. J. and Grigg R. W. (2004). Anthropogenic and natural stresses on selected coral reefs in Hawai̒i: a multidecade synthesis of impact and recovery. *Pacific Science*, **58**(2), 281–304, https://doi.org/10.1353/psc.2004.0016

Donigian A. S., Bicknell B. R. and Wolfram E. (2009). Modeling the contribution of copper from Brake Wear Debris to the San Francisco Bay. Phase 2. Prepared by AQUA TERRA Consultants for the Brake Pad Partnership.

Dorfman M. (2006). Testing the Waters: A Guide to Water Quality at Vacation Beaches, National Resource Defense Council, Washington D.C.

Drayna P., McLellan S. L., Simpson P., Li S. and Gorelick M. H. (2010). Association between rainfall and pediatric emergency department visits for acute gastrointestinal illness. *Environmental Health Perspectives*, **118**(10), 1439–1443, https://doi.org/10.1289/ehp.0901671

Dressing S. A., Meals D. W., Harcum J. B., Spooner J., Stribling J. B., Richards R. P., Millard C. J., Lanberg S. A. and O'Donnell J. G. (2016). Monitoring and evaluating nonpoint source watershed projects, US EPA Report 841-1-R-16-010, Washington D.C. https://www.epa.gov/sites/production/files/2016-06/documents/nps_monitoring_guide_may_2016-combined_plain.pdf (accessed 20 July 2019).

Dvonch J. T., Keeler G. J. and Marsik F. J. (2005). The influence of meteorological conditions on the wet deposition of mercury in southern Florida. *Journal of Applied Meteorology and Climatology*, **44**(9), 1421–1435, https://doi.org/10.1175/JAM2272.1

Eckley C. S., Eagles-Smith C., Tate M. T., Kowalski B., Danehy R., Johnson S. L. and Krabbenhoft D. P. (2018). Stream mercury export in response to contemporary timber harvesting methods (Pacific Coastal Mountains, Oregon, USA). *Environmental Science and Technology*, **52**(4), 1971–1980, https://doi.org/10.1021/acs.est.7b05197

Ehrhart B. J., Shannon R. D. and Jarrett A. R. (2002). Effects of construction site sedimentation basins on receiving stream ecosystems. *Transactions of the American Society of Agriculture and Biological Engineers*, **45**(3), 675–680.

Eriksen M., Mason S., Wilson S., Box C., Zellers A., Edwards W., Farley H. and Amato S. (2013). Microplastic pollution in the surface waters of the Laurentian Great Lakes. *Marine Pollution Bulletin*, **77**(1–2), 177–182, https://doi.org/10.1016/j.marpolbul.2013.10.007

Eshleman K. M., Miller-Marshall L. M. and Webb J. R. (1995). Long-term changes in episodic acidification of streams in Shenandoah National Park, Virginia (USA). *Water, Air, and Soil Pollution*, **85**(2), 517–522, https://doi.org/10.1007/BF00476881

Fahrenfeld N., Knowlton K., Krometis L. A., Hession W. C., Xia K., Lipscomb E., Libuit K., Green B. L. and Pruden A. (2014). Effect of manure application on abundance of antibiotic resistance genes and their attenuation rates in soil: field scale mass balance approach. *Environmental Science and Technology*, **48**, 2643–2650, https://doi.org/10.1021/es404988k

Faucette L. B., Cardoso-Gendreau F. A., Codling E., Sadeghi A. M., Pachepsky Y. A. and Shelton D. R. (2009). Storm water pollutant removal performance of compost filter socks. *Journal of Environmental Quality*, **38**, 1233–1239, https://doi.org/10.2134/jeq2008.0306

Faust D. R., Kröger R., Moore M. T. and Rush S. A. (2018). Management practices used in agricultural drainage ditches to reduce Gulf of Mexico hypoxia. *Bulletin of Environmental Contamination and Toxicology*, **100**(1), 32–40, https://doi.org/10.1007/s00128-017-2231-2

Fay L. and Shi X. (2012). Environmental impacts of chemicals for snow and ice control: state of the knowledge. *Water, Air, Soil & Pollution*, **223**(5), 2751–2770, https://doi.org/10.1007/s11270-011-1064-6

Fisher T. S., Hayward D. G., Stephens R. D. and Stenstrom M. K. (1999). Dioxins and furans urban runoff. *Journal of Environmental Engineering, ASCE*, **125**(2), 185–191, https://doi.org/10.1061/(ASCE)0733-9372(1999)125:2(185)

Fitzsimmons E. G. (2014). Tap Water Ban for Toledo Residents, The New York Times, https://www.nytimes.com/2014/08/04/us/toledo-faces-second-day-of-water-ban.html (accessed 20 July 2019).

Francy D., Helsel D. and Nally R. (2000). Occurrence and distribution of microbiological indicators in groundwater and stream water. *Water Environment Research*, **72**, 152–161, https://doi.org/10.2175/106143000X137220

Gabriel M. C., Williamson D. and Pitt R. (2002). Availability of Atmospherically Deposited Mercury to Runoff and Receiving Waters. Proceeding of the 2002 National TMDL Science and Policy Conference, November 2002. Water Environment Federation.

Gaffield S. J., Goo R. L., Richards L. A. and Jackson R. J. (2003). Public health effects of inadequately managed stormwater runoff. *American Journal of Public Health*, **93**, 1527–1533, https://doi.org/10.2105/AJPH.93.9.1527

Galli J. (1991). Thermal Impacts Associated with Urbanization and Stormwater Best Management Practices. Metropolitan Washington Council of Governments, Washington, DC, and Maryland Department of the Environment, Annapolis, MD.

Galli J. and Dubose R. (1990). Water Temperature and Freshwater Stream Biota: An Overview. Maryland Department of the Environment, Sediment and Stormwater Administration, Baltimore, MD.

Gammons C. H., Shope C. L. and Duaime T. E. (2005). A 24 h investigation of the hydrogeochemistry of baseflow and stormwater in an urban area impacted by mining: Butte, Montana. *Hydrological Processes*, **19**(14), 2737–2753, https://doi.org/10.1002/hyp.5783

Garner E., Benitez R., von Wagoner E., Sawyer R., Schaberg E., Hession W. C., Krometis L. H., Badgley B. D. and Pruden A. (2017). Stormwater loadings of antibiotic resistance genes in an urban stream. *Water Research*, **123**, 144–152, https://doi.org/10.1016/j.watres.2017.06.046

Gawlowicz S. (2016). Researchers estimate 10,000 metric tons of plastic enter Great Lakes every year. https://www.sciencedaily.com/releases/2016/12/161219151752.htm (accessed 31 July 2019).

Godwin K. S., Hafner S. D. and Buff M. F. (2005). Long-term trends in sodium and chloride in the Mohawk River, New York: the effect of fifty years of road salt application. *Environmental Pollution*, **124**(2), 273–281, https://doi.org/10.1016/S0269-7491(02)00481-5

Graham S., Schempp S. and Troell J. (2011). Regulating nonpoint source water pollution in a federal government: four case studies. *International Journal of Water Resources Development*, **27**(1), 53–69, https://doi.org/10.1080/07900627.2010.531378

Gray L. (2004). Changes in water quality and macroinvertebrate communities resulting from urban stormflows in the Provo River, Utah, U.S.A. *Hydrobiologia*, **518**(1–3), 33–46, https://doi.org/10.1023/B:HYDR.0000025055.15164.40

Guo J., Romanak K., Westenbroek S., Li A., Kreis O. G., Hites R. A. and Venier M. (2017). Updated polychlorinated biphenyl mass budget for Lake Michigan. *Environmental Science and Technology*, **51**(21) 12455–12465, https://doi.org/10.1021/acs.est.7b02904

Herb W. R., Janke B., Mohseni O. S. and Stefan H. G. (2008). Thermal pollution of streams by runoff from paved surfaces. *Hydrological Processes*, **22**(7), 987–999, https://doi.org/10.1002/hyp.6986

Hornbuckle K. C., Carlson D. L., Swackhamer D. L., Baker J. E. and Eisenreich S. J. (2006). Polychlorinated Biphenyls in the Great Lakes. *The Handbook of Environmental Chemistry*, **5**(N), 13–70.

Houtz E. F. and Sedlak D. L. (2012). Oxidative conversion as a means of detecting precursors to perfluoroalkyl acids in urban runoff. *Environmental Science and Technology*, **46**(17), 9342– 9349, https://doi.org/10.1021/es302274g

Hwang H., Fiala M. J., Park D. and Wade T. L. (2016). Review of pollutants in urban road dust and stormwater runoff: part I. Heavy metals released from vehicles. *International Journal of Urban Sciences*, **20**, 334–360, https://doi.org/10.1080/12265934.2016.1193041

Hwang H., Fiala M. J., Wade T. L. and Park D. (2018). Review of pollutants in urban road dust: part II. Organic contaminants from vehicles and road management. *International Journal of Urban Sciences*, **23**, 445–463, https://doi.org/10.1080/12265934.2018.1538811

Incardona J. P., Linbo T. L. and Scholz N. L. (2011). Cardiac toxicity of 5-ring polycyclic aromatic hydrocarbons is differentially dependent on the aryl hydrocarbon receptor 2 isoform during zebrafish development. *Toxicology and Applied Pharmacology*, **257**(2), 242–249, https://doi.org/10.1016/j.taap.2011.09.010

Janke B. D., Rinlay J. C. and Hobbie S. E. (2017). Trees and streets as drivers of urban stormwater nutrient pollution. *Environmental Science and Technology*, **51**, 9569–9579, https://doi.org/10.1021/acs.est.7b02225

Jeng H. W., Jin G. and Englande A. J. (2002). Fate of Indicator Organisms from Stormwater Runoff Into Brackish Recreational Waters. Proceeding of the Disinfection 2002 Conference, February 2002. Water Environment Federation, Alexandria, VA, USA.

Jiang S., Noble R. and Chui W. P. (2001). Human adenoviruses and coliphages in urban runoff-impacted coastal waters of southern California. *Applied and Environmental Microbiology*, **67**, 179–184, https://doi.org/10.1128/AEM.67.1.179-184.2001

Jones-Lee A. and Lee G. F. (2000). Development of TMDL Goals for Control of Organophosphate Pesticide-Caused Aquatic Life Toxicity in Urban Stormwater Runoff. Proceedings of the Water Environment Federation National 73rd Annual Conference, Anaheim, CA, USA, October 2000. Water Environment Federation, Alexandria, VA, USA.

Joy S. R., Bartelt-Hunt S. L., Snow D. D., Gilley J. E., Woodbury B. L., Parker D. B., Marx D. B. and Li X. (2013). Fate and transport of antimicrobials and antimicrobial resistance genes in soil and runoff following land application of swine manure slurry. *Environmental Science and Technology*, **47**, 12081–12088, https://doi.org/10.1021/es4026358

Karraker N. E., Gibbs J. P. and Vonesh J. R. (2008). Impacts of road deicing salt on the demography of vernal pool-breeding amphibians. *Ecological Applications*, **18**(3), 724–734, https://doi.org/10.1890/07-1644.1

Kaushal S. S., Likens G. E., Jaworski N. A., Pace M. L., Sides A. M., Seekell D., Belt K. T., Secor D. H. and Wingate R. L. (2010). Rising stream and river temperatures in the United States. *Frontiers in Ecology and the Environment*, **8**(9), 461–466, https://doi.org/10.1890/090037

Keener H. M., Faucette B. and Klingman M. H. (2007). Flow-through rates and evaluation of solids separation of compost filter socks versus silt fence in sediment control applications. *Journal of Environmental Quality*, **36**, 742–752, https://doi.org/10.2134/jeq2006.0359

Kerr J. M., DePinto J. V., McGrath D., Sowa S. P. and Swinton S. M. (2016). Sustainable management of Great Lakes watersheds dominated by agricultural land use. *Journal of Great Lakes Research*, **42**, 1252–1259, https://doi.org/10.1016/j.jglr.2016.10.001

Key-Log Economics (2019). Lake Erie, ecosystem services & economic benefits from phosphorous reductions. Available online at: http://www.keylogeconomics.com/lakeerievalue.html (accessed 9 September 2019).

Khairy M., Muir D., Teixeira C. and Lohmann R. (2015). Spatial distribution, air–water fugacity ratios and source apportionment of polychlorinated biphenyls in the lower Great Lakes Basin. *Environmental Science and Technology*, **49**(23) 13787–13797, https://doi.org/10.1021/acs.est.5b00186

Kim L-H., Kang J., Kayhanian M., Gil K.-I., Stenstrom M. K. and Zoh K.-D. (2006). Characteristics of litter waste in highway storm runoff. *Water Science and Technology*, **53**(2), 225–234, https://doi.org/10.2166/wst.2006.056

Kohlhepp G. W. and Hellenthal R. A. (1992). The effects of sediment deposition on insect populations and production in a northern Indiana stream. In: Proceedings of 1991 Midwest Pollution Control Biologists Meeting. Environmental Indicators: Measurement and Assessment Endpoints, T. P. Simon and W. S. Davis (eds.). EPA 905-R-92-003, Washington, DC, pp. 73–84.

Krug W. R. and Goddard G. L. (1986). Effects or urbanization on streamflow, sediment loads, and channel morphology in Pheasant Branch Basin near Middleton, Wisconsin, Report No. 85-4068, US Geological Survey, Denver, CO, USA.

Kuivila K. M., Hladik M. L., Ingersoll C. G., Kemble N. E., Moran P. W., Calhoun D. L., Nowell L. H. and Gilliom R. J. (2012). Occurrence and potential sources of pyrethroid insecticides in stream sediments from seven US metropolitan areas. *Environmental Science and Technology*, **46**, 4297–4303, https://doi.org/10.1021/es2044882

Lake Chaplain Basin Program (LCBP) (2018). 2018 State of the Lake and Ecosystems Indicators Report. Available at: https://sol.lcbp.org/en/clean-water/phosphorus/ (accessed 11 September 2019).

Lake Chaplain Basin Program (LCBP) (2019). *Nutrients*. Available at: https://www.lcbp.org/water-environment/water-quality/nutrients/ (accessed 23 August 2019).

Lee G., Bigham J. M. and Williams D. J. (2008). Metal release from bottom sediments of Ocoee Lake No.3, a primary catchment area for the Ducktown Mining District. *Journal of Environmental Quality*, **37**, 344–352, https://doi.org/10.2134/jeq2007.0223

Levine S. N., Lini A., Ostrofsky M. L., Burgess-Grant H., Lami A., Collyer-Gilles E., Reuter D., Schwarting-Miller L. and Kamman N. (2018). The relative roles of point and nonpoint phosphorus sources in the eutrophication of Lake Champlain as recorded in sediment cores. *Journal of Great Lakes Research*, **44**, 1043–1056, https://doi.org/10.1016/j.jglr.2018.06.005

Lewitus A. J., Brock L. M., Burke M. K., DeMattio K. A. and Wilde S. B. (2008). Lagoonal stormwater detention ponds as promoters of harmful algal blooms and eutrophication along the south Carolina coast. *Harmful Algae*, **8**(1), 60–65, https://doi.org/10.1016/j.hal.2008.08.012

Liang X., Park M.-H. and Stenstrom M. K. (2019), Socio-economic factors of high trash generation in the city of Los Angeles. *Water Science and Technology*, **80**, 408–417, https://doi.org/10.2166/wst.2019.283

Lifset R. J., Eckelman M. J., Harper E. M., Hausfather Z. and Urbina G. (2012). Metal lost and found: dissipative uses and releases of copper in the United States 1975–2000. *Science of the Total Environment*, **417–418**, 138–147, https://doi.org/10.1016/j.scitotenv.2011.09.075

Linbo T. L., Stehr C. M., Incardona J. P. and Scholz N. L. (2006). Dissolved copper triggers cell death in the peripheral mechanosensory system of larval fish. *Environmental Toxicology and Chemistry*, **25**, 597–603, https://doi.org/10.1897/05-241R.1

Mallin M. A., Burkholder J. M., McIver M. R., Shank G. C., Glasgow H. B., Touchette B. W. and Springer J. (1997). Comparative effects of poultry and swine waste lagoon spills on the quality of receiving streamwaters. *Journal of Environmental Quality*, **26**, 1622–1631, https://doi.org/10.2134/jeq1997.00472425002600060023x

Mason R., Reinfelder J. and Morel F. (1995). Bioaccumulation of mercury and methylmercury. *Water, Air, and Soil Pollution*, **80**, 915–921, https://doi.org/10.1007/BF01189744

Maxted J. and Shaver E. (1997). The use of retention basins to mitigate stormwater impacts on aquatic life. In: Effects of Water Development and Management on Aquatic Ecosystems, A. Roesner (ed.), American Society of Civil Engineers, New York, NY, 596pp.

McCoy J. A. and Johnson L. T. (2010). Boating Pollution Economics & Impacts. University of California, Cooperative Extension, County of San Diego (UCCE-SD) Fact Sheet 2010–2. Available at: http://ucanr.org/sites/coast/files/59476.pdf (accessed 12 July 2019).

McLellan S. L., Hollis E. J., Depas M. M., Dyke M. Van, Harris J. and Scopel C. O. (2007). Distribution and fate of *Escherichia coli* in Lake Michigan following contamination with urban stormwater and combined sewer overflows. *Journal of Great Lakes Research*, **33**, 566–580, https://doi.org/10.3394/0380-1330(2007)33[566:DAFOEC]2.0.CO;2

Meixner T. (2004). Wildfire impacts on water quality. Southwest Hydrology, 24–25.

Minnesota Pollution Control Agency (MPCA) (2017). Minnesota Stormwater Manual. Environmental impacts of road salt and other de-icing chemicals. Available at: https://stormwater.pca.state.mn.us/index.php/Environmental_impacts_of_road_salt_and_other_de-icing_chemicals (accessed 19 July 2019).

Moore S. L., Gregorio D., Carreon M., Weisberg S. B. and Leecaster M. K. (2001). Composition and distribution of beach debris in Orange County, California. *Marine Pollution Bulletin*, **42**, 241–245, https://doi.org/10.1016/S0025-326X(00)00148-X

Moore J., Braun D., Meals D., Potter F., Middleman M. and Howe E. (2015). Harvesting pollution: quantifying agricultural phosphorus inputs to Lake Champlain. *Proceedings of the Water Environment Federation*, **2015**(9), 3784–3795, https://doi.org/10.2175/193864715819555337

Munck I. A., Bennett C. M., Camilli K. S. and Nowak R. S. (2010). Long-term impact of de-icing salts on tree health in the Lake Tahoe Basin: environmental influences and interactions with insects and diseases. *Forest Ecology and Management*, **260**, 1218–1229, https://doi.org/10.1016/j.foreco.2010.07.015

Nason J., Bloomquist D. and Sprick M. (2012). Factors influencing dissolved copper concentrations in Oregon highway storm water runoff. *Journal of Environmental Engineering*, **138**, 734–742, https://doi.org/10.1061/(ASCE)EE.1943-7870.0000537

National Academies of Sciences, Engineering, and Medicine (NASEM) (2019). Improving the EPA Multi-Sector General Permit for Industrial Stormwater Discharges. The National Academies Press, Washington, DC, https://doi.org/10.17226/25355

National Research Council (NRC) (2009). Urban Stormwater Management in the United States. The National Academy Press, Washington DC, 598 pp.

Nelson E. J. and Booth D. B. (2002). Sediment sources in an urbanizing, mixed land-use watershed. *Journal of Hydrology*, **264**(1–4), 51–68, https://doi.org/10.1016/S0022-1694(02)00059-8

Nolan B. T., Ruddy B. C., Hitt K. J. and Helsel D. R. (1997). Risk of nitrate in groundwaters of the United States – a national perspective. *Environmental Science and Technology*, **31**, 2229–2236, https://doi.org/10.1021/es960818d

Norris L. A. and Webb W. L. (1989). Effects of Fire Retardant on Water Quality. In: Proceedings of the Symposium on Fire and Watershed Management, October 26–28, Sacramento, California. General Technical Report.

PSW-109. Berkeley, California: US Department of Agriculture, Forest Service, Pacific Southwest Forest and Range Experiment Station: 79–86, Publication Series: General Technical Report (GTR).

North Carolina. Division of Environmental Management (NCDEM) (1990). North Carolina Coastal Marinas: Water Quality Assessment, Report No. 90–01, NCDEP, Raleigh, NC, USA, 74pp.

Osterkamp W. R., Heilman P. and Lane L. J. (1998). Economic considerations of continental sediment-monitoring program. *International Journal of Sediment Research*, **13**(4), 12–24.

Parker J. K., McIntyre D. and Noble R. T. (2010). Characterizing fecal contamination in stormwater runoff in coastal north Carolina, USA. *Water Research*, **44**, 4186–4194, https://doi.org/10.1016/j.watres.2010.05.018

Pedersen J. A., Yeager M. A. and Suffet I. H. (2006). Organophosphorus insecticides in agricultural and residential runoff: field observations and implications for total maximum daily load development. *Environmental Science and Technology*, **40**, 2120–2127, https://doi.org/10.1021/es051677v

Pinkney A. E., Harshbarger J. C., May E. B. and Reichert W. L. (2004). Tumor prevalence and biomarkers of exposure and response in brown bullhead (*Ameriurus nebulosus*) from the Anacostia River, Washington, DC and Tuckahoe River, Maryland, USA. *Environmental Toxicology and Chemistry*, **23**, 638–647, https://doi.org/10.1897/03-77

Pirrone N., Cinnirella S., Feng X., Finkelman R. B., Friedli H. R., Leaner J., Mason R., Mukherjee A. B., Stracher G. B., Streets D. G. and Telmer K. (2010). Global mercury emissions to the atmosphere from anthropogenic and natural sources. *Atmospheric Chemistry, and Physics*, **10**, 5951–5964, https://doi.org/10.5194/acp-10-5951-2010

Pitt R. (2015). National Stormwater Quality Database (NSQD), Version 4.02. http://www.bmpdatabase.org/nsqd.html (accessed 18 July 2019).

Pitt R. (2018). National Stormwater Quality Database (NSQD), Version 4.02. http://www.bmpdatabase.org/Docs/NSQD_ver_4_brief_Feb_18_2018.pdf (accessed 18 July 2019).

Pitt R., Clark S., Parmer K. and Field R. (1996). Groundwater Contamination From Stormwater. Ann Arbor Press, Chelsea, MI, USA, 219pp.

Pruss A. (1998). Review of epidemiological studies on health effects from exposure to recreational water. *International Journal of Epidemiology*, **27**(1), 1–9, https://doi.org/10.1093/ije/27.1.1

Przepiora A., Hesterberg D., Parsons J. E., Gilliam J. W., Cassel D. K. and Faircloth W. (1998). Field evaluation of calcium sulfate as a chemical flocculant for sedimentation basins. *Journal of Environmental Quality*, **27**, 669–678, https://doi.org/10.2134/jeq1998.00472425002700030026x

Ribaudo M. (2009). Non-point pollution regulation approaches in the US. In: The Management of Water Quality and Irrigation Techniques, J. Albiac and A. Dinar (eds.), Earthscan, London, pp. 83–102.

Richards R. P., Baker D. B., Crumrine J. P. and Stearns A. M. (2010). Unusually large loads in 2007 from the Maumee and Sandusky Rivers, tributaries to Lake Erie. *Journal of Soil and Water Conservation*, **65**(6), 450–462, https://doi.org/10.2489/jswc.65.6.450

Roberts T., Serdar D., Maroncelli J. and Davies H. (2011). Control of Toxic Chemicals in the Puget Sound: Phase 3: Primary Sources of Selected Toxic Chemicals and Quantities Released in the Puget Sound Basin. Publication No. 11-03-024. Washington State Department of Ecology, Olympia, WA, USA. www.ecy.wa.gov/biblio/1103024.html (accessed 1 August 2019).

Rowden R. D., Libra R. D. and Hallberg G. R. (1995). Surface Water Monitoring in the Big Spring Basin 1986–1992: A Summary Review. Geological Survey Bureau Technical Information Series 33. Iowa Department of Natural Resources, Des Moines, IA, USA.

Ryan S. E. and Grant G. E. (1991). Downstream effects of timber harvesting on channel morphology in Elk River Basin, Oregon. *Journal of Environmental Quality*, **20**, 60–72, https://doi.org/10.2134/jeq1991.00472425002000010011x

Sabin L. D., Jeong H. L., Stolzenbach K. D. and Schiff K. C. (2005). Contribution of trace metals from atmospheric deposition to stormwater runoff in a small impervious urban catchment. *Water Research*, **39**, 3929–3937, https://doi.org/10.1016/j.watres.2005.07.003

Sandahl J. F., Baldwin D. H., Jenkins J. J. and Scholz N. L. (2007). A sensory system at the interface between urban stormwater runoff and salmon survival. *Environmental Science and Technology*, **41**, 2998–3004, https://doi.org/10.1021/es062287r

Schoonover J. E., Graeme Lockaby B. and Helms B. S. (2006). Impacts of land cover on stream hydrology in the West Georgia Piedmont, USA. *Journal of Environmental Quality*, **35**, 2123–2131, https://doi.org/10.2134/jeq2006.0113

Schroeder W. and Munthe J. (1998). Atmospheric mercury – an overview. *Atmospheric Environment*, **32**, 809–822, https://doi.org/10.1016/S1352-2310(97)00293-8

Schueler T. (1987). Controlling Urban Runoff: A Practical Manual for Planning and Designing Urban BMPs. Metropolitan Washington Council of Governments, Washington, DC.

Scott T. A. (2017). Flexible, collaborative, and meaningful? The case of the US coastal nonpoint pollution control program. *Journal of Environmental Planning and Management*, **61**, 272–290, https://doi.org/10.1080/09640568.2017.1301896

Selvakumar A. and Borst M. (2006). Variation of microorganism concentration in urban stormwater runoff with land use and seasons. *Journal of Water and Health*, **4**, 109–124, https://doi.org/10.2166/wh.2006.0009

Sheavly S. (2007). National Marine Debris Monitoring Program: Final program deport, data analysis and summary, Prepared for USEPA by Ocean Conservancy, grant number X83053401–02, 76pp. Available at: http://www.portalasporta.it/dati_plastica/NMDMP_Report_April_2008.pdf (accessed 1 August 2019).

Shimabuku M., Diringer S. and Cooley H. (2018). Stormwater Capture in California: Innovative Policies and Funding Opportunities. Pacific Institute, Oakland, CA, USA. Available at: https://pacinst.org/wp-content/uploads/2018/07/Pacific-Institute-Stormwater-Capture-in-California.pdf (accessed 10 August 2019).

Silver P., Rupprecht S. M. and Stauffer M. F. (2009). Temperature-dependent effects of road deicing salt on chironomid larvae. *Wetlands*, **29**, 942–951, https://doi.org/10.1672/08-212.1

Smith K. P., Sorenson J. R. and Granato G. E. (2018). Characterization of stormwater runoff from bridge decks in eastern Massachusetts, 2014–16, US Geological Survey Scientific Investigations Report 2018–5033, 73 pp. Available at: https://doi.org/10.3133/sir20185033 (accessed 10 August 2019).

Sonoda K., Walker C. E. and Yeakley J. A. (2001). Near-stream land use effects on stream water nutrient distribution in an urbanizing watershed. *Journal of American Water Resources Association*, **37**, 1517–1532, https://doi.org/10.1111/j.1752-1688.2001.tb03657.x

Stewart J. and Skousen J. (2003). Water quality changes in a polluted stream over a twenty-five-year period. *Journal of Environmental Quality*, **32**, 654–661, https://doi.org/10.2134/jeq2003.6540

Suarez M. P., Rifai H. S., Schimek J. and Bloom M. (2006). Dioxin in stormwater runoff in Houston, Texas. *Journal of Environmental Engineering*, **132**, 1633–1643, https://doi.org/10.1061/(ASCE)0733-9372(2006)132:12(1633)

Sutton R., Mason S. A., Stanek S. K., Willis-Norton E., Wren I. F. and Box C. (2016). Microplastic contamination in the San Francisco Bay, California, USA. *Marine Pollution Bulletin*, **109**, 230–235, https://doi.org/10.1016/j.marpolbul.2016.05.077

Swadener L. M., Javeherian M. H., Jones E. G. and Bartelt-Hunt S. L. (2014). Chemical characterization and toxicity of bridge deck runoff and impacts to receiving water quality. *Journal of Environmental Engineering*, **140**, 04014010, https://doi.org/10.1061/(ASCE)EE.1943-7870.0000818

Trahan N. A. and Peterson C. M. (2008). Impacts of Magnesium Chloride-Based Deicers on Roadside Vegetation. In: Proceedings of 6th Intl. Symposium on Snow Removal and Ice Control Technology. Transportation Research Circular E-C126. SNOW08-050. Transportation Research Board, Washington, DC, pp. 171–186.

Traister E. and Anisfeld S. C. (2006). Variability of indicator bacteria at different time scales in the upper Hoosic rRiver watershed. *Environmental Science and Technology*, **40**, 4990–4995, https://doi.org/10.1021/es0601437

Turner R. E. and Rabalais N. N. (2019). The Gulf of Mexico. In: World Sea: An Environmental Evaluation, Volume I: Europe, the Americas and West Africa, C. Sheppard (ed.), 2nd edn, Academic, Oxford, UK, pp. 445–464.

Turner R. E., Rabalais N. N., Alexander R. B., McIsaac G. and Howarth R. W. (2007). Characterization of nutrient, organic carbon, and sediment loads and concentrations from the Mississippi river into the northern Gulf of Mexico. *Estuaries and Coasts*, **30**, 773–790, https://doi.org/10.1007/BF02841333

United States Department of Agriculture (USDA) (2011). Nationwide aerial application of fire retardant on national forest system land, Final Environmental Impact Statement, Forest Service, Fire and Aviation Management, Washington, DC.

United States Environmental Protection Agency (US EPA) (1983). Final report of the nationwide urban runoff program, US Environmental Protection Agency, Water Planning Division, Washington, D.C.

US EPA (1986). Ambient Water Quality Criteria for Bacteria—1986. EPA 440-5-84-002. US Environmental Protection Agency, Office of Water, Washington, D.C.

US EPA (1999). Preliminary data summary of urban storm water best management practices.

US EPA (2000a). Abandoned Mine Site Characterization and Cleanup Handbook. EPA 910-B-00- 001. US EPA Region 10, Seattle, Washington.

US EPA (2000b). 3rd edn, Guidance for Assessing Chemical Contaminant Data for use in Fish Advisories Volume 1: Fish Sampling and Analysis. EPA 823-B-00-007. US Environmental Protection Agency, Office of Science and Technology, Washington, D.C.

US EPA (2001). National management measures guidance to control nonpoint source pollution from marinas and recreational boating, US EPA Report, EPA 841-B-01-005, Office of Water.

US EPA (2003). National management measures for the control of nonpoint pollution from agriculture, US EPA Report, EPA-841-B-03-004, Office of Water, Washington, D.C.

US EPA (2005a). National management measures for the control of nonpoint pollution from forestry, US EPA Report, EPA-841-B-05-001, Office of Water, Washington, D.C.

US EPA (2005b). National management measures for the control of nonpoint pollution from urban areas, US EPA Report, EPA-841-B-05-004, Office of Water, Washington, D.C.

US EPA (2005c). Protecting water quality from agricultural runoff, EPA 841-F-05-001. https://www.epa.gov/sites/production/files/2015-09/documents/ag_runoff_fact_sheet.pdf (accessed 15 August 2019).

US EPA (2007). National Management Measures for the Control of Nonpoint Pollution From Hydromodification. EPA 841-B-07-002. Office of Water, Washington, DC.

US EPA (2017). The national water quality inventory: report to Congress, EPA 841-R-16-011, Office of Water, Washington, D.C.

US EPA (2019a). National summary of state information, http://ofmpub.epa.gov/waters10/attains_nation_cy.control (accessed 15 July 2019).

US EPA (2019b). Polluted runoff: Nonpoint source (NPS) pollution, https://www.epa.gov/nps (accessed 12 July 2019).

US EPA (2019c). Types of nonpoint source pollution, https://www.epa.gov/nps/types-nonpoint-source-pollution (accessed 12 Jul 2019).

US EPA (2019d). Introduction to the Clean Water Act, https://cfpub.epa.gov/watertrain/pdf/modules/IntrotoCWA.pdf (accessed 12 July 2019).

US EPA (2019e). Nutrient Pollution, https://www.epa.gov/nutrientpollution (accessed 12 July 2019).

WA DOE (Washington State Department of Ecology) (2014). 2012 Stormwater Management Manual for Western Washington as Amended in December 2014. Publication No. 14-10-055. Washington State Department of Ecology, Water Quality Program, https://fortress.wa.gov/ecy/publications/SummaryPages/1410055.html (accessed 10 July 2019).

Waschbusch R. J. (1999). Evaluation of the effectiveness of an urban stormwater treatment unit in Madison, Wisconsin, 1996-97, US Geological Survey Open-File Report 99-4195, Middleton, WI, USA.

Water Environment Federation Stormwater Institute (WEF) (2015). Rainfall to Results: the future of stormwater, WEF, Alexandria, VA, USA. Available at: https://wefstormwaterinstitute.org/wp-content/uploads/2015/09/Rainfall-to-Results.pdf (accessed 20 July 2019).

Weston D. P. and Lydy M. J. (2012). Stormwater input of pyrethroid insecticides to an urban river. *Environmental Toxicology and Chemistry*, **31**(7), 1579–1586, https://doi.org/10.1002/etc.1847

Weston D. P., Holmes R. W. and Lydy M. J. (2009). Residential runoff as a source of pyrethroid pesticides to urban creeks. *Environmental Pollution*, **157**(1), 287–294, https://doi.org/10.1016/j.envpol.2008.06.037

Whitman R. L., Nevers M. B. and Byappanahalli M. N. (2006). Examination of the watershed-wide distribution of *Escherichia coli* along southern Lake Michigan: an integrated approach. *Applied and Environmental Microbiology*, **72**, 7301–7310, https://doi.org/10.1128/AEM.00454-06

Wolfand J. M., Seller C., Bell C. D., Cho Y., Oetjen K., Hogue T. S. and Luthy R. G. (2019). Occurrence of urban-use pesticides and management with enhanced stormwater control measures at the watershed scale, *Environmental Science and Technology*, **53**, 3634–3644, https://doi.org/10.1021/acs.est.8b05833

Writer J. H. and Murphy S. F. (2012). Wildfire effects on source-water quality – lessons from Fourmile Canyon fire, Colorado, and implications for drinking-water treatment, US Geological Survey Fact Sheet 2012–3095. https://pubs.usgs.gov/fs/2012/3095/FS12-3095.pdf (accessed 20 August 2019).

Writer J. H., Brown G. K., Taylor H. E., Schoenfuss H. L., Jahns N. D., Bartell S. E., Kiesling R. L., Ferrey M. L. and Barber L. B. (2010). Anthropogenic tracers, endocrine active chemicals, and endocrine disruption in Minnesota lakes not impacted by point-source discharges. *Science of the Total Environment*, **409**, 100–111, https://doi.org/10.1016/j.scitotenv.2010.07.018

Young S., Juhl A. and O'Mullan G. D. (2013). Antibiotic-resistant bacteria in the Hudson River Estuary linked to wet weather sewage contamination. *Journal of Water and Health*, **11**, 297–310, https://doi.org/10.2166/wh.2013.131

Zannetti P. (1990). Dry and wet deposition. In: Air Pollution Modeling. Springer, Boston, MA, USA, pp. 249–262.

doi: 10.2166/9781789061123_139

Chapter 9

Rural diffuse pollution in Scotland: impacts and management

F. Napier[1], I. Sime[2], M. Aitken[1] and K. Kortland[3]

[1]Scottish Environment Protection Agency, Strathallan House, Castle Business Park, Stirling FK9 4TZ, UK
[2]NatureScot, Great Glen House, Leachkin Road, Inverness IV3 8NW, UK
[3]Forestry and Land Scotland, Smithton, Inverness IV2 7NL, UK

Keywords: Diffuse pollution; farming; forestry; land management; mitigation; water quality

9.1 INTRODUCTION

The last few decades have seen environmental protection move from fragmented, often narrowly focused management to a more holistic consideration of whole ecosystems. For the aquatic environment, this has meant a shift in perspective and scale from individual waterbodies to whole catchment (and even basin) management. In Europe, the introduction of Directive 2000/60/EC, the Water Framework Directive (WFD), in 2000 heralded a paradigm shift in water management for Member States, with a move away from water quality assessment based on chemical parameters towards whole ecosystem assessment measured by ecological parameters, and management based on hydro-geographical rather than political boundaries.

The over-arching purpose of the WFD is to improve water quality and prevent any deterioration from good status through a strategy of river basin management. For each river basin district, a management plan has been established and is updated every six years in an iterative process of determining current status (through a programme of monitoring and classification), identifying water bodies which are at risk of failing the WFD objectives, setting objectives for each waterbody and identifying the measures required to achieve these objectives.

An important component of WFD is that it places on member states a specific requirement to control diffuse pollution. Diffuse pollution is by nature a complex problem, reflecting a wide range of potential pollutant sources and pathways together with high temporal and spatial variability means that changes occur on a daily, weekly and annual basis. All of these factors, coupled with the individual characteristics of catchments and the water environment mean that pollution from diffuse sources can be challenging to remedy and improvements difficult to monitor.

This chapter reports on diffuse pollution impacts in Scotland, and how the challenges of managing and reducing these impacts are being addressed.

9.2 WATER REGULATION AND DIFFUSE POLLUTION

Legislation relating to the water environment in Scotland stretches to a list of over fifty various regulations, directions, acts and orders. The key legislation relevant to regulating activities with the potential to cause diffuse pollution is discussed here. The European Commission (EC) WFD requires water bodies to meet 'good' ecological status or potential. This WFD legislation became law in Scotland as the Water Environment and Water Services (Scotland) Act 2003 (WEWS). In Scotland, the Water Environment (Controlled Activities) (Scotland) Regulations 2011 (generally referred to as CAR) and their further amendments apply regulatory controls over activities which may affect Scotland's water environment. The regulations cover rivers, lochs, transitional waters (estuaries), coastal waters, groundwater, and groundwater-dependent wetlands.

Statutory objectives are set for Scotland's waters by the river basin management planning (RBMP) process required by WFD. RBMP plans are produced every six years by Scotland's environmental regulator (SEPA) on behalf of Scottish Government. The plans cover actions for all responsible authorities in Scotland and assess and report: the current status of the water environment; pressures affecting the water environment where it is in less than good condition; the actions required to protect and improve the water environment (referred to as 'measures'); a summary of objectives or outcomes following implementation of measures. RBMP is a cyclical process of characterisation, objective setting, monitoring and reporting.

Activities likely to cause rural diffuse pollution are specifically regulated under the CAR regulations and are referred to as Diffuse Pollution General Binding Rules (DP GBRs). The DP GBRs aim to reduce the risk of diffuse pollution from rural land-use activities and help to protect and improve water quality. The GBRs focus on land and water run-off management and are split into specific activities, each explaining the steps required to protect the water environment. Examples of the activities which the GBRs cover include fertiliser, manure and slurry storage and application; keeping livestock; land cultivation; pesticide application and so on. The DP GBRs were derived from already established and accepted codes of good agricultural practice, and present a statutory baseline designed to provide a general level of environmental protection and contribute significantly to water quality improvements. The provisions of the CAR regulations and DP GBRs addressing pollution from diffuse sources, are also important for achieving compliance with EU Directives protecting recreational waters, shellfisheries, and controlling dangerous substances.

The Nitrates Directive (91/676/EEC) requires Member States to identify areas where the concentrations of nitrate in groundwater exceed, or are likely to exceed, the levels set in the Directive. These areas are designated as Nitrate Vulnerable Zones (NVZs). The Action Programme for Nitrate Vulnerable Zones (Scotland) Regulations 2008 (as amended), meet Scotland's legal and environmental obligations for NVZs. These regulations set out what additional actions (in addition to GBR compliance) farmers are required to take to comply with the NVZ rules.

9.3 DIFFUSE POLLUTION IMPACTS

Scotland has more than 125 000 km of rivers and streams, and over 25 500 lochs, containing 90% of the UK's surface freshwater. The country's freshwater resources are an important part of the landscape and the economy, providing water for industry and agriculture, habitats for wildlife and providing recreation and tourism amenity opportunities. Many individuals, companies and communities rely on groundwater for drinking water, agriculture and industry. Groundwater also feeds wetlands and river flows during dry spells and is vital to the maintenance of their rich ecology and biodiversity (Critchlow-Watton *et al.*, 2014).

Some of Scotland's most important socio-economic industries depend on the high quality of the country's rivers. Scotch whisky for example is the UK's largest food and drink sector and accounts for 80% of Scotland's food and drink exports, worth over £4billion per year. Finfish aquaculture is another important sector, contributing £620 million annually to Scottish economy (Critchlow-Watton *et al.*, 2014).

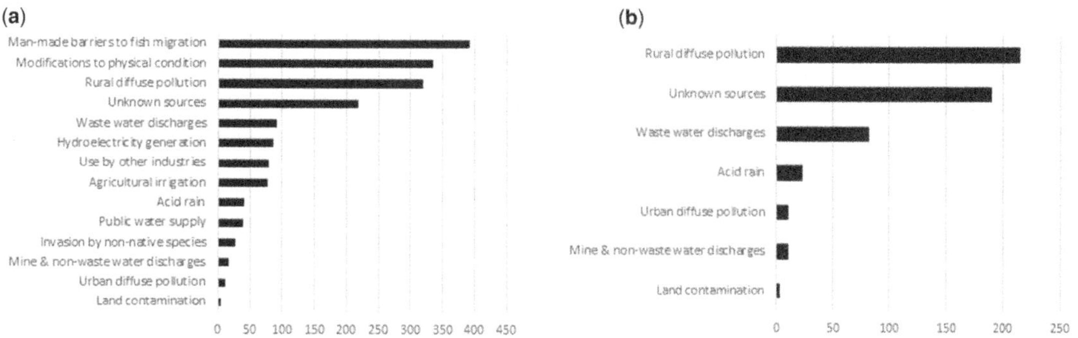

Figure 9.1 (a) Number of surface waters with pressures affecting overall condition; (b) number of surface waters with pressures affecting water quality (from SEPA, 2015).

The rural land-use sectors have a significant influence on Scotland's environment (SEPA, 2019). Scotland's 2015 River Basin Management Plan (RBMP) identified rural diffuse pollution as the single largest water quality pressure causing downgrades in Scotland's waters, reporting 246 water bodies failing water quality objectives, and 70 impacted protected areas (SEPA, 2015).

The main pollutants involved are nutrients, faecal pathogens, eroded soil, and pesticides, from a range of agricultural, urban green space, forestry and mining land uses.

9.3.1 Current status

The 2015 RBMP reported the main pressures currently affecting the overall condition of surface waters (rivers and lochs) in Scotland as man-made barriers to fish migration, physical changes to the beds and banks and rural diffuse pollution, with rural diffuse pollution identified as the largest single pressure on water quality (see Figure 9.1).

Groundwater is a hugely valuable water resource for Scotland. It provides 73 per cent of Scotland's private water and 5 per cent of public water supplies, for at least 330 000 people in total (Ó Dochartaigh *et al.*, 2015). Groundwater for public water supply is abstracted from approximately 100 boreholes and springs across Scotland. More than 4000 boreholes, and some large springs, are used for large private, industrial or agricultural supplies; and many more–approximately 20 000–boreholes, small springs and wells provide private water supplies for at least 80 000 people (Ó Dochartaigh *et al.*, 2015).

Groundwater is also very important for Scotland's economy and provides almost all of the water sold by the bottled water industry (Ó Dochartaigh *et al.*, 2015; SEWeb, 2011)). It also underpins

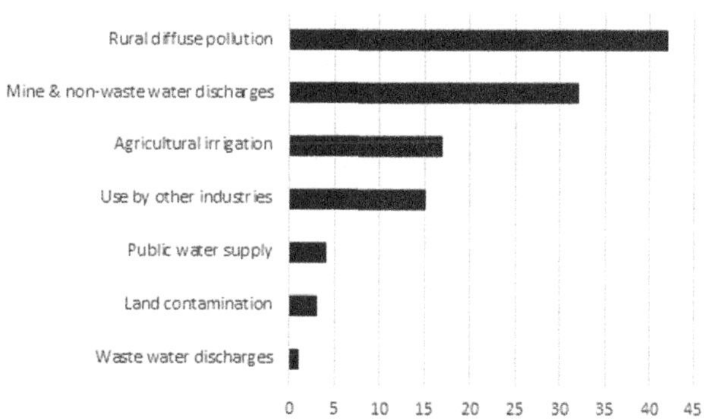

Figure 9.2 Number of groundwaters with pressures affecting overall condition (from SEPA, 2015).

Scotland's whisky exports (Scotland's leading export after oil and gas), providing 70 per cent of the water bottled by the whisky distilling industry. It is widely used in other industries, such as brewing and fish farming, and for agricultural and recreational (including golf course) irrigation.

According to the latest RBMP assessment (see Figure 9.2), rural diffuse pollution is currently the largest source of groundwater impairment, with approximately 10% of Scotland's 408 groundwater bodies impacted by this pressure (see case study 4.3 below).

Scotland has 49 estuaries assessed under WFD, covering an area of approximately 1000 km^2. While the majority (85%) of these are at good or high status, three are currently at moderate due to diffuse nutrient inputs from agriculture (SEWeb, 2014; see case study 9.4.3 below).

9.4 CASE STUDIES

9.4.1 Diffuse pollution priority catchments

In Scotland, the Diffuse Pollution Management Advisory Group (DPMAG) is driving actions to improve the water environment by taking a partnership approach to reducing rural diffuse pollution risks (FWS, 2021a). This group is made up of a range of organisations from organisations who have rural diffuse pollution interests. A national coordinated approach is underway to work with farmers, forestry and recreational businesses to improve and protect water quality by reducing diffuse pollution risks. Actions include:

- A national campaign to prevent water bodies from deterioration in status and make improvement where they are not far from a status boundary.
- A targeted approach in identified 'Priority Catchments' (see Figure 9.3) where the extent of diffuse pollution problems on the water environment requires a more focused approach.

The 'Priority Catchment' approach has been shown to be highly effective at managing and reducing diffuse pollution risks (Aitken & Field, 2015). It is risk-based, flexible and integrated and ensures that regulation is proportionate for farm businesses and focuses efforts on priority issues. A key need is to provide information, advice and guidance where appropriate, and to promote funding for diffuse pollution mitigation measures. A sound scientific and evidence base was, and is, required to support the mitigation of diffuse pollution, both in terms of stakeholder buy-in and to accurately target measures to achieve the most cost-effective improvements. This includes research on land and soil management, sources of pollutants, their transport pathways and processes, as well as their impact and farm management practices.

Farmers and land managers need advice and tools to enable the right environmental improvements to be made. As well as holding workshops and open days about diffuse pollution with partner organisations, SEPA carries out farm visits which focus on measures that land managers can reduce the risks of diffuse pollution from their land (FWS, 2021b). Findings are explained to farmers in a way that enables them to make informed decisions, to improve their business and the environment. The farmer receive a letter, map-based report and mitigation advice, and a date for follow up visit if required. Considerable work has been completed in the catchments to put measures in place to mitigate the effects of diffuse pollution. These measures are a mixture of self-funded work, Scottish Government funded measures and local project initiatives that focus on multiple benefits.

9.4.2 Loch of strathbeg

The Loch of Strathbeg nature reserve features the UK's largest dune loch, and is protected by Ramsar, Site of Special Scientific Interest (SSSI) and Special Protection Area (SPA) status under the European 'Birds' Directive (2009/147/EC). The shallow loch itself is 616 ha in extent and, along with surrounding wetlands, grassland and woodland, supports thousands of wildfowl over winter, and protects a wealth of other wetland wildlife. A large part is a Royal Society for the Protection of Birds (RSPB) reserve.

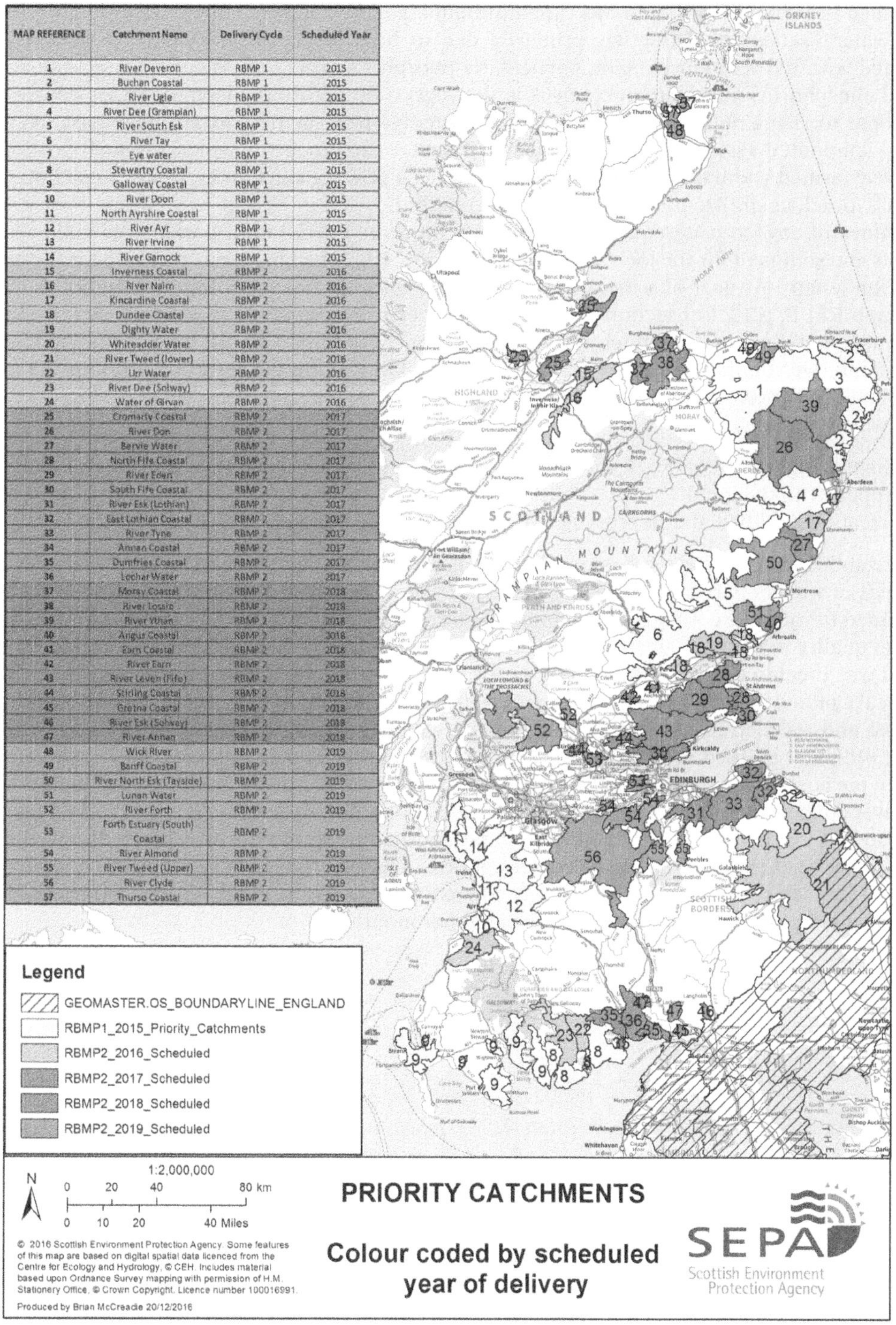

MAP REFERENCE	Catchment Name	Delivery Cycle	Scheduled Year
1	River Deveron	RBMP 1	2015
2	Buchan Coastal	RBMP 1	2015
3	River Ugie	RBMP 1	2015
4	River Dee (Grampian)	RBMP 1	2015
5	River South Esk	RBMP 1	2015
6	River Tay	RBMP 1	2015
7	Eye water	RBMP 1	2015
8	Stewartry Coastal	RBMP 1	2015
9	Galloway Coastal	RBMP 1	2015
10	River Doon	RBMP 1	2015
11	North Ayrshire Coastal	RBMP 1	2015
12	River Ayr	RBMP 1	2015
13	River Irvine	RBMP 1	2015
14	River Garnock	RBMP 1	2015
15	Inverness Coastal	RBMP 2	2016
16	River Nairn	RBMP 2	2016
17	Kincardine Coastal	RBMP 2	2016
18	Dundee Coastal	RBMP 2	2016
19	Dighty Water	RBMP 2	2016
20	Whiteadder Water	RBMP 2	2016
21	River Tweed (lower)	RBMP 2	2016
22	Urr Water	RBMP 2	2016
23	River Dee (Solway)	RBMP 2	2016
24	Water of Girvan	RBMP 2	2016
25	Cromarty Coastal	RBMP 2	2017
26	River Don	RBMP 2	2017
27	Bervie Water	RBMP 2	2017
28	North Fife Coastal	RBMP 2	2017
29	River Eden	RBMP 2	2017
30	South Fife Coastal	RBMP 2	2017
31	River Esk (Lothian)	RBMP 2	2017
32	East Lothian Coastal	RBMP 2	2017
33	River Tyne	RBMP 2	2017
34	Annan Coastal	RBMP 2	2017
35	Dumfries Coastal	RBMP 2	2017
36	Lochar Water	RBMP 2	2017
37	Moray Coastal	RBMP 2	2018
38	River Lossie	RBMP 2	2018
39	River Ythan	RBMP 2	2018
40	Angus Coastal	RBMP 2	2018
41	Earn Coastal	RBMP 2	2018
42	River Earn	RBMP 2	2018
43	River Leven (Fife)	RBMP 2	2018
44	Stirling Coastal	RBMP 2	2018
45	Gretna Coastal	RBMP 2	2018
46	River Esk (Solway)	RBMP 2	2018
47	River Annan	RBMP 2	2018
48	Wick River	RBMP 2	2019
49	Banff Coastal	RBMP 2	2019
50	River North Esk (Tayside)	RBMP 2	2019
51	Lunan Water	RBMP 2	2019
52	River Forth	RBMP 2	2019
53	Forth Estuary (South) Coastal	RBMP 2	2019
54	River Almond	RBMP 2	2019
55	River Tweed (Upper)	RBMP 2	2019
56	River Clyde	RBMP 2	2019
57	Thurso Coastal	RBMP 2	2019

Figure 9.3 Scotland's Priority Catchments (from FWS, 2021b).

The loch is currently failing to meet the minimum 'good ecological status' water quality required under Water Framework Directive, primarily due to high levels of chlorophyll A believed to be associated with nutrient enrichment, particularly phosphorus (the loch is currently graded as 'poor' for total phosphorus). Based upon changes in the vegetation within the loch, there are concerns that in response to rising nutrient levels the loch will change from being a macrophyte-based system to an algal-dominated system. The wider Buchan Coastal catchment in which the loch is situated is intensively farmed and has been identified by SEPA as a priority catchment, where a catchment-wide approach to reduce diffuse pollution impacts is required.

The Burn of Savoch watercourse feeds directly into Loch of Strathbeg and is the main source of nutrients and sediment for the loch. Until the summer of 2007 the burn was canalised for the majority of its 2 km length. As part of a major restoration project undertaken by Scottish National Heritage (SNH) and RSPB, a 1.2 km section of the burn has been re-meandered to create a new 1.8 km long shallow channel. This new channel flows through a purpose built 1 ha silt trap and along the whole length it can overtop its banks into a 23 ha reed bed. On leaving the reed bed it rejoins its original channel that has been modified to remove the bunds prior to entering the loch.

The work had two aims:

(a) improving water quality in the Burn of Savoch, and hence reducing sediment and phosphorus inputs into the loch;
(b) creating reed bed and improving habitat for breeding waders.

The combination of these measures has resulted in many more birds using these areas, with thousands of waders particularly benefitting every year. The expanded reed bed is used regularly by the bearded tit, one of Scotland's rarest breeding birds.

Water quality in the burn upstream and downstream of the wetland has been monitored since the wetland was created to assess the impact of the new wetland creation on water quality in the burn and reveal a mixed picture. Storm flows are currently being effectively attenuated by the wetland, with a delayed and reduced response observed downstream compared to upstream. This has resulted in reduced sediment and sediment-associated phosphorus in the burn downstream of the wetland, a key intended purpose of the restoration work. Nitrate concentrations in the burn are also being reduced, with lower concentrations measured downstream compared to upstream during both low and storm flows. However, it is evident that as the wetland and habitat continues to mature, the wetland has now become a net exporter of dissolved phosphorus. Dissolved oxygen concentration is also being negatively impacted, and is greatly reduced downstream of the wetland (Figure 9.4). Both of these phenomena are particularly marked over the summer months.

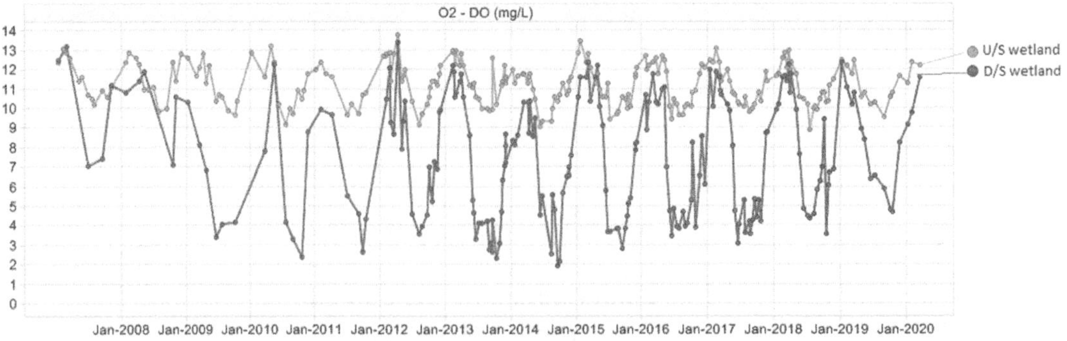

Figure 9.4 Concentrations of dissolved oxygen (DO) measured in Burn of Savoch upstream and downstream of wetland. (*Source*:SEPA).

It seems likely that elevated P downstream of the wetland is associated with the low dissolved oxygen (DO) periods in summer, as the wetland processes the influent contaminants. Low DO would favour release of soluble P from sediments. The reedbed treatment would probably be an unqualified success if it was for 'polishing' incoming flows from farmland; it will only achieve maximum benefits if it is the last stage in a catchment-wide program of measures starting at source on the farms to reduce sediment, P and BOD/COD load.

9.4.3 Ythan catchment

The Ythan catchment is in North Aberdeenshire. Like much of north-east Scotland, the catchment is mainly rural, and land use is predominantly mixed farming with most of the catchment consisting of either arable or grazing land with small areas of forestry. The lower 9 km of the river is tidal, and the Ythan estuary is an important wildlife reserve, attracting RAMSAR, SSSI and SPA/ SPC protected area designations (NatureScot, 2021).

Thousands of people within the catchment use private groundwater (GW) supplies (springs, wells and boreholes) as their source of drinking water, and any impact on GW primarily present risks for private water supplies. Groundwater also provides the majority of the flow to the river Ythan, especially in dry weather periods, and therefore to the Ythan estuary (SEPA, 2011).

The River Ythan catchment was designated a Nitrate Vulnerable Zone (NVZ) in 2000 (Scottish Government, 2021) due to nitrate impacts in groundwater and the estuary. The catchment is also one of SEPA's diffuse pollution 'Priority Catchments' (see section 9.4.1).

The Ythan estuary has been affected by eutrophication for many years, with frequent occurrences of algal mats (see Figure 9.5) threatening changes to the ecology of the estuary (MacGarvin 2001; Marine Scotland, 2020). It is currently at moderate status because of high nutrient inputs from agriculture (SEPA, 2014); typically 2000 tonnes of nitrogen leave the catchment and enter the estuary every year, with over 90% of the nitrate inputs to the catchment coming from agriculture (Aitken & Field, 2015).

Under the WFD, excessive growth of opportunistic macroalgae in estuaries is assessed by the coverage and density, expressed as a number from 1 to 0, referred to as the Ecological Quality Ratio (EQR). Macroalgal bloom (MAB) surveys carried out by SEPA in the Ythan estuary show that the overall EQR in the estuary has improved from moderate to good since 2009, which is encouraging. However the latest 2019 survey found that the total area covered with MAB species has increased since 2011, from 11 to 24 ha (Marine Scotland, 2020), and the classification for dissolved inorganic nitrogen as 'bad' compared to 'moderate' in 2017. Much work is happening in the catchment to reduce nutrient inputs to the water environment, but it is clear that significant and lasting improvements will take some time to establish.

In this intensively farmed catchment pesticides are an issue too. Intensive monitoring carried out by SEPA in the catchment showed chemical evidence of regular low-level pesticide inputs in surface water across the catchment. For surface waters, the analysis of ecological data showed evidence of pesticide impacts in surface water throughout a 30 year dataset. Approximately two-thirds of the ecological sites in the catchment showed distinct impacts on the invertebrate fauna at some point since the year 2000. Impacts tended to be sporadic, acute and generally short term. Recovery was observed at most ecological sites where impacts have been found in the past, and there was an overall improvement in the catchment over the past three decades (Aitken & Field, 2015).

Emelogu *et al.* (2013) also identified pesticides in streams associated with agricultural activities (comparing autosampling with silicone rubber passive sampling). Several of the pesticides measured were banned or have severely restricted uses within Europe as set down by various EC Regulations and Directives for persistent priority pollutants (PPPs) including diuron, isoproturon, chlorpyrifos, monolinuron, atrazine, trifluralin, dichlorvos, diazinon, fenuron, diflufenican, mevinphos and parathion ethyl (e.g., EC, 1979, 1991, 2009). The persistence of these substances makes attribution to sources difficult; they may reflect legacy uses, or use in contravention of regulations perhaps from old, stored

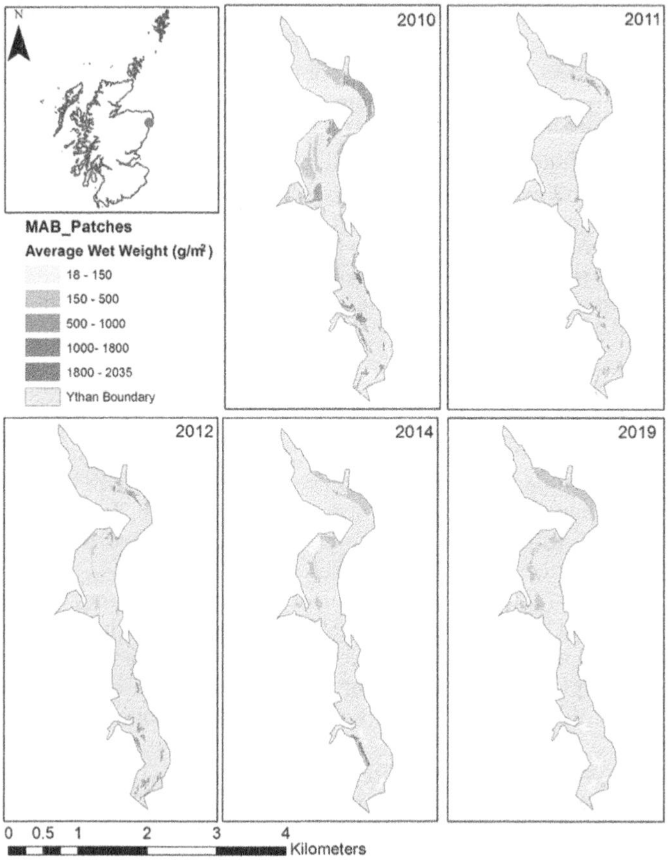

Figure 9.5 Macroalgal bloom (MAB) coverage 2010–2019 (from Marine Scotland, 2020).

material. Persistent pollutants can be transported to waterbodies by adsorption on soil particles and, if soluble, directly in runoff or leached from soil. Emelogu *et al.* (2013) noted that the fate of pesticides in the environment is complex and cannot be exclusively predicted, and reported that the relatively long persistence in soil-water after the last application of chemicals such as isoproturon, atrazine, trifluralin, and diuron has been demonstrated in both controlled conditions and in some field studies.

In summary, long-term monitoring such as undertaken in the Ythan, indicates that long timescales are needed to achieve/show improvement and reduction of risks to environmental or human health. Downstream impacts can be significant for the ecology of estuaries as well as compliance with health standards for rural water supplies. Diffuse source contamination needs to be addressed at source.

9.4.4 Freshwater pearl mussels

Freshwater pearl mussel (*Margaritifera margaritifera*) is a critically endangered species in Europe, protected internationally, but with a steadily declining range owing to a range of pressures, particularly pollution (IUCN, 2011). It is a remarkable mollusc that can live for more than a century, has very demanding water quality requirements and requires the presence of salmonids to act as a host species during the first year of its lifecycle (Skinner *et al.*, 2003). The number of freshwater pearl mussel populations has declined significantly since the beginning of the 20th century. In central and southern Europe, one estimate suggested that the decline in known populations is as high as 95% (Bauer, 1988). It has been acknowledged that this decline is continuing (Geist, 2010) and the best remaining populations are in countries along the north Atlantic coast (Scotland, Ireland) and in Fenno-Scandinavia and Russia (Lopes-Lima *et al.*, 2017). In Scotland, the decline is continuing and diffuse pollution has been identified as the main threat to many populations (JNCC, 2015; Watt *et al.*, 2015). Indeed, rivers with sustainable freshwater pearl mussel populations are characterized by extremely low levels of turbidity and suspended solids, exhibiting only minor peaks of very short duration that occur during periods of heavy rainfall or snowmelt (Degerman *et al.*, 2009).

In Scotland, efforts over the last 20 years to reduce diffuse pollution affecting the conservation of freshwater pearl mussel have mainly been prioritised to catchments that are designated for their national and international importance.

One of those rivers is a small, isolated catchment in Argyll, western Scotland, where the catchment is dominated by a coniferous forest plantation. The forest is managed by Forestry and Land Scotland, who oversee and implement a Land Management Plan that seeks to restore and conserve the freshwater pearl mussel population, as well as maintain a productive forest. In 1997 and 2004 the resident freshwater pearl mussel population was surveyed and, in both cases found to be dominated by old pearl mussels. Each survey recorded only a single juvenile, indicating the population was struggling to successfully recruit and likely to be unsustainable in the long term (Cosgrove & Young, 1998). The principle pressure in the catchment was the plantation forest, established decades previously. The Land Management Plan has sought to remove mature conifers from around the river and its tributaries, block drains discharging into watercourses, create buffer strips, establish native riparian woodlands and restore riparian wetlands and mires where possible. This work has been ongoing since 2003 and was also accelerated during the recent Pearls in Peril LIFE + project (2012–2017). These efforts have sought to reduce the amount of nutrients and, particularly, fine sediments entering the watercourse and threatening the freshwater pearl mussels.

The most recent survey of the pearl mussel population was in 2014 (Watt *et al.*, 2015). While the population remains small and dominated by old, large adult mussels, for the first time several juvenile pearl mussels were recorded. In comparison to the results of 2004 (Figure 9.6), smaller (younger) juveniles were also recorded. Outside of the fixed survey sites (the 2004 and 2014 surveys visited the

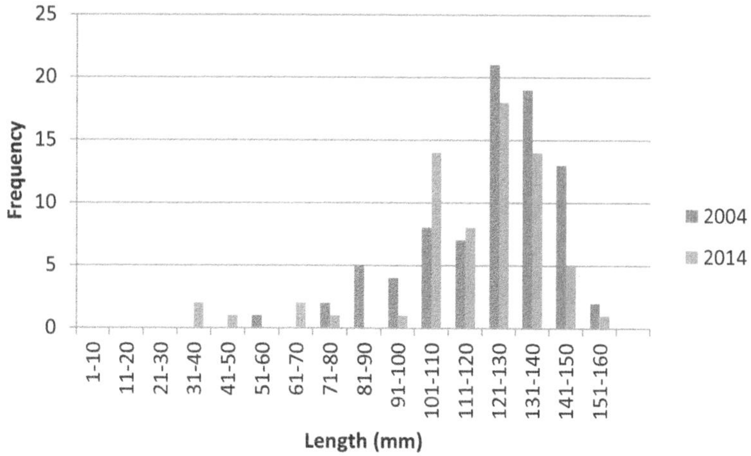

Figure 9.6 The size–frequency of freshwater pearl mussels from a river in Argyll, western Scotland, recorded in 2004 and 2014, indicating increased recruitment.

same, established survey locations) a very small pearl mussel at 10 mm long was also recorded in 2014, showing the first evidence of ongoing, successful recruitment in decades. These results, and actions by Forestry and Land Scotland since, have meant that the status of the river was deemed to be 'unfavourable recovering' due to the efforts to reduced diffuse source pollution. While the population has by no means fully recovered it is on a clear trajectory towards recovering. This is believed to be the first time such an outcome has been achieved for a freshwater pearl mussel population in Europe in the last 20 years.

These efforts to reduce diffuse pollution have not only benefited the conservation of freshwater pearl mussels in the river. The establishment of native riparian woodlands, particularly hazelwood, is also designed to increase habitat for Laborian lichens as well as leading to an increased local population of the hazel gloves fungus (*Hypocreopsis rhododendri*).

Another river where diffuse pollution has been identified as a particular pressure for freshwater pearl mussel is the much larger River South Esk on the east coast of Scotland. The freshwater pearl mussel population in the River South Esk is of such importance that the river has been designated as a Special Area of Conservation (SAC) under the Habitats Directive (Council Directive 92/43/EEC), for the conservation of the species. A key pressure on the River South Esk has been diffuse pollution, such that the river was identified as a 'priority catchment' by SEPA due to the risk posed to the conservation status of the SAC.

The River South Esk rises in the foothills of the Cairngorm mountains, where the catchment is mountainous and dominated by rough grazing and some forestry. In the lower reaches, the river passes over the Highland boundary fault and at lower altitude, flows through a number of settlements with the landuse including more intensive agriculture.

Since 2009, SEPA has led work to characterise and address diffuse pollution sources (see section 4.1), particularly in the lower catchment. This has involved extensive 'river walks', and advice to land managers to reduce diffuse pollution, particularly fine sediment, entering watercourses. This work has extended into the upper catchment, alongside extensive restoration works as part of the Pearls in Peril LIFE+ project (2012–2017) to improve riparian habitat and reduce diffuse pollution.

Since the river was designated as a SAC, the river has been monitored on three separate occasions. For this monitoring, the river was divided into three monitoring units. One was in the upper reaches in the foothills of the Cairngorms. The other two are in the lower reaches. The results have been summarised in Table 9.1, with the results compared to the favourable condition targets for the population parameters for the species described by JNCC (2015). These targets are:

(1) Population density: $\geq$ 5mussels per m^2 within sample tra nsects.
(2) Percentage juveniles: at least 20% of the population $\leq$ 65 mm in length (65 mm is widely taken to be the size at which pearl mussel reach sexual maturity).
(3) Presence of very young juveniles: at least one pearl mussel $\leq$ 30 mm in length.

The monitoring results are complicated but show that recruitment of juveniles in the upper catchment appears to have stopped over a period of 11 years to 2013. In the lower catchment recruitment has been maintained and, in places, increased markedly, to an extent that in 'lower catchment (a)' all the

Table 9.1 Population favourable condition targets for freshwater pearl mussel in the river south Esk 2002–2013 (those targets that have been met in italics).

	2002		2009		2013	
	Density	% juvs (<30 mm)	Density	% juvs (<30 mm)	Density	% juvs (<30 mm)
Upper catchment	0.5	7.1 (0)	0.3	2.3 (0)	0.4	0.0 (0)
Lower catchment (a)	***8.9***	3.7 (0)	***10***	4.5 (***15***)	***6.9***	***20.6 (21)***
Lower catchment (b)	1.7	***20.3*** (3)	1.2	17.3 (***4***)	1.6	16.9 (**1**)

Population Favourable Condition Targets were being met for the first time by 2013. While acknowledging that the most recent monitoring results from 2013 were collected only four years after the start of SEPA's efforts to address diffuse pollution, they show great encouragement for the recovery of freshwater pearl mussels in the lower reaches of the catchment. It is anticipated that, with ongoing efforts to reduce diffuse pollution, the next survey will record further improvements. In the upper catchment, the latest results were collected as restoration efforts commenced. The results show an alarming decline in the status and indicate the threat that fine, diffuse source, sediment poses to this species. It is hoped that restoration efforts since 2012 will yield benefits. A sediment fingerprinting project is also underway at present to better identify fine sediment sources in the upper catchment and allow them to be better targeted. Sediment is a significant diffuse pollutant mobilised by weather conditions and a characteristic of high flows, but landscape measures and managment practices can effectively manage the risks to the rivers.

9.5 CONCLUSIONS

Rural diffuse pollution is a significant cause of failure to meet good ecological status in Scotland. Important rural sources of diffuse pollution include nutrients, suspended solids (soil runoff e.g.), faecal pathogens and pesticides. Improving and protecting water quality will bring a range of benefits, including helping to:

- protect drinking water supplies and avoid increased purification treatment costs
- safeguard and support the expansion of businesses that depend on a high-quality water environment such as tourism, fisheries, aquaculture and whisky production
- maintain and expand opportunities for a wide range of recreational activities
- contribute to the social well-being and regeneration of areas in which a high-quality water environment is, or has the potential to be, an important community amenity
- protect and enhance wildlife including rare and endangered species.

Active engagement and knowledge exchange with farmers and land managers at a catchment scale is required to tackle diffuse pollution pressures. There is a need for an integrated approach to deliver environmental, economic and social objectives.

Where improvements have been achieved the following factors have been important:

- pollution prevention and control requires sector engagement with significant commitment of staff time and resources
- following regulations and best practices for soil, manure, nutrient and pesticide management; this can involve control at source and often will achieve cost savings for the farmer
- landscape interventions–appropriate for each land use–at source (in field and at steadings, on forest tracks and roads as well as forestry plantations) are key components of control strategies
- partnership working to ensure consistent messages from within the sectors (e.g., across agriculture and forestry, across environment and other government agencies) to deliver effective action across catchments
- appropriate monitoring is the key to effective control.

REFERENCES

Aitken M. and Field S. (2015). Practical Action on Farm to Reduce Rural Diffuse Pollution. (International Fertiliser Society Proceedings No 777). International Fertiliser Society, Colchester, UK, pp. 25.

Bauer G. (1988). Threats to the freshwater pearl mussel *Margaritifera margaritifera* L. in central Europe. *Biological Conservation*, **45**, 239–253, https://doi.org/10.1016/0006-3207(88)90056-0 (accessed 20 March 2022).

Cosgrove P. J. and Young M. R. (1998). The Status of the Freshwater Pearl Mussel M. margaritifera (L.) in Scotland. Confidential report to Scottish Natural Heritage, Edinburgh.

Critchlow-Watton N., Dobbie K. E., Bell R., Campbell S. D. G., Hinze D., Motion A., Robertson K., Russell M., Simpson J., Thomson D. and Towers W. (eds.) (2014). Scotland's State of the Environment Report, 2014. Scotland's Environment Web. http://www.environment.scotland.gov.uk/ (accessed 23 September 2021).

Degerman E., Alexanderson S., Bergengren J., Henrikson L., Johansson B.-E., Larsen B. M. and Söderberg H. (2009). Restoration of Freshwater Pearl Mussel Streams. WWF, Solna, Sweden.

Dochartaigh B. É. O., Macdonald A. M., Fitzsimons V. and Ward R. (2015). Scotland's Aquifers and Groundwater Bodies. British Geological Survey Open Report, OR/15/028.

EC (European community). (1979). Directive 79/117/EEC of 21 December 1978 on prohibiting the placing on the market and use of plant protection products containing certain active substances. *Off J Eur Commun*, **L 33**, 36.

EC (European community). (1991). Directive 91/414/EEC of 15 July 1991 concerning the placing of plant protection products on the market. *Off J Eur Commun*, **L 230**, 1–154.

EC (European community). (2009). Regulation 1107/2009/EC of the European Parliament and of the Council concerning the placing of plant protection products on the market and repealing Council Directives 79/117/EEC and 91/414/EEC. *Off J Eur Commun*, **L 309**, 1–50.

Emelogu E. S., Pollard P., Robinson C. D., Webster L., McKenzie C., Napier F., Steven L. and Moffat C. F. (2013). Identification of selected organic contaminants in streams associated with agricultural activities and comparison between autosampling and silicone rubber passive sampling. *Science of the Total Environment*, **445–446**, 261–272, https://doi.org/10.1016/j.scitotenv.2012.12.053

Farming and Water Scotland (FWS) (2021a). Diffuse Pollution. https://www.farmingandwaterscotland.org/diffuse-pollution/ (accessed 24 September 2021).

Farming and Water Scotland (FWS) (2021b). Scotland's Water & Scotland's Priority Catchments. https://www.farmingandwaterscotland.org/funding-grants-and-resources/scotlands-water-scotlands-priority-catchments/ (accessed 24 September 2021).

Geist J. (2010). Strategies for the conservation of endangered freshwater pearl mussels (*Margaritifera margaritifera* L.): A synthesis of conservation genetics and ecology. *Hydrobiologia*, **644**, 69–88, https://doi.org/10.1007/s10750-010-0190-2

IUCN (2011). Species on the edge of survival. [online] Available at: https://www.iucn.org/content/species-edge-survival-0 (accessed 20 March 2022).

JNCC–Joint Nature Conservation Committee (2015). Common Standards Monitoring Guidance for Freshwater Fauna, Version October 2015, JNCC, Freshwater, ISSN 1743-8160.

Lopes-Lima M., Sousa R., Geist J., Aldridge D. C., Araujo R.*et al*. (2017). Conservation status of freshwater mussels in Europe: State of the art and future challenges. *Biological Reviews*, **92**(1), 572–607.

MacGarvin M. (2001). Out of Sight–Out of Mind: Marine Eutrophication in the United Kingdom. Modus vivendi for WWF-UK, Glenlivet, Scotland, UK.

Marine Scotland (2020). Case Study: Opportunistic Macroalgal Blooms (Ythan Estuary). http://marine.gov.scot/sma/assessment/case-study-opportunistic-macroalgal-blooms-ythan-estuary (accessed 24 September 2021).

NatureScot (2021). Forvie NNR–About the Reserve. https://www.nature.scot/enjoying-outdoors/scotlands-national-nature-reserves/forvie-nnr/forvie-nnr-about-reserve (accessed 24 September 2021).

Scottish Government (2021). Nitrate Vulnerable Zones. https://www.gov.scot/policies/agriculture-and-the-environment/nvz/ (accessed 24 September 2021).

SEPA (2015). The river basin management plan for the Scotland river basin district: 2015–2027, SEPA website. https://www.sepa.org.uk/media/163445/the-river-basin-management-plan-for-the-scotland-river-basin-district-2015-2027.pdfotherinfo (accessed 23 September 2021).

SEPA (2019). Significant Water Management Issues for Scotland, SEPA website. https://www.sepa.org.uk/media/490770/191219_swmi_final.pdf (accessed 23 September 2021).

SEPA Ythan Project Group (2011). Nitrates and Pesticides Impacts and Pressures in the Ythan catchment. Volume 2: Technical Report.

SEWeb (2011). Groundwater, Scotland's Environment Web. https://www.environment.gov.scot/media/1230/water-groundwater.pdf (accessed 23 September 2021).

SEWeb (2014). Estuaries, Scotland's Environment Web. https://www.environment.gov.scot/media/1175/water-estuaries.pdf# (accessed 20 March 2022).

Skinner A., Young M. and Hastie L. (2003). Ecology of the Freshwater Pearl Mussel. Conserving Natura 2000 Rivers Ecology Series No. 2 English Nature, Peterborough. ISBN 1857167031.

Watt J., Cosgrove P. J. and Hastie L. C. (2015). A National Freshwater Pearl Mussel (Margaritifera margaritifera L.) Survey of Scotland. (Scottish Natural Heritage Commissioned Report No. 901). https://www.nature.scot/sites/default/files/2017-07/Publication%202015%20-%20SNH%20Commissioned%20Report%20901%20-%20A%20national%20freshwater%20pearl%20mussel%20%28Margaritifera%20margaritifera%2C-%20L.%29%20survey%20of%20Scotland.pdf (accessed 24 September 2021).

IWA PUBLISHING

doi: 10.2166/9781789061123_151

Chapter 10

Implications of land use and land-use changes to water quality in Korea: status, management and policy development

F. K. Geronimo[1] and Lehyung Kim[1]

Department of Civil and Environmental Engineering, Kongju National University, Cheonan City, Republic of Korea. E-mail: fkgeronimo@kongju.ac.kr, leehyung@kongju.ac.kr

Keywords: Agricultural area; land use and land use changes (LULUC); livestock; policy development; urban landuse

10.1 INTRODUCTION

According to UN-Water, land use and land-use changes (LULUC) refers to the conversion or transition of the earth's original surface by humans which is an important determinant of local, regional or continental-scale water cycles (UN-Water, 2018). The modification of the land surface was driven by humans to improve livelihoods. However, nowadays, the rate, extent, and intensity of LULUC are far greater than it was thousands of years ago (Hassan *et al.*, 2016). The 20th century stands out not only as the most dynamic period of anthropogenic ecosystem transformation of the past 300 years but also as the period during which the terrestrial biosphere transitioned from a primarily wild and seminatural state to a primarily used state (Ellis *et al.*, 2010). The amount of urban, mixed settlements, and populated areas are continuously increasing while the wildlands and natural landscapes continuously decreased. The transition of the inherent landscape to an extremely developed environment or urbanization increases the impervious surfaces covered by building rooftops, landscaping, and transportation land use such as roads, highways, parking lots, and sidewalks. The development of urban surfaces prevents infiltration, thereby increasing the runoff directed to stormwater networks and deteriorating the water resources in highly urbanized areas (Qiu, 2013). Land-use changes affect both the hydrologic cycle and water quality through the significant pollutant loads that it diverts to surface water bodies globally (Honkonen & Rantalainen, 2013). Particularly, land-use changes directly affected the increase in surface imperviousness, increased stormwater runoff, urban flooding and disasters, increased level of non-point source (NPS) pollution, and intensified effects of climate changes (Kong *et al.*, 2017).

Urban environmental problems began when the population in urban areas increased. However, economic development and improved living standards have increased citizens' desire to live in a safer and clean urban environment. As urban environmental problems were caused by the increase in population density and ecosystem degradation, policy-makers, and experts introduced an excellent system to prevent urban flooding and created settlements with increased water treatment facilities and green spaces to maintain a clean environment. Conventional drainage systems were designed to

Editor(s): Puangrat Kajitvichyanukul and Brian D'Arcy

rapidly eliminate floodwaters and prevent damage to properties; however, rapid drainage of water can lead to the urban heat island phenomenon due to the distortion of the natural water cycle. Moreover, discharging large amounts of NPS pollutants into the natural water bodies can result in algal blooms and ultimately, water quality deterioration.

In order to mitigate the negative effects of land-use changes to water bodies, most developed countries have enacted different policies to manage stormwater in their respective countries. The policies enacted to control and manage stormwater in some developed countries were summarized in Table 10.1. Priority legislations have increased together with a dedicated budget from different developed countries (Liptan, 2017). The identification of problems brought about by non-point source pollution

Table 10.1 Summary of stormwater regulations in different countries.

Country	Law/Act	Brief Description of Law/Act
United Kingdom	Flood & Water Management Act 2010 (F&WMA 2010)	Local Authorities will have responsibility for surface water management in their area. They will produce Strategic Water Management Plans (SWMP). Local Authorities will have to adopt a wide range of SUDS There will be National Standards for Sustainable Drainage Systems – which will include standards for Stormwater Treatment.
Japan	Water Circulation Basic Act	'Water circulation,' as referred to in the Act, means the cycle of water evaporation and of rain falling, running down over the ground surface, sinking into the ground, and reaching the sea. Rivers are the centers of circulation. (Water Circulation Basic Act, art. 2.) The Act recognizes that proper water circulation is important for the environment, people's daily life, and industrial production, and that water is the valuable property of the nation. Therefore, maintaining proper water circulation is important and water circulation through rivers must be managed comprehensively
USA	The Clean Water Act, the Clean Water Action Plan, and the Coastal Zone Management Act	These three laws support the national pollutant discharge elimination system of the USA for stormwater discharges from different sources including construction activities, industrial activities, municipal sources, and transportation sources
Germany	Wasserhaushaltsgesetz (Water Management Law)	Basic law for every water management in Germany but every federal state has its own 'water law' and additional regulations, defining the aspects of diffuse and non-point source control.
France	Order of 21 July 2015	Development of centralized treatment systems and decentralized sanitation facilities, with the exception of decentralized treatment facilities receiving a gross load of organic pollution less than or equal to 1.2 kg/d of BOD_5
Australia	Environment Protection Act 1993 and Environment Protection (Water Quality) Policy 2015 (the Water Quality Policy)	Water Quality Policy is second-level legislation and offers more specific protection for the state's waters. It prohibits the pollution of the stormwater system and our natural waters. The policy has general obligations which every person, business, and industry must comply with as well as specific obligations for particular activities.
Korea	Water Quality and Ecosystem Conservation Act	This act contained the stormwater runoff pollution level management from several business establishments which should follow the discharge standards set by the Ministry of Environment. In addition, this act also included setting control areas identified by the Ministry of Environment as areas which degrades the rivers, lakes and marshes ecosystem and are harmful to public health.

or diffuse pollution was only recognized after having proper management measures for point source pollution. Also, existing stormwater management policies have evolved from focusing solely on peak discharge control to the inclusion of stormwater quality management as well. Except for Germany, all developing countries listed have policies requiring the need for installing management systems such as low impact development (LID), sustainable drainage systems (SuDS), green infrastructures (GI), water-sensitive urban designs (WSUDS), and best management practices (BMP).

The introduction of the total pollutant load management system (TPLMS) in Article 4 of the 'Water Quality and Ecosystem Protection Act' in Korea has been applied since 2004. This was enacted due to the observation that despite the proper management of point source pollution, the concentrations of biochemical oxygen demand (BOD), chemical oxygen demand (COD), total organic carbon (TOC), total nitrogen (TN), and dissolved nitrogen (dissolved N) in the four major rivers of Korea did not show any improvements or even worsen as exhibited in Figure 10.1. Particularly, the dissolved N and nitrate (NO_3-N), which are readily available for phytoplankton uptake and affected algal bloom, were found to be increasing after 2015. While the phosphorus management in Korea has seen stable improvement since 2000, the TN/TP ratio, which is one of the factors being considered in evaluating the health of rivers, in the major rivers was found to be increasing. Considering that phosphorus management has reached its limit in watershed management for the four major rivers, nitrogen management should be an important viewpoint for managing eutrophication causing algal bloom. In addition, identification

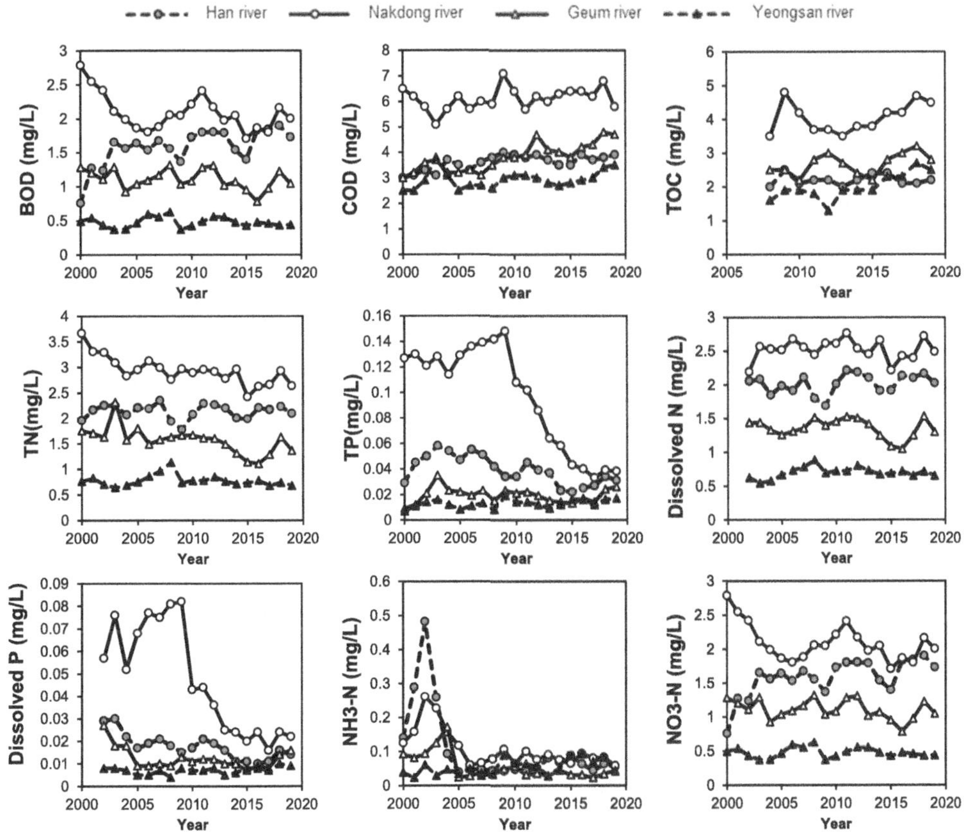

Figure 10.1 Annual water quality changes in four major rivers of Korea.

and monitoring of the effects of NPS pollution in large scale catchments are still undergoing challenges. The act was revised in 2012 to extend it to a third phase until 2020 to include other variables such as water-environment, weather, economic, and technologies that affect water quality. During the first phase of NPS management in Korea, policies were developed to properly manage NPS pollution. During the second phase, development, testing, and design of different technologies, such as LID and GI, were implemented. The third phase focused on the continuous development of LID and GI for cost-effective treatment.

10.2 URBAN, AGRICULTURAL AND LIVESTOCK LANDUSES

10.2.1 History, status and management strategies for urban land uses in Korea

Korea's urbanization rate has been steadily increasing since 1970s, reaching almost 80% with the observation of a rapid increase from the year 2000. Urban development in Korea began in 1960s alongside industrialization and economic development policies. Ulsan city was considered as the first 'new town' with a population of 150 000; however, the term 'new town' was only officially coined in the year 1970s. The creation of new industrial cities in 1970s was encouraged to promote heavy and chemical industries. During this period, Changwon city, having a population of 300 000, was designated as the first 'new town.' Since then, the creation of new towns was expanded to Daedeok Research Institute town, Yeocheon Industrial city, Gumi Industrial complex, Seoul Gangnam new town, and Gwacheon-Banwol new town. In the 1980s development of 'new town in towns' was promoted in Mokdong and Sanggye-dong. During this decade, two million houses were constructed to allow the creation of the first new towns in other areas including Bundang, Ilsan, Pyeongchon, Sanbon, and the middle eastern part of Korea. In the 1990s, in response to the criticism of the simultaneous development of large-scale new towns and the development of small-scale decentralized housing sites, semi-agricultural and forest land development started, allowing counties and nearby large cities to be developed. This scheme resulted in reckless development which in turn damaged the natural ecosystem. Urban planning and development were promoted in 2000 to replace the negative image of new towns and small decentralized development. During this decade, the second phase of the new town development included Seongnam, Pangyo, Hwaseong, Dongtan, Gimpo, Han River, Paju, Unjeong, Suwon, Gwanggyo, Yangju, Wirye, Godeok, Incheon, Asan, Daejeon, and Duan.

The expansion of impermeable areas due to urban development distorts the natural water circulation, causing various problems such as increased stormwater runoff, generation of non-point pollutants, depleted groundwater levels, and increased urban drought and urban flooding frequency. According to the Ministry of Environment, Seoul city, which has about 54% imperviousness rate, was identified to have the highest imperviousness rate among the metropolitan cities, followed by Busan, Gwangju, Daegu, Incheon, and Daejeon. As shown in Figure 10.2, the imperviousness rate of metropolitan cities, excepting Ulsan and Sejong, exceeded 20%. Considering the national provincial level, Gyeonggi-do's imperviousness rate is 22.7% wherein eight of the local governments, excluding Seoul and Mokpo, which were located in Gyeonggi-do, were included in the top ten cities with the highest imperviousness rate in Korea. Particularly, the imperviousness rate of Bucheon city, Suwon city, Gwangmyeong city, Osan city, and Anyang city were 61.7%, 49.3%, 43.9%, 39%, and 39%, respectively.

Managing stormwater has a deep-rooted history dating back to 3000 BC termed as urban drainage (Burian & Edwards, 2002). Through the years, different terms have been associated with urban stormwater management (Fletcher *et al.*, 2015). Most of these terms were from developed countries that identified the need to control NPS pollution to properly manage surface and groundwater quality. The utilization of nature-based solutions (NBS) including LID technologies, a multi-beneficial stormwater management approach that connects the ecosystem with urban revitalization, has been extensively practiced to address the potential risks of mismanaged urban stormwater runoff. As such, the management of urban drainage has become significantly complex over the past few decades, shifting from focused approaches to a multi-beneficial approach where several objectives

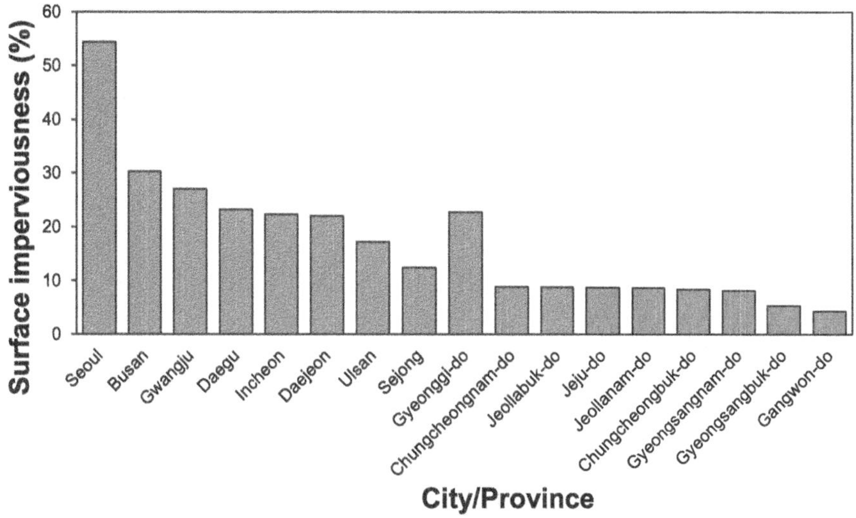

Figure 10.2 Surface imperviousness of metropolitan cities and provinces in Korea.

drive the design and decision-making processes (Fletcher *et al.*, 2015). LID technologies and GI were proven to be effective in mitigating the detrimental effects of urban stormwater runoff. The operations and treatment of these technologies occur via specific flow and environmental process pathways. Since stormwater runoff has high uncertainties due to environmental factors, an appropriate design approach requires identifying the treatment mechanisms occurring inside each system. LID practices including decentralized technologies control stormwater from the source to mimic the pre-developed water cycle through mechanisms of microscale stormwater storage, increased infiltration, and lengthening of the flow path and runoff time (Damodaram *et al.*, 2010). In addition, understanding the mechanisms affecting the fate of pollutants in a stormwater treatment system still needs investigation for optimum design and performance of stormwater treatment systems. The use of GI is a multi-beneficial stormwater management approach that connects the ecosystem with urban revitalization considering political and socio-economic constraints (Keeley *et al.*, 2013). Fluctuating flow and environmental conditions affect the operation and treatment performance of GI (Davis *et al.*, 2010).

Different manuals have been established in different countries to guide designers in applying appropriate stormwater treatment technologies in a catchment. These manuals have been used for the implementation of policies including climate action, water health, and disaster risk management (Greenbelt Foundation and Green Infrastructure Ontario Coalition, 2017). The commonly used technologies used as substitutes to conventional social infrastructures in Korea included infiltration planters, bioretention facilities, green roofs, tree box filters, rain gardens, permeable pavements, and constructed wetlands summarized in Figure 10.3. Developed countries have different focus and purposes for utilizing stormwater management technologies since stormwater has been identified as a key cause of pollution and the declining health of waterways (Melbourne Water, 2013). Policy and government support were key drivers for the full application of these techniques in several developed countries. Treatment levels of stormwater technologies vary depending on the size of contaminants in stormwater which may be addressed by a specific type of stormwater treatment technology. However, the application of these stormwater treatment techniques has been well known not only due to their treatment function but also for their capability of delivering a broad range of ecosystem services (Natural England, 2009). The application of these technologies varies on the scale of catchment being drained. Source controls include pervious pavements and green roofs while site controls include

Figure 10.3 Transition of conventional social infrastructures to green infrastructures (Modified from Kim, 2019a).

infiltration systems, bioretention, rain gardens, and tree box filters. Regional-scale treatment usually employs wetlands and retention ponds. As a result of the application of these nature-based solutions, creation of water circulation cities were enacted wherein cities including Gimhae, Andong, Gwangju, Ulsan and Daejeon were selected as Demo cities in 2016.

10.2.2 Agricultural and livestock industries in Korea

In 2016, World Bank reported that 17.4% of the total land area of Korea is dedicated to agricultural activities where high organic fertilizers are being used. Specifically, high nutrient concentration in water bodies leads to excessive algal growth and affects the level of organics since nitrogen (N) is essential to the production of plant and animal tissue while phosphorus (P) is vital for converting sunlight into usable energy and essential to cellular growth and reproduction. In general, the outflow of nutrients discharged from sewage and wastewater has greatly decreased since the introduction of total phosphorus treatment facilities in the 2000s. However, the occurrence of green algae and fish kill is increasing every year, and the cause was identified to be the agricultural and livestock NPS. As of 2018, rice and paddy fields accounted for 11% and 8%, respectively, of the total land area in Korea. The impact on water quality caused by the excessive use of fertilizers is increasing due to the increase in annual crop production per unit area. Increasing the amount of nutrients such as compost and fertilizer in agricultural areas and advanced agricultural technologies such as facility cultivation and water cultivation have contributed to the increase in land efficiency of cultivated land, but surplus nutrients have a great influence on water quality and aquatic ecology (Kim, 2019b). Figure 10.4 shows

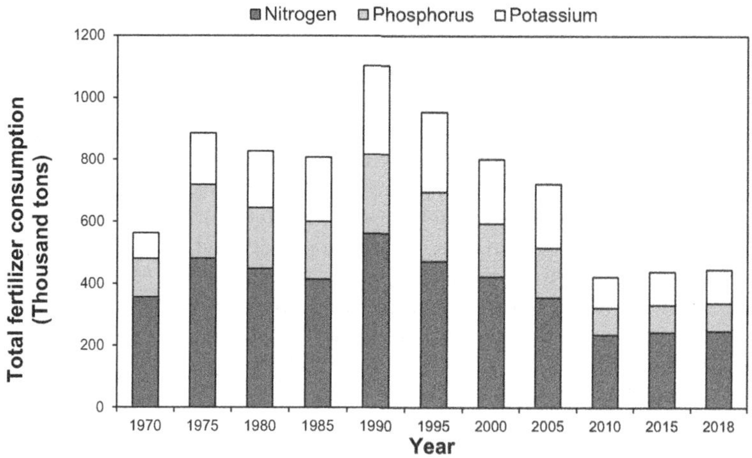

Figure 10.4 Fertilizer consumption in Korea.

the trend of annual chemical fertilizer consumption in Korea since 1970. The total consumption of chemical fertilizers reached its peak in the 1990s, then declined rapidly, and has stabilized at about 430 000 tons per year since 2010. Since 2005, nitrogen fertilizer consumption has been about 250 000 tons per year while the phosphorus-type fertilizers have been maintained at 100 000 tons per year. Potassium fertilizers, on the other hand, have been maintained at around 110 000 tons per year. Fertilizer consumption per unit area peaked at a maximum of 457 kg/ha in the 1990s and is currently maintained at around 250 kg/ha as exhibited in Figure 10.5. The decrease in total fertilizer consumption is also due to the development of agricultural technology. However, the decrease in cultivated land and increased use of organic compost made from livestock manure were identified as the main reasons for the decrease in total fertilizer consumption. The rice cultivation area, which

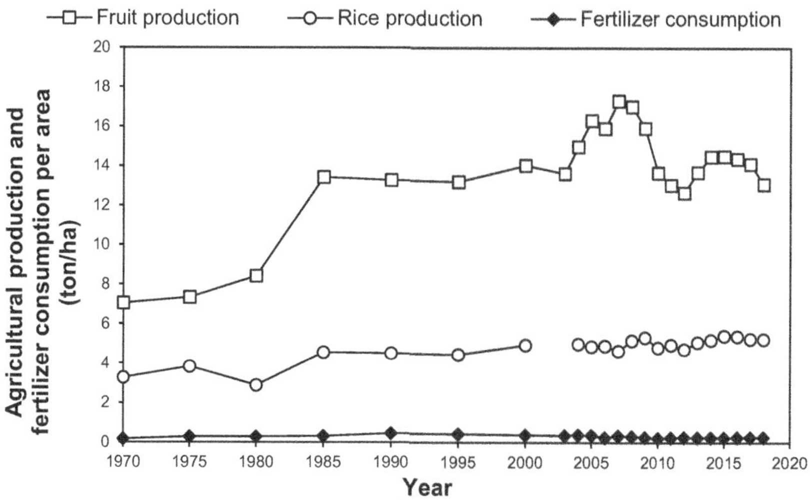

Figure 10.5 Agricultural production and fertilizer consumption per area in Korea.

occupies the largest area among cultivated land in agricultural areas, has been maintained at around 1.2 million ha, but has decreased since 1990, and is currently being cultivated at about 800 000 ha with the rice production of about 3.9 million tons. However, the country's fruit cultivation area has been steadily increasing, reaching about 2.8 million ha in the late 2000s, and about 200 000 tons of fruit are being produced annually. Considering the production area and production volume of fruits over the past 20 years, apples and peaches have increased while pears and grapes have decreased. In the last 50 years, vegetable production has been increasing significantly. These changes in the cultivation area and production volume of each crop in the agricultural area were greatly affected by urbanization, rural aging, and changes in the cultivation patterns which had great effects on the water environment and water quality in the watershed.

In Korea, the growth in the number of livestock animals has intensified since 1970 to meet the food demand related to economic and population growth. As exhibited in Figure 10.6, the livestock industry has increased significantly. In particular, the number of pigs was about 1.2 million in 1970, but now exceeds 11 million, showing a growth rate of about 9 times over 50 years. The number of Korean cattle has also steadily increased and now exceeds 3 million, and the number of dairy cattle has steadily increased to more than 400 000. The number of chickens raised rapidly increased from about 30 million in the 1970s to about 180 million. Looking at the characteristics of the livestock industry, although the number of livestock areas is greatly decreasing, the number of reared animals is increasing significantly, posing a serious impact on the water environment (Kim, 2019c; Ministry of Agriculture, Forestry and Livestock, 2019). In the past, livestock manure was managed through purification treatment, resource conversion, and marine discharge. Due to the total ban on ocean dumping of organic wastes including animal manure in the late 2000s, there has been no ocean emission since 2012. As of 2018, about 70.5% of livestock manure was managed as resources, about 23.7% is consigned, and about 5.6% is purified. As a result of the government's policy of recycling livestock manure, about 130 000m^3/day of compost and liquid manure was used in agricultural areas. Composts originating from animal manure is recognized as an organic fertilizer utilized on most farmlands except for water supply protected areas. However, due to the fertilization issues concerning the maturity of organic compost, problems such as the occurrence of green algae and surface and groundwater contamination occurred. According to the OECD Agricultural Environment Efficiency Report (2013), the nitrogen (N) concentration in the cultivated soil in Korea was the highest among OECD countries, while Japan had

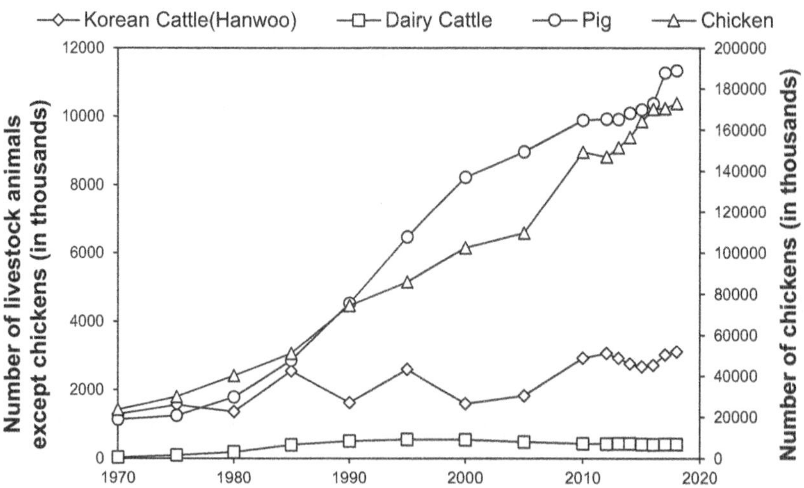

Figure 10.6 Changes in livestock animal count in Korea from 1970 to 2018.

the highest phosphorus (P) concentration. This high nutrient distribution in the soil resulted from the use of organic compost from livestock manure and excessive use of inorganic fertilizers. In particular, organic composts originating from domestic livestock manure are discharged to an extent that exceeds the annual total nutrient distribution required by domestic farmland. It is, therefore, necessary to implement strong measures against non-point pollution in agricultural and livestock areas.

10.3 IMPACT OF LULUC ON WATER QUALITY IN LARGE CATCHMENTS: CASE OF JEONJU STREAM AND SEONGJU ICHEON STREAM

Two streams were monitored automatically to identify the effects of land use to water quality in Korea. Specifically, Jeonju stream is the model watershed to evaluate the effects of increasing urban areas to nearby water bodies while Seongju Icheon stream is the model watershed to assess the effects of increasing agricultural areas to nearby water bodies. The 16.35 km Jeonju stream in Jeonju city, Jeollabukdo, Korea is composed of 64%, 12%, 9% and 6% forest area, agricultural area, urban area and paddy field, respectively. The monitoring stations in Jeonju stream were demonstrated in Figure 10.7. Jeonju stream has a drainage area of 455.3 km^2. On the other hand, Seongju Icheon stream in Sungju County, Gyeongsang Province in South Korea covers a watershed area of 127.23 km^2 with a reach of 23.0 kilometres. The whole catchment of Seongju Icheon stream was mainly composed of 67%, 14%, 5% and 3% forest area, agricultural area, paddy field and urban area, respectively as exhibited in Figure 10.8. The catchment of Seongju Icheon stream has an industrial complex, sewage treatment plant, and wastewater treatment plant which discharges to the stream.

The monitored rainfall events in Jeonju stream were ranging from 0.4 to 117 mm with mean ± standard deviation of 37.88 ± 29.13 mm. 61% of monitored rainfall events were conducted in summer season while

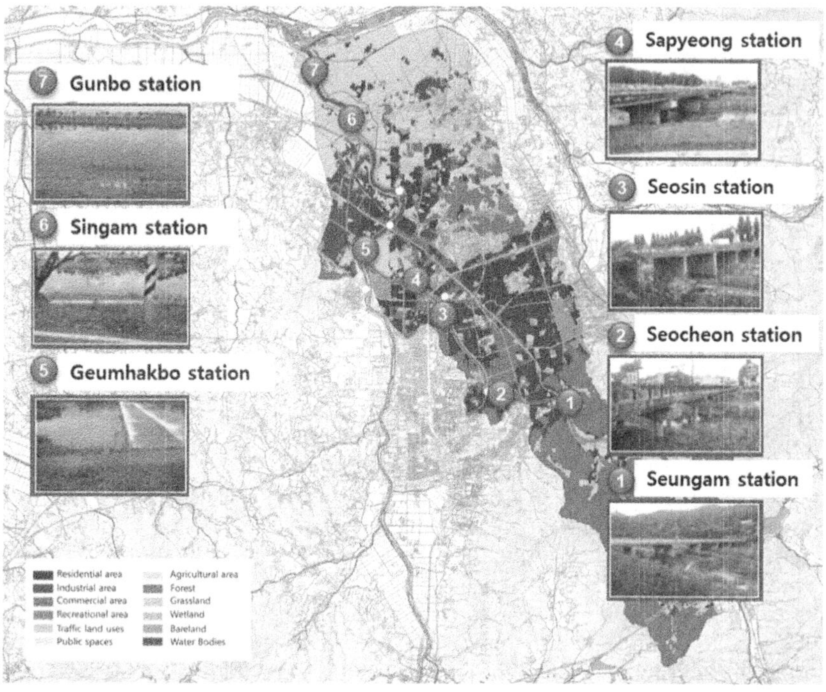

Figure 10.7 Land use map of Jeonju stream.

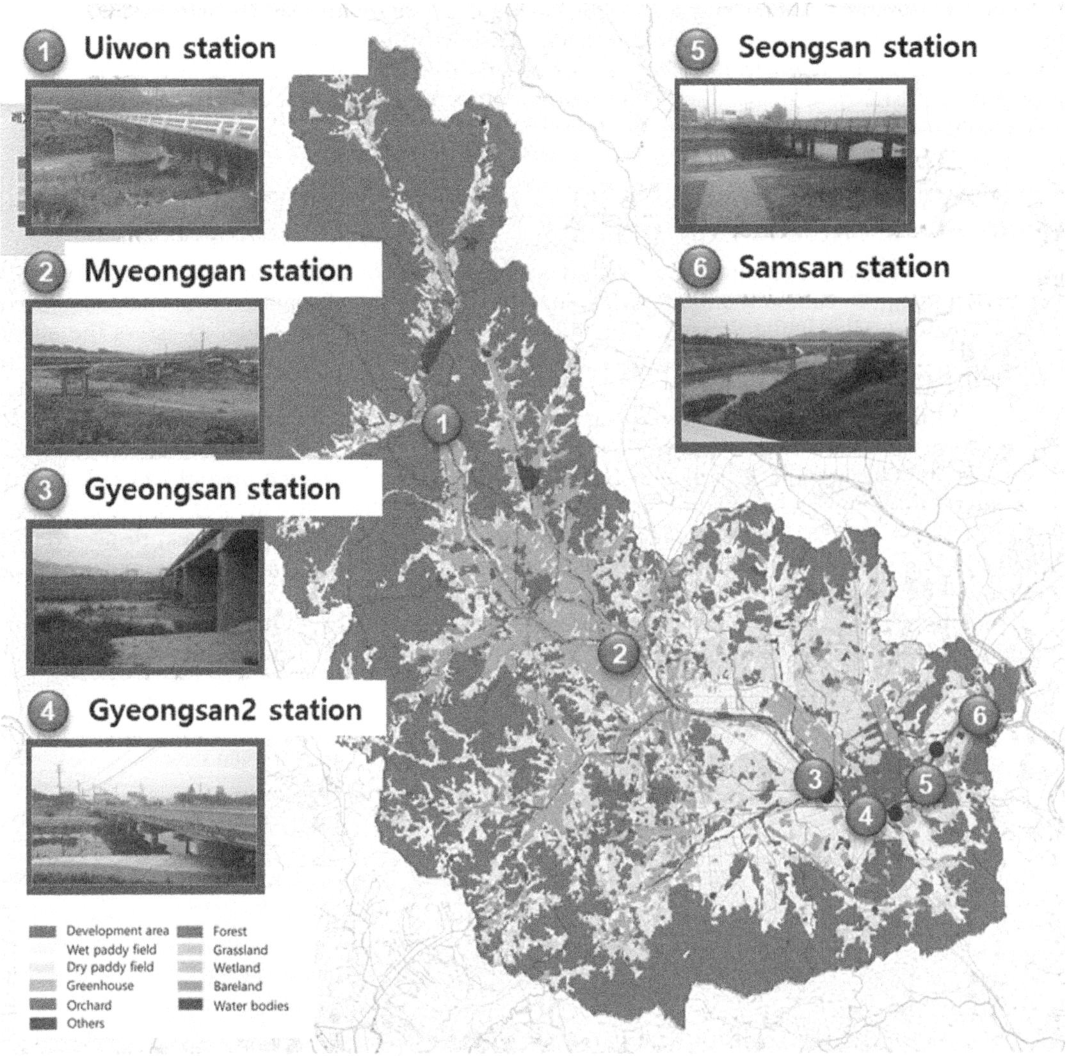

Figure 10.8 Land use map of Seongju Icheon stream.

35% and 4% of the monitored rainfall events occurred in fall and winter season, respectively. From 2015 to 2017 36% to 64% of the rainfall occurred during summer season whereas 13% to 30% and 8% to 14% occurred in autumn and winter, respectively. The antecedent dry days (ADD) of the monitored rainfall events ranges from 0.3 to 19 days with mean ± standard deviation of 5.43 ± 4.47 days. The longest ADD was observed in the rainfall event monitored during fall season whereas the shortest occurred in summer season which was considered as the wettest season in Korea. The duration of rainfall events monitored ranges from 0.3 to 56.9 hours with mean ± standard deviation of 13.5 ± 11.9 hours while the rainfall intensity was observed to be 0.26 to 27.25 mm/hr mean ± standard deviation of 4.78 ± 5.49 mm/hr.

The monitored rainfall events in Seongju Icheon stream were slightly greater compared to Jeonju stream ranging from 3.58 to 146 mm with mean ± standard deviation of 41.26 ± 36.64 mm. Among the monitored rainfall events, 66% were conducted in summer season while the remaining 29% and

5% of the monitored rainfall events were during fall and winter season, respectively. From 2015 to 2017 39% to 60% of the rainfall occurred during summer season whereas 20% to 34% and 5% to 11% occurred in autumn and winter, respectively. The antecedent dry days (ADD) of the monitored rainfall events in Seongju Icheon stream were in similar with that of Jeonju stream which ranges from 0.4 to 19.3 days with mean $\pm$ standard deviation of 4.56 ± 3.45 days. Similarly, the longest ADD was also observed in a rainfall event monitored during fall season whereas the shortest occurred in summer season. The duration of rainfall events monitored ranges from 0.3 to 56.9 hours with mean $\pm$ standard deviation of 13.5 ± 11.9 hours while the rainfall intensity was observed to be 0.26 to 27.25 mm/hr mean $\pm$ standard deviation of 4.78 ± 5.49 mm/hr.

The pollutant load in the rivers and streams during rainfall depends on several factors, for example, watershed characteristics such as watershed area and land use, and rainfall characteristics including rainfall depth, rainfall intensity and antecedent dry days. Most of the pollutant load in the upstream station of Jeonju stream were low due to the land use of its watershed area which was mainly composed of forest areas. Sudden increase in pollutant loads were observed in Seocheon station due to the residential and urban land uses in the watershed area (Figure 10.9a). After Seosin station, pollutant load decreased due to the inflow of one of the big tributaries of Jeonju stream thereby resulting to dilution effect. While it was apparent that the pollutant load usually showed an increasing trend from Seungam to Seocheon, one of the most notable recorded increase was the event in 21 August 2017. This event has a recorded rainfall depth of 68 to 72 mm. The sudden increase in flow resulted to a sudden and too much increase in pollutant load per unit area in Seocheon station. Another notable increase was observed in 26 June 2017 when the SS load rose up to greater than 1 600 000 kg/km^2. This rainfall event has depth ranging from 66.3 to 99.5 mm with an average rainfall intensity of 6.85 to 10.12 mm/hr. The pollutant load tend to increase as the stormwater runoff from urban land uses were discharged to the rivers however the pollutant load tend to decrease in Geumhakcheon station, Singam station and Gunbo Station due to dilution effect of the nearby wastewater treatment facilities.

Another reason for the decrease in pollutant load was probably due to sedimentation due to long residence time in the river and relatively large catchment area. In addition, the low flow rate in Singam station may have been due to the bridge in the area prolonging residence time of pollutants while increasing the tendency for sedimentation due to gravity. On the other hand, most of the pollutant load per unit area in the Seongju Icheon stream from Uiwon station to Samsan station has relatively same trends with peak values usually observed in Myeonggan station and Seongsan station which have drainage area of 82 and 125 km^2, respectively (Figure 10.9b). The pollutant load increase in Seongsan station was due to a nearby wastewater treatment facility which has by-pass flow during rainfall events. In 2017, the rainfall which occurred on 11 September was the most notable because of the stable increase of BOD and SS load per unit area from Uiwon station to Samsan station. Due to the dilution effect that may have been caused by the discharge of the nearby wastewater treatment facility, TN and TP load per unit area were decrease from Seongsan station to Samsan station. For rivers and streams with watershed areas mostly composed of rural and agricultural land uses, the pollutant load does not increase rapidly during rainfall compared to urban land uses since the natural state of the area was still preserved. Naturally, the pollutant load per unit area in rivers and streams increases in areas where urban land uses were located and decreases when there is no additional pollutant load input from non-point source pollution due to its self-purification capabilities.

The BOD contribution due to non-point sources to Jeonju stream ranges from 30% to 98% during rainfall events as shown in Figure 10.10a. The SS contribution ratio due to non-point source pollution during rainfall ranges from 22% to 99% with an average of 87% in the downstream of Jeonju stream, Gunbo station. On the other hand, The TP contribution to Jeonju stream ranges from 34% to 96%. The TN contribution ratio due to non-point source pollution during rainfall ranges from 14% to 97% with an average of 74% in the downstream of Jeonju stream, Gunbo station. Tributaries of Jeonju stream contributed 20% to 97% of TN due to non-point source pollution during rainfall. It was identified that

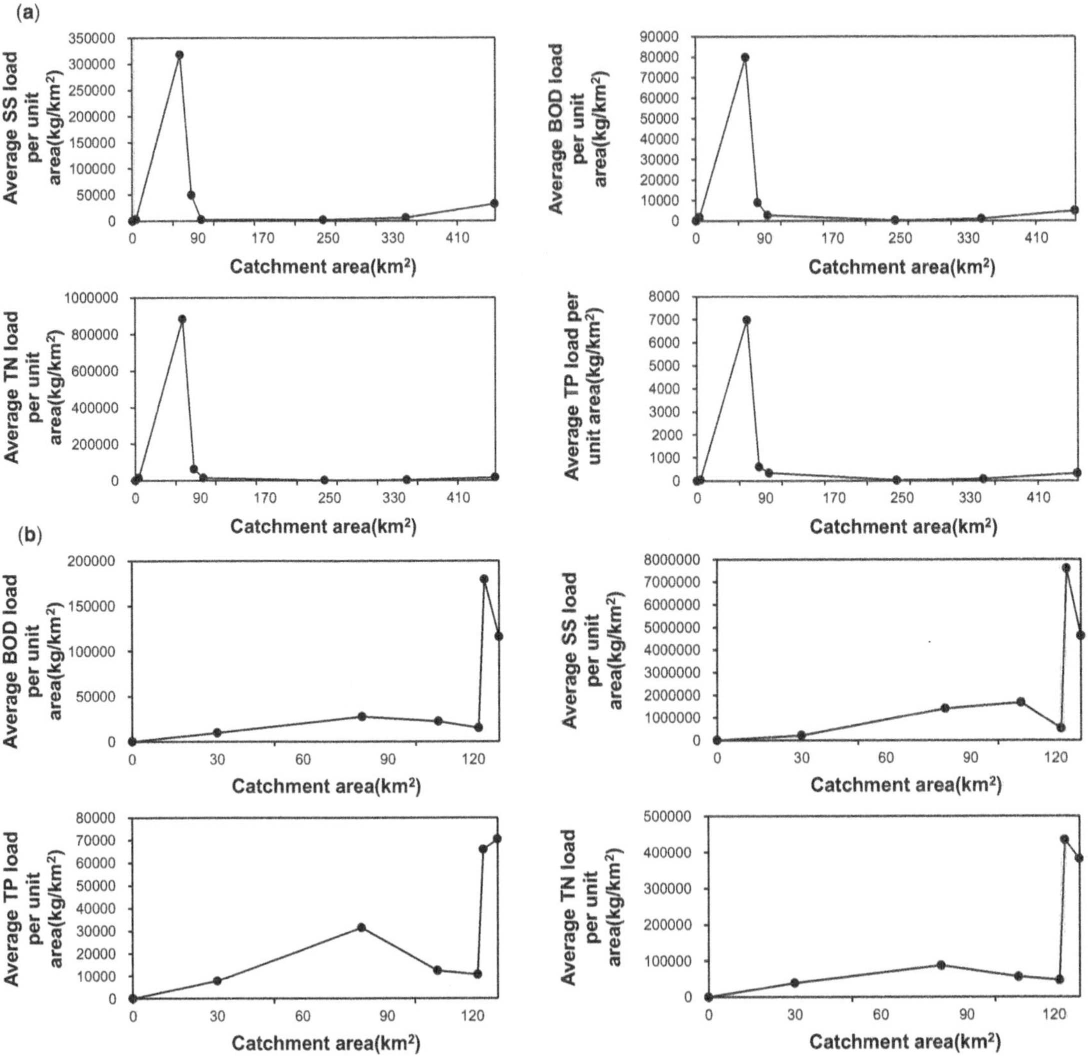

Figure 10.9 Variation of pollutant load per unit area with respect to catchment area in (a) Jeonju stream and (b) Seongju Icheon stream.

the tributaries of Jeonju stream contributed higher non-point source pollutants. This was attributed to the industrial complexes influencing water quality in the tributaries. In Geonsancheon, the relatively large amount of SS was due to the on-going construction. These findings also suggested that various contaminants were adsorbed in SS concentration thereby increasing its total concentration. BOD, SS, TN and TP contribution ratio was found to have an increasing trend from Seungam station to Sapyeong station after which, the contribution rate was found to have variable trend attributed to the dilution effect brought about by the flow from the tributaries.

The contribution rate of BOD due to nonpoint pollution sources during rainfall events varied but ranges from 36% to 96% in Seongju Icheon stream according to Figure 10.10b. Generally, there was

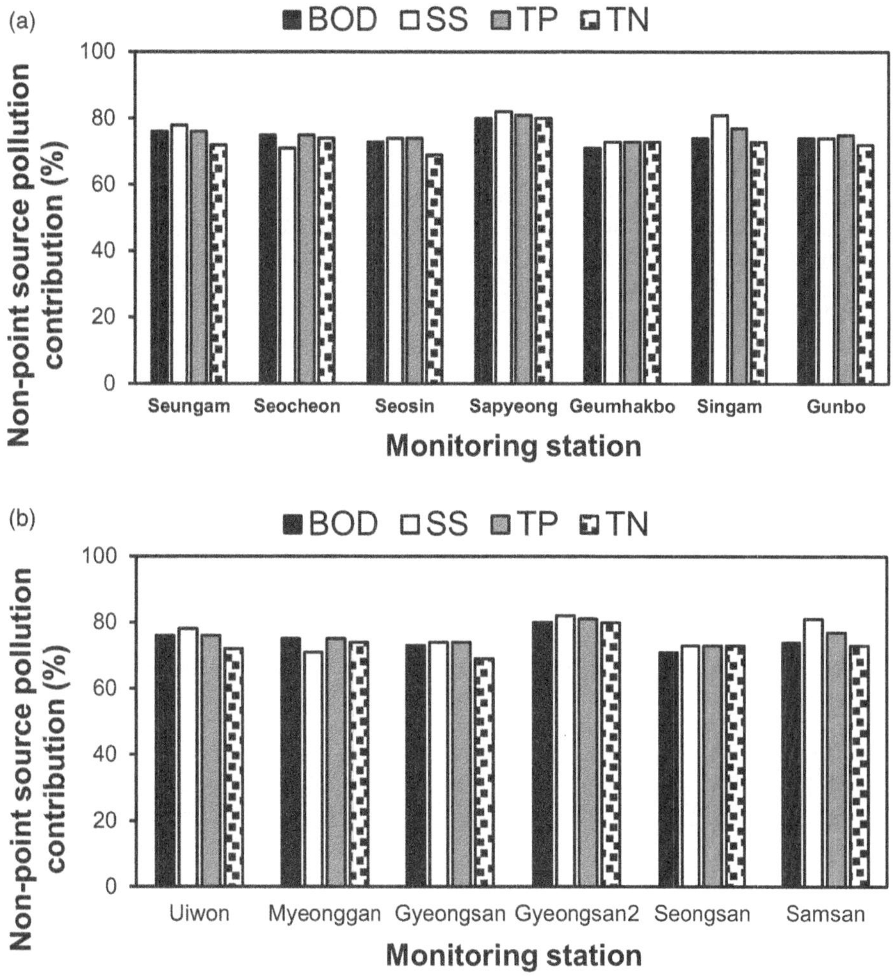

Figure 10.10 Non-point source pollution contribution per monitoring station of (a) Jeonju stream and (b) Seongju Icheon stream.

in increasing trend of contribution ratio from upstream to downstream except for the fluctuations in Gyeongsan station and Gyeongsan2 station. SS, TP and TN concentration ranges were 33% to 100%, 25% to 98% and 23% to 93%, respectively. The average non-point source contribution ratio observed in Samsan station were 81%, 77%, 74%, 73% and 73% for SS, TP, BOD, TN and TOC, respectively. Apparently, 92 to 95% occurrence frequency of the non-point source contribution ratio was greater than 50% implying that the stormwater runoff has great influence on water quality deterioration of the river. However, the mainstream of the river was observed to have a higher tendency to have greater nonpoint pollution contribution due to the urban area located midstream to downstream of the rivers compared to upstream, which were composed of forests. Another cause of higher non-point contribution rate was the discharge from the tributaries receiving stormwater runoff from industrial watershed. The relatively high non-point source contribution ratio of Seongju Icheon stream was mainly due to agricultural land uses.

10.4 OTHER POLLUTANTS OF EMERGING CONCERNS IN DIFFERENT LAND USES

Impervious land uses have been identified as a source of different pollutants being washed off with stormwater runoff during rainfall events. Summarized in Figure 10.11 are the comparison of heavy metals event mean concentration (EMC) in stormwater runoff from different impervious areas. Motor vehicle emissions, drips of crankcase oil, vehicle tire wear, asphalt road surfaces, domestic fire emissions, spillage or deliberate dumping of waste oil, body paint, wheel balance weights, commercial car washing facilities and corrosion of roofing materials are all diffuse sources of PAH and heavy metal contaminants in urban environments (Brown & Peake *et al.*, 2006; Müller *et al.*, 2020). Only the Zn EMC of roads amounting to 0.1 ± 0.08 mg/L was found to be significantly greater than parking lot by 21% ($p < 0.05$). For Cd EMC, road contained 0.1 ± 0.07 mg/L was found to be significantly greater by 27% and 33% compared to combined road and parking lot, and road, respectively ($p < 0.05$). No significant difference was found for Pb EMC from all impervious catchment types ($p > 0.05$). TZn EMC from roofing, amounting to 0.2 ± 0.11 mg/L, was found to be significantly less than road and combined road and parking lot by 64% and 55%, respectively ($p < 0.001$). Consequently, TCd EMC of roads, amounting to 0.1 ± 0.09 mg/L, was found to be significantly greater than the combined road and parking lot amounting to 0.1 ± 0.08 mg/L ($p < 0.05$), while the combined road and parking lot was found to be significantly greater than roof amounting to 0.1 ± 0.05 mg/L ($p < 0.01$). Lastly, TPb EMC from roofs amounting to 0.3 ± 0.22 mg/L was also found to be significantly greater than parking lot and combined road and parking lot by 1.6 and 2.3 times, respectively ($p < 0.05$). TPb EMC of road amounting to 0.3 ± 0.3 mg/L was also found to be significantly greater than parking lots, amounting to 0.1 ± 0.1 mg/L ($p < 0.01$). It was also found that Zn was mostly particulate bound for road and combined road and parking lot. On the other hand, Cd from all the impervious land uses was found to be in the dissolved phase, while for Pb only roofing was found to be in the particulate phase. These findings were in conjunction with the findings of other studies that a considerable ratio of heavy metals from urban stormwater runoff was in dissolved phase (Ladislas *et al.*, 2012; Li *et al.*, 2012; Rangsivek & Jekel, 2005). These findings were especially useful in selecting the media to be used in stormwater LID

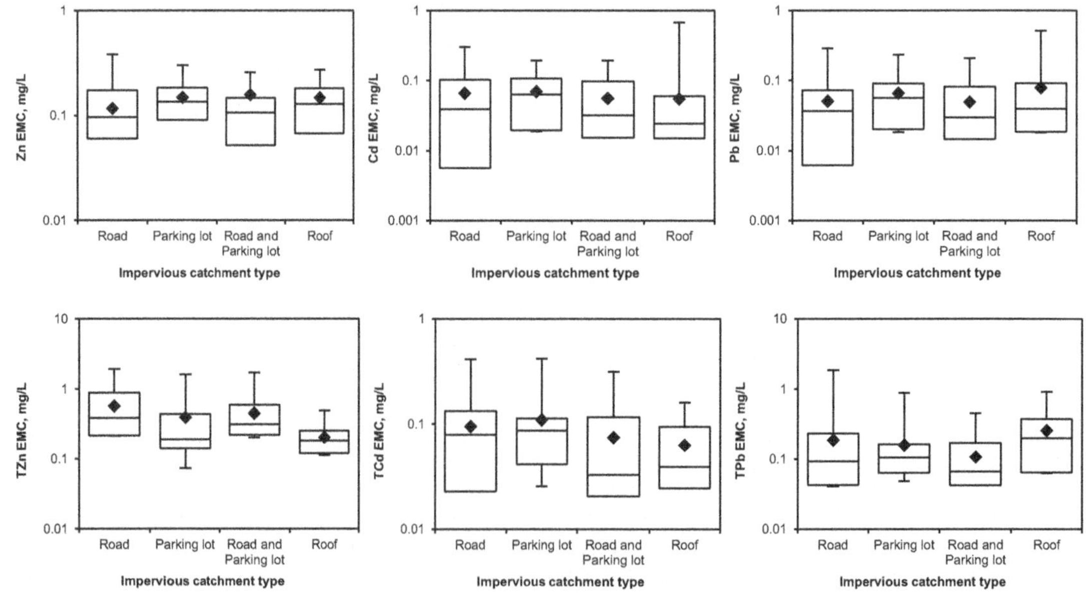

Figure 10.11 Comparison of heavy metal event mean concentrations (EMC) from different impervious land uses.

technologies. Specifically, filter materials with high pH, low organic matter, and large surface areas such as sand may be used for adsorbing dissolved metals (Søberg *et al.*, 2019). However, this finding just considered the physico-chemical removal mechanism via filter media disregarding the contribution of microorganisms and plants through which biosorption and phytoremediation, respectively of heavy metals may occur. This will imply that a higher possibility of clogging that decreases infiltration and filtration may be expected from filter media since the proper microbial environment was not considered (Seki *et al.*, 1998). Therefore, selection of filter media that considered both physico-chemical and biological treatment mechanisms should be considered for a better LID performance.

Nakdong river, one of the four major rivers in Korea, is receiving effluent from agricultural areas and industrial complexes that causing emerging concerns for different types of pollutants emitted by these land uses in the south eastern part of Korea (Lee *et al.*, 2017). Table 10.2 shows the concentrations of polycyclic hydrocarbons (PAHs) and polychlorinated biphenyls (PCBs) in Nakdong river estuary. Polychlorinated biphenyls (PCBs) are members of the group of halogenated aromatic

Table 10.2 Concentrations of polycyclic hydrocarbons (PAHs) and polychlorinated biphenyls (PCBs) in sediments from barrier islands in the Nakdong River Estuary, Korea (Adapted from: Lee *et al.*, 2017).

Compound	Range	Mean	S.D.	Compound	Range	Mean	S.D.
	(ng g^{-1} dry wt)	(ng g^{-1} dry wt)	(ng g^{-1} dry wt)		(pg g^{-1} dry wt)	(pg g^{-1} dry wt)	(pg g^{-1} dry wt)
Naphthalene	ND*	–	–	PCB (#8)	ND*–86.0	86.00	0.00
Acenaphthylene	ND*–3.1	1.28	0.88	PCB (#18)	ND*–1.0	1.00	0.00
Acenaphthene	ND–3.7	2.61	0.75	PCB (#52)	ND*–33.0	33.00	0.00
Fluorene	0.13–12.5	4.96	3.84	PCB (#44)	ND*–2.0	2.00	0.00
Phenanthrene	1.86–68.4	31.19	21.56	PCB (#101)	ND*–3.0	2.00	0.63
Anthracene	0.27–4.0	2.01	1.16	PCB (#87)	ND*	–	–
Fluoranthene	2.17–41.3	15.18	11.68	PCB (#110)	ND*–7.0	4.00	3.00
Pyrene	1.80–55.8	15.95	15.57	PCB (#118)	ND*–4.0	2.50	1.50
Benzo[a] anthracene	0.28–5.4	1.82	1.4	PCB (#153)	ND*–2.0	2.00	0.00
Chrysene	0.59–10.5	3.23	2.79	PCB (#105)	13.00–120.0	50.10	40.59
Benzo[b] fluoranthene	ND*–8.6	3.01	2.68	PCB (#138)	ND*	–	–
Benzo[k] fluoranthene	ND–2.2	1.01	0.62	PCB (#187)	ND*	–	–
Benzo[a]pyrene	ND*	–	–	PCB (#128)	ND*	–	–
Indeno(12,3-CD) pyrene	ND*	–	–	PCB (#180)	ND*–120.0	120.00	0.00
Dibenz(a,h) anthracene	ND*	–	–	PCB (#170)	ND*–298.0	260.33	33.81
Benzo[ghi] perylene	ND*	–	–	PCB (#194)	ND*	–	–
				PCB (#205)	ND*–431.0	268.17	78.56
				PCB (#206)	ND*–172.0	172.00	0.00
				PCB (#209)	ND*–1.0	1.00	0.00
ΣPAHs	10.76–167.73	65.22	45.71	ΣPCBs	197.0–754.0	371.60	178.24

hydrocarbons (HAHs) and consist of 209 isomers and congeners with different numbers and positions of chlorine atoms substituted on the biphenyl moiety (Giesy & Kannan, 1998). Natural causes and human activities have been proven to affect the generation and release of PAH into the environment causing carcinogenic and mutagenic effects to different organisms including human beings (Moon *et al.*, 2003). These PAHs and PCBs in sediments will eventually be released to bodies of water when not managed properly. In Korea, however, currently there are still insufficient exposure data to assess the harmful effects of both pollutant groups.

10.5 POLICIES AND GOVERNANCE

In Korea, the management of stormwater runoff began in 1963, wherein the water quality conservation act was established by the Ministry of Environment (MOE). Due to the increasing initiatives of the government to restore and improve the water quality of the four major rivers of Korea, the 'Four Major River Restoration Project' was established under the supervision of MOE between 1998 to 2000. The recent World Economic Forum's World Crisis Report (WEF, 2020) identified climate change, biodiversity reduction, natural disasters, environmental disasters caused by development and industrialization, and water crises as the top challenges. At a time when international warnings and country-specific responses are raised due to climate change and development, the establishment of the Fourth phase of 'Comprehensive Non-point Source Pollution Management Measures for the Four Major Rivers' is very relevant in Korea. Since NPS pollution is caused by various land uses, cooperation between different government agencies was deemed essential and highly encouraged. For this reason, the first (2004) and the second phase (2006) of 'Comprehensive Non-point Source Pollution Management Measures for the Four Major Rivers' was jointly established by related ministries led by the office of the prime minister. This legislation also allowed the strengthening of the implementation of the Total Maximum Daily Load (TMDL) program. In addition, the importance of legislation on the management of policies, pilot projects, treatment technologies, and designs and maintenance standards were emphasized, studied, and derived in the first and second phases. The third phase which started in 2012 emphasized the importance of management not just by the national government but of the local government as well. This encouraged the full project implementation with the local government as the lead agencies with the support of the national government. During the third phase, the legal measures and civil participation were also improved.

For the fourth phase, through the leadership of the office of the prime minister in partnership with the related government agencies, the government identified four major targets including urban areas, agricultural areas, forest areas, and management. During this phase, it is envisioned to have a 5% reduction in TP emission by 2025 or from 52.7 million tons/day to 20.1 million tons/day. Specifically, in urban areas, the imperviousness rate of areas with 25% and above imperviousness rate will be reduced from 40.3% imperviousness rate target in 2017 to 37.6% by 2030. For agricultural and livestock land uses, the NPS management areas were set to be increased from a total of nine areas in 2020 to 20 areas by 2025. For forest areas, the goal was to expand the management areas from 37.9 km^2 in 2020 to 47.2 km^2 by 2025. Lastly, a form of management encouraging citizen participation was planned to increase from seven areas in 2020 to 50 areas by 2025.

In the fourth phase, the management of urban areas will cover imperviousness rate reduction and water circulation rate improvement. There will also be a mandatory application of LID technologies, especially for urban development. The government will also expand its support and subsidies, not only to the local government but also to private companies as well. Lastly, a more proactive and comprehensive combined sewer overflow management will be implemented.

Since Korea is one of the countries with the highest agricultural and livestock activities, the fourth phase included management of pollution in livestock industries prior to wastewater discharge. Specifically, farmers will have self-sustaining farm villages with minimal NPS pollutant production. Water environment management will be more focused on the needs of a specific basin rather than regions and branches which will entail a more detailed approach for integrated water management. Lastly, the management strategies for NPS reduction will be implemented through different joint projects of the Ministry of Environment and Ministry of Agriculture, Food, and Rural Affairs.

Water quality and environmental problems in forest areas are usually caused by soil erosion from arable land and forest waste management process. Until now, the management of NPS pollution in forest areas has been a government-level land purchase policy for areas where erosion-by-products are commonly observed. However, a limited budg*et al.*so entails a limitation on the efficient management of soil erosion in forest areas, which accounted for more than 70% of the land area in Korea. In the fourth phase, the main policy is to prevent the occurrence of non-point pollution in advance by setting up a forest waterfront management area, and the management of illegal agricultural areas through cooperation among the different ministries.

It is important to establish management-based policies to address non-point pollution efficiently. In the third measure, non-point pollution management in areas where NPS pollutants are intensively discharged will shift from an application and support methods to designation and management methods. The NPS pollution management entity, which was previously under different ministries, will be integrated into the National Non-point Pollution Management Center. NPS pollutants subject to management will also be expanded to include emerging pollutants and microplastics apart from the typical water quality indicators (i.e., BOD, TP, and SS). The national non-point pollution measuring network established and operated on a trial basis shall be integrated into the water quality measuring network by reflecting the watershed-centered integrated management policy.

REFERENCES

Brown J. N. and Peake B. M. (2006). Sources of heavy metals and polycyclic aromatic hydrocarbons in urban stormwater runoff. *Science of the Total Environment*, **359**(1–3), 145–155, https://doi.org/10.1016/j.scitotenv.2005.05.016

Burian S. J. and Edwards F. G. (2002). Historical perspectives of urban drainage. In: Global Solutions for Urban Drainage (Proceedings of 9th International Conference on Urban Drainage), E. W. Strecker and W. C. Huber (eds.), ASCE, Reston, VA, USA, pp. 1–16.

Damodaram C., Giacomoni M. H., Prakash Khedun C., Holmes H., Ryan A., Saour W. and Zechman E. M. (2010). Simulation of combined best management practices and low impact development for sustainable stormwater management 1. *Journal of the American Water Resources Association*, **46**, 907–918, https://doi.org/10.1111/j.1752-1688.2010.00462.x

Davis A. P., Traver R. G. and Hunt W. F. (2010). Improving urban stormwater quality: applying fundamental principles. *Journal of Contemporary Water Research & Education*, **146**, 3–10, https://doi.org/10.1111/j.1936-704X.2010.00387.x

Ellis E. C., Klein Goldewijk K., Siebert S., Lightman D. and Ramankutty N. (2010). Anthropogenic transformation of the biomes, 1700 to 2000. *Global Ecology and Biogeography*, **19**, 589–606.

Fletcher T. D., Shuster W., Hunt W. F., Ashley R., Butler D., Arthur S., Trowsdale S., Barraud S., Semadeni-Davies A., Betrand-Krajewski J. L., Mikkelsen P. S., Rivard G., Uhl M., Dagenais D. and Viklander M. (2015). SUDS, LID, BMPs, WSUD and more – the evolution and application of terminology surrounding urban drainage. *Urban Water*, **12**(7), 525–542, https://doi.org/10.1080/1573062X.2014.916314

Giesy J. P. and Kannan K. (1998). Dioxin-like and non-dioxin-like toxic effects of polychlorinated biphenyls (PCBs): implications for risk assessment. *Critical Reviews in Toxicology*, **28**(6), 511–569, https://doi.org/10.1080/10408449891344263

Greenbelt Foundation and Green Infrastructure Ontario Coalition. (2017). A Green Infrastructure Guide for Small Cities, Towns and Rural Communities, Ontario, Canada. Retrieved from: https://d3n8a8pro7vhmx.cloudfront.net/greenbelt/pages/5202/attachments/original/1504021812/Green_Infrastructure_Final.pdf?1504021812 (accessed 24 August 2020).

Hassan Z., Shabbir R., Ahmad S. S., Malik A. H., Aziz N., Butt A. and Erum S. (2016). Dynamics of land use and land cover change (LULCC) using geospatial techniques: a case study of Islamabad Pakistan. *SpringerPlus*, **5**(1), 812, https://doi.org/10.1186/s40064-016-2414-z

Honkonen O. and Rantalainen A. L. (2013). Impact of urbanization on the concentrations and distribution of organic contaminants in boreal lake sediments. *Environmental Monitoring and Assessment*, **185**, 1437–1449, https://doi.org/10.1007/s10661-012-2643-8

Keeley M., Koburger A., Dolowitz D. P., Medearis D., Nickel D. and Shuster W. (2013). Perspectives on the use of green infrastructure for stormwater management in Cleveland and Milwaukee. *Environmental Management*, **51**, 1093–1108, https://doi.org/10.1007/s00267-013-0032-x

Kim L. H. (2019a). Advancement of Customized Design Technology for Green Infrastructure Facilities. Report for Korea Environmental Industry & Technology Institute, Seoul, Korea (In Korean).

Kim L. H. (2019b). Types and Management of Agricultural Non-Point Pollution. Seminar material for Chungnam Research Institute, Chungnam, Korea (In Korean).

Kim L. H. (2019c). Management of Agricultural Land in Catchment Area for Reducing Algal Bloom in Daecheong Lake. Geumgang Water System Management Committee Report (in Korean).

Kong F., Ban Y., Yin H., James P. and Dronova I. (2017). Modeling stormwater management at the city district level in response to changes in land use and low impact development. *Environmental Modelling & Software*, **95**, 132–142, https://doi.org/10.1016/j.envsoft.2017.06.021

Ladislas S., El-Mufleh A., Gérente C., Chazarenc F., Andrès Y. and Béchet B. (2012). Potential of aquatic macrophytes as bioindicators of heavy metal pollution in urban stormwater runoff. *Water, Air, & Soil Pollution*, **223**, 877–888, https://doi.org/10.1007/s11270-011-0909-3

Lee J. H., Woo H. J., Jeong K. S., Kang J. W., Choi J. U., Jeong E. J., Park K. S. and Lee D. H. (2017). Spatial distribution of polycyclic aromatic hydrocarbon and polychlorinated biphenyl sources in the Nakdong River Estuary, South Korea. *Journal of Environmental Science and Health, A*, **52**, 1173–1183, https://doi.org/10.1080/10934529.2017.1356207

Li W., Shen Z., Tian T., Liu R. and Qiu J. (2012). Temporal variation of heavy metal pollution in urban stormwater runoff. *Frontiers of Environmental Science & Engineering*, **6**, 692–700, https://doi.org/10.1007/s11783-012-0444-5

Liptan T. W. (2017). Sustainable Stormwater Management: A Landscape-Driven Approach to Planning and Design. Timber Press, Portland, OR, USA.

Melbourne Water. (2013). Water Sensitive Urban Design Guidelines. Retrieved from https://www.melbournewater.com.au/sites/default/files/South-Eastern-councils-WSUD-guidelines.pdf (accessed August 1, 2020).

Ministry of Agriculture, Forestry and Livestock. (2019). 2019 Major Statistics of Agriculture, Food and Rural Affairs. Ministry of Agriculture, Food and Rural Affairs. Registration number 11-1543000-000128-10 (In Korean). https://www.mafra.go.kr/mafra/293/subview.do?enc=Zm5jdDF8QEB8JTJGYmJzJTJGbWFmcmElMkY2OCUyRjMyOTQ5MiUyRmFydGNsVmlldy5kbyUzRg%3D%3D (accessed 1 August 2020).

Moon S. H., Lee M. G. and Kam S. K. (2003). Distribution and origin of polycyclic aromatic hydrocarbons (PAHs) in surface sediments inside Hallim Harbor of Jeju Island, Korea. *Journal of Environmental Science International*, **12**, 1145–1157, https://doi.org/10.5322/JES.2003.12.11.1145

Müller A., Österlund H., Marsalek J. and Viklander M. (2020). The pollution conveyed by urban runoff: a review of sources. *Science of the Total Environment*, **709**, 136125, https://doi.org/10.1016/j.scitotenv.2019.136125

Natural England. (2009). Green Infrastructure Guidance. Natural England, York, UK. Available online at: http://publications.naturalengland.org.uk/publication/35033 (accessed August 1, 2020).

Qiu Z. (2013). Comparative assessment of stormwater and nonpoint source pollution best management practices in suburban watershed management. *Water*, **5**, 280–291, https://doi.org/10.3390/w5010280

Rangsivek R. and Jekel M. R. (2005). Removal of dissolved metals by zero-valent iron (ZVI): kinetics, equilibria, processes and implications for stormwater runoff treatment. *Water Research*, **39**, 4153–4163, https://doi.org/10.1016/j.watres.2005.07.040

Seki K., Miyazaki T. and Nakano M. (1998). Effects of microorganisms on hydraulic conductivity decrease in infiltration. *European Journal of Soil Science*, **49**(2), 231–236, https://doi.org/10.1046/j.1365-2389.1998.00152.x

Søberg L. C., Winston R., Viklander M. and Blecken G. T. (2019). Dissolved metal adsorption capacities and fractionation in filter materials for use in stormwater bioretention facilities. *Water Research X*, **4**, 100032, https://doi.org/10.1016/j.wroa.2019.100032

WEF. (2020). The Global Risks Report 2020. World Economic Forum, Geneva, Switzerland.

WWAP (United Nations World Water Assessment Programme)/UN-Water. (2018). The United Nations World Water Development Report 2018: Nature-Based Solutions for Water. UNESCO, Paris.

doi: 10.2166/9781789061123_169

Chapter 11

Protected areas and the diffuse pollution problem: rivers of the Kruger National Park, South Africa

E. S. Riddell[1,2], L. Boyd[3], R. M. Petersen[1] and R. G. M. Heath[3]

[1]South African National Parks, Skukuza, Kruger National Park, South Africa
[2]Centre for Water Resources Research, University of KwaZulu-Natal, Pietermaritzburg, South Africa
[3]Golder Associates Africa (Pty) Ltd, Midrand, South Africa

Keywords: Aquatic ecosystems; diffuse pollution; Inkomati; Kruger National Park; Limpopo; Mozambique; protected areas; rivers; South Africa; water quality

11.1 INTRODUCTION

National parks and game reserves, typically large tracts of land set aside for the protection of landscapes and ecosystems began in the late 19th century in North America and soon followed in many parts of Africa during the colonial period when there was a surge in the transformation of pristine landscapes through development in the form of towns, industry, farms, mining and forestry plantations. Parts of what is now known as the Kruger National Park (KNP) in South Africa were set aside as early as the 1890s, including the Sabie (established 1898) and Shingwedzi Game Reserves (established 1903), before being joined to form the KNP in 1926. These protected areas were established to protect their abundant wildlife from hunting and to retain natural spaces for the benefit and enjoyment of society, and for the most part are now maintained through tourism revenues. Typically, these parks, including the KNP, have gone through different periods of management, sometimes including various degrees of ecosystem manipulation such as prescribed burning or artificial water provision for game.

Ostensibly these protected areas are situated within one or several watersheds. The KNP itself straddles the downstream reaches of several large watersheds in international river basins, and these river systems are key aesthetic and biodiversity assets of the protected area estate. Being situated downstream in these river basins means impacts on the water resource can be significant and in a multitude of forms, including depleted river flows due to upstream abstraction for irrigated agriculture, domestic use, industry and mining. Such impacts were recognised as being detrimental to park ecosystems during the 1980s (Joubert, 1986), and this provided the impetus for the impactful Kruger National Park Rivers Research Program, which ran through three phases during the 1990s (Breen *et al.*, 2000; Venter & Deacon, 1995). The outcomes of this multidisciplinary research endeavour continue to guide and inform the KNP's freshwater management program, notably through promoting ecosystem thresholds of potential concern employed through a management philosophy of strategic adaptive management (McLoughlin *et al.*, 2011; Pollard *et al.*, 2011). This has gone on to influence the

Editor(s): Puangrat Kajitvichyanukul and Brian D'Arcy

management of other ecosystems facets over the years. In the water space this has been accompanied by a co-ordinated participatory approach leading to some notable successes in river flow management for the KNP in recent years, to achieve environmental flow targets given provision for in South Africa's National Water Act of 1998 (Act 36 of 1998) (e.g., Riddell *et al.*, 2014) and put to the test during the severe *El Niño* induced drought of 2015–16 (Biggs *et al.*, 2017; Harwood *et al.*, 2017).

Despite these documented improvements in the operational component of river management leading to improved provision of the quantity component of environmental flows towards the KNP, corrective action on water quality remains elusive. This is particularly the case for diffuse sources, those arising from weather-driven mobilisation of pollutants from both land and atmospheric sources into the freshwater environment. These impacts in the water column also present a significant risk to the viability of the aquatic environment in these protected areas.

The conceptualisation and characterisation of diffuse pollution, and hence corresponding landscape and other interventions, was undertaken by Novotny *et al.* (1994). Of the typical diffuse source pollutants, identified as sediment, nutrients, pesticides, persistent toxic pollutants including toxic metals, and faecal pathogens, USEPA identified sediment as the most important pollutant of rivers and lakes in USA (USEPA, 1994). The principal sources of sediment were identified as being from arable and livestock farming, including rangeland. Other important sources were forestry, urban and industrial development, and mining.

Impacts on flagship aquatic ecosystems in general from poor land and water resources management can be significant. For example, diffuse pollution from land use activities in Queensland, notably through nitrogen and pesticide input from intensive cropping and sediment load, has increased from the exploitation of grazing land capacity and has been recognised as a major threat to the Great Barrier Reef in Australia (Kroon *et al.*, 2016). Similarly, Boin *et al.* (2019) cite circumstances afflicting the world's largest wetland system, the Pantanal in Brazil, where excessive mobilisation of agricultural pesticides and increased sediment delivery to the wetlands from eroding upland drainages were the principal water quality problems amongst a multitude of environmental stressors on this other magnificently productive and diverse ecosystem. Of course there are also those that advocate strongly for protected areas to be considered in the integrated planning of land and water for the ecosystem services that they provide, including the ameliorating effect on water quality through assimilative buffering provided by contiguous and pristine riparian corridors (e.g., Pittock *et al.*, 2008).

These issues are pertinent to the catchments of the rivers draining into and through the KNP. This is given all the more credence due to continued and well reported instances of nutrient enrichment, eutrophication, salinisation, acid mine drainage, and sedimentation in South African river systems. This is pressured through the national developmental pathway concomitant with insufficient monitoring and understanding of process and impact. This has been highlighted as an urgent issue requiring the implementation of catchment-based diffuse pollution strategies including land use regulation with an ambitious target for completion by 2023, as advocated in the new National Water and Sanitation Master Plan (DWS, 2018d).

This review should give the reader a sense of the KNP's place within the broader drainage landscape with an overview of the historical and contemporary diffuse pollution impacts on these water resources. This then provides the means to synthesise these effects with recommendations to be taken forward for future management of this prized aquatic biodiversity, benefits for the conservation estate at large, and ecosystem services to society downstream.

11.1.1 The Kruger National Park landscape

The KNP itself is a 20 000 km^2 protected area with a north-south orientation straddling the border between South Africa, Mozambique and Zimbabwe (Figure 11.1). It forms part of the 35 000 km^2 Great Limpopo Transfrontier Park (GLTP), jointly managed by these three countries, and is situated in the lowland reaches of two transboundary rivers basins, the Limpopo and Incomati. Geologically the KNP is underlain by igneous rocks, granite and basalt, whilst its upstream catchments namely the

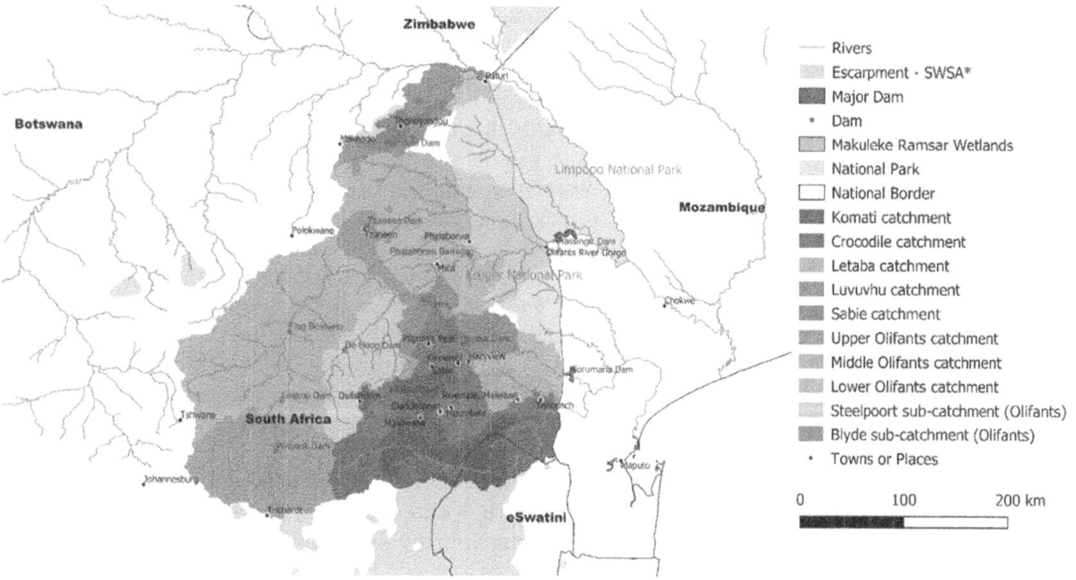

Figure 11.1 Map of the Kruger National Park and its contributing catchment areas (*delineated Strategic Water Source Areas).

Crocodile, Sabie, Olifants, Letaba, and Luvuvhu arise on basement igneous rocks in their headwaters of the South African Highveld plateau before traversing sedimentary rocks as they descend the escarpment of the northern Drakensberg range into the lowveld and then beyond through the KNP to the Mozambique coastal plain. Rainfall varies from approximately 2000 mm per annum along the escarpment down to 500 mm in the eastern lowveld region. The KNP itself is classed as a semi-arid sub-tropical savanna system with gently undulating relief typical of the southern African interior with mixed trees and grasses. Its riparian zones are relatively broad with well developed channels (Figure 11.2), often in braiding or anastomosing forms with a wide range of habitats of water, sand,

Figure 11.2 Typical river setting within the lowveld arid savanna in the Kruger National Park, the Luvuvhu River at Pafuri during winter low flows (left, source: E. Riddell), Olifants River during low flows at Phalaborwa (right, source: E. Riddell).

rock, reeds, shrubs and trees, the proportions of which are largely dependent on the spatial hierarchy of the drainage network. Some 600 km of perennial river reach exists within the KNP itself (O'Keeffe & Rogers, 2003).

These river systems offer some of the most diverse aquatic fauna in South Africa, particularly the Sabie and Luvuvhu rivers, which host some 46 fish species (Pienaar, 1968) representative of the Zambezian and Southern icthyofauna regions. A number of these catchments also form part of the National Freshwater Ecosystem Priority Areas (Nel *et al.*, 2011) designated for protective efforts to safeguard sensitive species and their support areas, such as spawning and recruitment sites. These rivers also house large populations of Nile crocodile (*Crocodylus niloticus*) and hippopotamus (*Hippopotamus amphibious*). According to the South African River Ecostatus Monitoring Program (http://www.dwa.gov.za/iwqs/rhp/naehmp.aspx), these rivers have a range of ecostatuses, with the Sabie and Luvuvhu being close to pristine at an A/B class, whilst the Crocodile, Olifants and Luvuvhu have a moderately impacted ecostatus of C (DWS, 2016a). Fortunately most of the tributaries that arise within the KNP itself are close to pristine; however there are some tributaries outside the KNP, notably the Ga-Selati River that flows into the Olifants River at the park's western boundary which is severely compromised with an 'E' ecostatus.

Rains fall predominantly during the austral summer between November to March, and are therefore strongly seasonal, giving rise to relatively turbid water conditions during mid-summer 'high flows', which transform to much clearer, tranquil waters during the winter dry-season low flows from April to October.

11.1.2 Crocodile River

The Crocodile River catchment, a major tributary of the international Incomati Basin, is 10 446 km^2 in extent and originates near the town of Dullstroom, an area famous for its trout farming. From here it flows due east into the Kwena Dam and thereafter through forestry plantations and fruit orchards towards the Mpumalanga provincial capital, Mbombela. The White River sub-catchment to the east of Mbombela is also an intensive fruit cropping area with several smaller dams. The river then passes through the Crocodile River gorge on the outlying area of the Mpumalanga Drakensberg escarpment before reaching the lowveld savanna region which gives way to extensive citrus and sugar cultivation on its southern banks and the KNP on its northern banks before joining the Komati River at the border with Mozambique. The catchment is over-allocated in terms of its available yield (ICMA, 2010), a factor that was noted as a concern as far back as the 1970s (van der Zel, 1977), and this compelled the increase in storage in that system in the form of the Kwena Dam, commissioned in the mid-1980s. The flows in the Crocodile River are regulated by releases for agriculture from this dam.

The Crocodile River itself suffers from notable pollution from point sources, including elevated salts, particularly observed in the Elands tributary through the mobilisation of chloride from return flows from the paper mill at Ngodwana. This has received a lot of attention in recent history (e.g., Ferreira *et al.* 2009; ICMA, 2010). The Joint Incomati Basin Study (JIBS, 2001) assessed the entire transboundary Incomati basin in terms of water quality and identified the Komati River, the main tributary in the basin, and south of the Crocodile catchment as already having showed worrying signs in terms of increasing electrical conductivity and pH over time. The same study suggested that the Crocodile River followed a significant but less severe trend and that these effects became more pronounced during dry-season (winter) low flow months. The Inkomati Water Availability Assessment Study (DWAF, 2009) performed a similar analysis using the national Resource Quality Services historical data on water quality in the Inkomati. This reiterated that the main water quality issues were being felt in the lower end of the river systems towards the KNP. It was notable that Ashton *et al.* (1995) suggested impacts associated with a combination of excessive low flows and loss of environmental capacity to assimilate chemical constituents due to reduced dilution capacity resulting from irrigation withdrawals. This meant that the river system could not sufficiently assimilate constituents such as aluminium, iron and manganese that emanated from upstream agroforestry areas, nor the sediments or dissolved salts arising from irrigated agricultural lands at the lower end of the catchment. Overall the DWAF (2009) study showed that there was a significant increase in salinity from the Kwena Dam downstream to Tenbosch towards the eastern boundary of the KNP. This was reiterated with selected operational water flow volume targets linked to releases from the Kwena Dam to downstream target gauges to dilute the system and reduce TDS concentrations significantly (down from peaks of 700 mg/l) for the benefit not only of the environment but also irrigation users themselves (Deksissa *et al.*, 2003). Similar problematic trends were observed for nitrates, nitrites, sodium, sulphate and phosphate. These effects are reflective of the significant impacts of land use activities in this catchment, with contributions from the Elands and Kaap tributaries beings singled out in particular for their impacts on the water quality of the main stem of the Crocodile River. Phosphate was seen to increase particularly east of Mbombela as the river traversed large citrus agriculture. Sulphate was also seen as a particular problem emanating from the extensive and abandoned mining operations in the Kaap River. In addition to the ecological impacts, the elevated salinity in the lower reaches of the Crocodile has also come in for investigation by the irrigated sugar industry (van der Laan *et al.*, 2012).

Manganese is a notable pollutant in the Gladdespruit River near Mbombela arising from some of the quarrying in this sub-catchment, leading to elevated loads in the main stem of the Crocodile River (DWA, 2013a). This was also highlighted as arising from upstream forestry areas (Ashton *et al.*, 1995), and noted all the way down the system to KNP's eastern boundary, and worryingly also in fish tissues sampled within the Crocodile River at KNP (Roux *et al.*, 1994).

Moreover Soko and Gyedu-Ababio (2015) noted that fish were particularly sensitive to elevated TDS in both wet summer and dry winter seasons in the lower Crocodile River in the vicinity of the KNP, which has resulted in a significant drop in fish species diversity, with only species tolerant of impacted conditions remaining, such as the hardy Barb species *Barbus viviparous.*

Overall, the synergistic effects of point and diffuse source pollution over the last few decades emanating from expanding agriculture and rangelands, forestry, rapid urbanisation, increased industrialisation, dysfunctional wastewater treatment works (WWTW) and the development of mines have resulted in a decline in ecosystem health in the Crocodile catchment. As a result the Crocodile River is currently in Class C/D, a moderately to largely modified condition (DWA *et al.*, 2013b). The results from fish, macroinvertebrates and diatom surveys over the past few years indicate that the river is showing signs of severe stress as bio-monitoring within the KNP has revealed below the expected fish species diversity and the long-term average bio-monitoring scores. The abundance of diatom species *Melosira* and *Nitzschia paleacea* collected from annual surveys in the KNP further reflect that the Crocodile River is heavily polluted. Fish tissues examined in work by Heath (1999) and Heath and Claassen (1999) also revealed varying concentrations of metals (arsenic, cadmium, copper, iron, manganese, nickel and zinc) and pesticides (DDT, DDE, DDD) bio-accumulated in 11 species. Worryingly in 2019, in-stream hippo die-off near Riverside, Malelane was observed. To date, the exact cause of the die-off has not been established although initial results suggest possible pesticide or herbicide contamination. Microbial pollution in the form of *Escherichia coli* (*E. coli)* emanating from poor sanitation infrastructure and dysfunctional WWTW poses another significant threat to human health as *E. coli* levels are intolerably high along many stretches of this river and others.

11.1.3 Sabie River

The Sabie catchment is a relatively small system at 7096 km^2 arising in the high-altitude grasslands and forested gorges of the Drakensberg escarpment to the west of the old gold mining towns of Sabie and Pilgrim's Rest. It is also a significant tributary of the Incomati Basin which it joins downstream of the KNP, in Mozambique. Before descending into the lowveld region occupied by the KNP the Sabie River flows through expansive plantation forestry predominantly of blue gum (*eucalyptus*) and pine (*pinus*). Just upstream of the town of Hazyview the landscape gives way to soft-fruit orchards including extensive banana plantations in the Kiepersol area. To the east of Hazyview the Sabie catchment moves into the undulating lowveld terrain that is typical of the KNP landscape. In this region there is extensive peri-urban settlement, and smallholder agricultural production in the form of government irrigation schemes. Importantly, the Sabie catchment has the largest proportion of its catchment within the confines of the KNP and adjacent protected areas, when compared to the other perennial catchments of the KNP.

The Sabie River is a flagship river for the KNP in respect of ecological protection due to its rich aquatic biodiversity (O'Keeffe & Rogers, 2003) and high ecostatus. The Sabie River has not suffered the pollution extremes that have affected the other perennial rivers significantly in recent decades and is considered to be in a largely natural, un-impacted state. This is important considering that approximately a hundred years ago it was severely impacted to the point of sterilisation due to cyanide leaching from upstream gold mines in the Pilgrim's Rest area (Pienaar, 1968; Van Vuuren, 2017). Currently, known threats include poor wastewater treatment in small urban areas and continued changes in land use practices upstream of the KNP (Stolz, 2018). Another concern is the impact that the planned raising of the dam wall of the Corumana manmade lake in Mozambique, will have on the Sabie River gorge reach in the KNP. The synergistic effects of these upstream stressors on the Sabie River are however still poorly understood compared to other river systems.

Importantly, in the present day the Sabie River provides good dilution capacity in its contribution to the Incomati system in Mozambique (Saraiva-Okello, 2019). Nevertheless, the Sabie River system is still beset with challenges, such as poor sanitation management in smaller communities, adding organic and microbiological contaminants which impact on the waters of the catchment. Organic contaminants are specifically noted as a potential risk during large rainfall events (DWAF, 2009). Irrigation of the extensive banana plantations and numerous fruit orchards, particularly in the Kiepersol-Hazyview area is also expected to negatively influence Sabie River water quality and quantity (ICMA, 2010). Both these effects were highlighted by DWA (2013) when setting the catchments' Resource Quality Objectives, with high algal growth becoming apparent. The same report highlighted an emerging problem of increasing suspended solids load in the Sabie River due to catchment clearing, urbanised but unsurfaced road networks, and in some parts of the landscape over-grazing of indigenous pasture. Furthermore, a recent ecostatus study of the entire catchment highlighted the negative effects of agro-forestry practices in the upstream catchment on the biotic responses in the aquatic environment (IUCMA, 2017). This was also revealed in an ecological impacts study undertaken by Riddell *et al.* (2019) where an apparent negative aquatic biota response to variable dissolved mineral solid loads was noted.

The Sand River is the main sub-catchment of the Sabie and is densely populated with peri-urban communities, a situation that arose from the forced resettlement period during the apartheid era in South Africa. With continued population growth, in some cases by 1000% since the 1960s, this situation has resulted in an increased waste output and organic pollution into the water courses. Concurrently, overgrazing by livestock has caused extensive erosion of the riverbanks and in-stream sedimentation (DWAF, 2009). This has led to near complete loss of species dependent on mixed bedrock habitat due to smothering of the macro-channel floor with sandy deposits leading to homogenisation of habitats in the lower reaches of this tributary (IUCMA, 2017).

11.1.4 Olifants River

The Olifants River originates at Trichardt, east of Johannesburg, and flows through the Kruger National Park into Mozambique at the Massingir Dam; thereafter it joins the Limpopo River and provisions water to the extensive Chokwe Irrigation Scheme before discharging into the Indian Ocean. The Olifants River flows through various sub-catchments with diverse water users. In addition, the entire Olifants River basin is highly stressed in terms of being over-allocated with further pressure through a growing population and development. Recently DWS (2017), as part of a national review to manage water quality, emphasised that in this catchment diffuse sources of pollutants are the dominant factor impacting on water quality and its ability to provision assimilative services.

The Upper Olifants sub-catchment covers an area of 11 461 km^2 and is the 'headwaters' of the Olifants River basin. It has the highest density of open cast and underground coal mines in South Africa, all at various stages of construction, operation and closure with coal for export through the coastal export town of Richards Bay and for direct use within the catchment itself in a multitude of active coal fired power stations. These activities have grown substantially over the past 20 years. The activities of the mines, power stations and associated industries, as well as the two largest urban centres in the area, contribute to a very high load of diffuse source pollution, resulting in a lack of any further assimilative capacity for future salt loads. Whilst point-source effects through extensive dysfunction in wastewater treatment works are leading to hypertrophic conditions in the upper Olifants, the increased concentration, particularly from sulphates, indicates the residual impacts of mining on catchment runoff (Dabrowski & de Klerk, 2013). The two largest dams in the area are the Witbank and Middelburg dams, both of which receive high loads of total dissolved solids (TDS) and sulphates, as well as nutrients from the urban wastewater treatment works. This impacts the dilution capacity for downstream ecosystems. Overall the DWS (2017) study showed that compliance to water quality planning limits (limits set based on, amongst others, specific water user guidelines, land use and present state of water quality) for total dissolved salt, in each of the management units (disaggregated sub-catchments) was high. Almost 70% of the areas were non-compliant for TDS and orthophosphate concentrations were almost 100% non-compliant. The same study recommended that in order to improve the situation and achieve the water quality planning limits, many of the sub-catchments in the upper Olifants require salinity load to be removed from the system. This was highlighted by Ashton and Dabrowski (2011) who did an overview study on the water quality in the Olifants catchment. The water quality impacts on the ecological of the Olifants River were also highlighted by Heath *et al.* (2010).

Loskop Dam drains the Upper Olifants sub-catchment into the Middle Olifants sub-catchment, the largest sub-catchment with an area of 22 791 km^2. Immediately downstream of this is the second largest irrigation scheme in South Africa that needs to comply with export quality requirements, including the Global GAP, before it flows through areas with extensive subsistence agriculture, limited platinum mining, light manufacturing and tourism (DWS, 2018a). A large concern in this sub-catchment is the impact of pit latrines on the groundwater resources. The Flag Boshielo Dam is the largest dam in this sub-catchment and supplies downstream users which include the platinum and chrome mines, agriculture, domestic use and the transfer of water from the Olifants to the expanding city of Polokwane in the adjacent Limpopo River basin. In this stretch of river DWS (2017) noted close to 80% non-compliance to the national water quality planning limits for TDS and orthophosphate throughout the sub-catchment. Pesticide use and potential contamination is a concern that will require further studies to assess the extent of pollution, as the current state is still poorly understood (DWS, 2018a).

The Steelpoort sub-catchment, which includes the newly commissioned De Hoop Dam, covers a smaller area, 7136 km^2. Major water users in the catchment that also contribute to diffuse source pollution include seven controlled irrigation schemes as well as several major platinum and chrome mining operations, and expansive peri-urban areas in close proximity to the mines. These all contribute diffuse source pollution, specifically, metals and nutrients. The Lower Olifants covers an area of 12 154 km^2 representing the catchment of the Olifants River between the Steelpoort confluence and the Mozambique border. This sub-catchment also showed relatively high non-compliance for TDS at 45% (DWS, 2017). Metal contamination from the mines and industries needs to be better monitored in a catchment context as it is currently specific to each mine and not well integrated (DWS, 2018b). As for the Middle Olifants sub-catchments, pesticide use in the irrigated areas and potential contamination is a concern that will require further studies to assess the extent of pollution, as the current state is still poorly understood (DWS, 2018b).

Two significant towns in the Lower Olifants River catchment are Phalaborwa and Hoedspruit. The mines and industry in Phalaborwa contribute high loads of diffuse pollution to the Ga-Selati River which confluences with the Olifants River at the KNP border. The Ga-Selati has also been known to experience severe point-source pollution over the years, the most recent being in 2013 (Marr *et al.*, 2017). The DWS (2017) study suggested that data was limited for this region of the Olifants River, however, where there was data in the Phalaborwa area specifically, non-compliance for TDS and orthophosphate was high (DWS, 2018c).

While turbidity and TDS had been highlighted as a concern in the Olifants system as far back as the 1980s (Wepener *et al.*, 1999) the impacts of this on the biotic response have been quite broad, including that those conditions had supposedly led to the loss of six fish species in this river system (Russell and Rogers, 1991, cited in Wepener *et al.*, 1999). This led to knock-on effects on predator–prey interactions in this system, resulting in reduced numbers of the rare Pel's Fishing Owl (*Scotopeli peli*) and the top aquatic predator, the Tiger Fish (*Hydrocynus vittatus*) (Moore *et al.*, 1991, cited in Wepener *et al.* 1999). Unfortunately, this sediment-related problem continues to the present day with an array of research efforts to determine its impact, most notably that of Gerber *et al.* (2015a) who note that whilst the sediment loads are high in the Olifants River basin, the contaminants they bring with them in the form of metals and organochlorine are actually below the levels of contaminated sediments from the global record. Wepener *et al.* (1999) noted chromium, manganese, lead and zinc as metals with a significant impact on the aquatic ecosystem – with very high concentrations of copper and manganese arriving from the Selati River. Another impact on the water column arises more locally to the KNP, in the form of fine wind-blown dust from mine tailings at the Phalaborwa mining complex (Wepener *et al.*, 1999), However, it is noted that in recent years a dust suppression process has been introduced to ameliorate these impacts.

The turbidity and sedimentation problems add to numerous stressors having adverse ecological impacts, including potential losses in biodiversity and ecosystem processes of the Olifants River. A notable example of this as a causal mechanism, was the large die-off of Nile crocodile in the Olifants River Gorge in 2007–8 due to the disease *pansteatitis*. It was believed that one of the drivers for the hardening of the fatty tissue in the crocodiles, rendering the animals lethargic with the result that they slowly starve, was due to bioaccumulation of metals bound to the silts deposited in this gorge area (Ashton, 2010; Ferreira & Pienaar, 2011). While the debate on the precise mechanism has moved to the role of multiple stressors including flow alterations, trophic accumulation and alien fish species in the food chain, and coincidental eutrophic conditions, the underlying cause is still believed to be due to the sedimentation risk in the gorge area bringing with it metals, pesticides and endocrine-disrupting chemicals (EDC), as referenced by numerous scientists including Woodborne *et al.* (2012), Bouwman *et al.* (2014) and Du Preez *et al.* (2016). Further signals of a compromised Olifants River ecosystem within the KNP have extended to fish (Du Preez *et al.*, 1997; Huchzermeyer, 2012), macro-invertebrates (Kemp *et al.*, 2014; Wolmarans *et al.*, 2014) and bird communities (Myburgh & Botha, 2009). Worryingly, communities that rely on the Olifants River for their food face the risk of contamination from metals such as lead, antimony and chromium found at dangerously elevated levels in fish along the length of the river (Jooste *et al.*, 2014, 2015).

More recently there is a new threat of increased turbidity by the uncontrolled sand mining that is taking place illegally in the Olifants and Selati rivers on the borders of the KNP. The removal of the stabilising reed beds by this illegal sand mining has enhanced the sediment load entering the KNP.

11.1.5 Letaba River

The Letaba River, a major sub-catchment of the Olifants River basin, covers a total area of 13 400 km^2. Upper portions are located within the Kruger to Canyons Biosphere Reserve and several smaller protected areas along the escarpment. The Groot Letaba River joins the Olifants River just upstream of the border at the Olifants River Gorge into Mozambique. Upstream of the KNP, the economy of the catchment area is based on commercial forestry, tea, sub-tropical fruits, summer crops, vegetables and livestock farming, with ecotourism regarded as a core industry of the catchment. In the upstream region, releases of very good quality water from the Tzaneen Dam are noted. However downstream of this area, large urban settlements (and to a lesser extent irrigation) impact considerably on any available assimilative capacity with respect to nutrients.

Diffuse pollutants in the Letaba River appear to be less well studied, however as noted in planning documents they have been primarily linked to commercial agriculture in the upstream tropical-fruit producing area near the town of Tzaneen, where agricultural return flows and runoff result in fertiliser input to the system, elevating the salt and nutrient concentrations. Secondary diffuse sources come from the large grazing lands which increase turbidity and microbial activity in the river system (DWAF, 2006) impacting the KNP river systems, a situation that is not unique to the Letaba River system only. Unfortunately, this river system also has a TDS problem with 86% non-compliance against water quality planning limits and 43% non-compliance for orthophosphate (DWS, 2017).

Overall, the challenges in the Letaba around water quality and diffuse pollution arise from its limited dilution capacity due to its distinct hydrological regime dominated by low flow conditions (see e.g., Vlok & Engelbrecht, 2000). A decline in aquatic health, especially fish and macro invertebrate communities, is attributed to altered flow regimes where the hydrology of the river has changed from a perennial to a largely ephemeral river with no-flow situations occurring for prolonged periods in the KNP (Venter & Deacon, 1995). Although, comparatively speaking, the Letaba River is a less polluted system, recent diatom community surveys in the Kruger Park indicate the presence of *Cocconeis placentula* which are commonly associated with meso-eutrophic waters and abundant *Gomphonema parvulum*, and *Nitzschia* species which are indicative of polluted waters. There has also been evidence of metal and pesticide bioaccumulation in fish and possible carcinogenic threats to human health (Heath & Claassen, 1999), and the sources need to be further studied and better understood. Interestingly, whilst the Letaba is a sediment-rich river system with a very sandy macro-channel, in comparison to the neighbouring Olifants River it appears to have better sediment quality in terms of being less contaminated by metals and other contaminants during the early 2010s when compared to studies in the early 1990s (Gerber *et al.*, 2015a). This suggests some improvement in catchment condition, although the Letaba River's geomorphology is such that it does contain coarser sediments, meaning that contaminants such as metals are less able to bind to particles in the water column than would otherwise impact silt-rich systems such as the Olifants.

11.1.6 Luvuvhu River

The Luvuvhu River starts close to the town of Makhado and is the smallest of the KNP's perennial river systems at 5952 km^2. It follows a north-easterly route along the foothills of the Soutpansberg Mountains and is joined by its major tributary the Mutale River in the Parfuri region of the KNP. It is a flagship river system due to its high biodiversity as well as, through overbank recharge water replenishing the rich floodplain pans of the Makuleke Wetlands, a Ramsar site of international importance, at its confluence with the Limpopo River at Pafuri. The western part of the Luvuvhu catchment is dominated by commercial tropical-fruit production upstream of the regional centre, Thohoyandou, thereafter flowing through a predominantly rural but relatively densely populated setting of the Venda area with subsistence and smallholder agriculture and extensive grazing shrub lands.

Whilst this river exhibits high aquatic biodiversity and is considered to be in a generally healthy state, diffuse pollution sources are a cause for concern. Gerber *et al.* (2015a) examined the contaminants in sediments throughout the reach of river within the KNP which showed that organochlorines (OCPs) increased downstream, probably having arisen from diffuse agricultural inputs from pesticide application in the upstream catchment. These, along with other contaminants are coupled to the finer sediments, such as silts deposited within the Luvuvhu floodplain at its confluence with the Limpopo River during back water flooding. This is all the more concerning in that the levels of OCPs in the Luvuvhu River were higher than in the Olifants River.

The toxicant dichlorodiphenyl-trichloroethane (DDT) is a broad-spectrum insecticide currently banned in most countries around the world. However, in South Africa, DDT is still permitted for use and is still used to control the spread of malaria. Specifically in the Limpopo Province spraying of DDT has been ongoing for almost 56 years (Brink, 2009). Although the Luvuvhu is one of the least polluted rivers in the Limpopo catchment the negative effects of metals and DDT exposure to human health and aquatic biota are worrisome. Studies undertaken by various scientists (Bornman *et al.*, 2009; Brink, 2009; Gerber *et al.*, 2016a, 2016b, 2017; Heath & Claassen, 1999) looking at the bioaccumulation of pesticides in various fish species, noted that the Luvuvhu River displayed the highest DDT concentration levels ever recorded in South African freshwater systems and in many cases the concentrations were higher than most contaminated areas from around the world. This is of concern as it now seems to be having an impact on rare and vulnerable eggs of piscivorous bird species and their viability (Bouwman *et al.*, 2013). Gerber *et al.* (2015b) also noted that other agricultural residues, notably ammonium and orthophosphates, were present in the sediment and if mobilised could result in nutrient enrichment in the system.

11.2 REGIONAL DIFFUSE SOURCES

While the focus of this review has been on constituent impact of supposed diffuse pollution problems at catchment level, there are other diffuse pollution problems that afflict the entire region around the KNP. The most notable is that of atmospheric deposition particularly of nitrogen and sulphur in aerosols. This was highlighted by Griffin *et al.* (2014) in their historical water quality trend analysis, with the Olifants catchment showing some worrying peaks of sulphate during the 1980s in particular. Lorentz *et al.* (2013) undertook a study for the national energy utility, ESKOM, who own and operate the large network of coal-fired power stations largely with the Olifants River headwaters. This study found that acid emissions and acid deposition were found to be increasing exponentially and in particular they noted that loads of sulphate and base cations were also increasing exponentially as revealed in data from the DWS water quality record, including in the Olifants and Sabie River catchments. It was also noted that dominant soils in the Olifants basin offered some buffering capacity to the deposition effect. However, the quantification of the proportional contribution of atmospheric deposition on the deteriorating water quality trend was a challenge.

One conclusion from the study was particularly concerning in that the mass balance in the Sabie River system could not be reconciled by atmospheric deposition alone, suggesting a source of unknown origin. Josipovic *et al.* (2011) suggested that the environmental effect of wet acid deposition east of the escarpment area was quite limited, and that the most severe effects of the deposition on environmental processes in soil and water were more pronounced in the Highveld region close to the source of the airborne pollution at the stations themselves. Interestingly, Gerber *et al.* (2015a) noted that airborne contaminants, particularly the metals cadmium, chromium, copper and zinc resulting from combustion activities and agriculture (fertiliser/pesticides) were present in high concentrations in both the Olifants and Luvuvhu River systems.

Another impact emerging as a serious threat to the KNP rivers systems that could be considered a diffuse source, is that of extensive sand mining in all catchments (see Figure 11.3), where entire macro-channels are denuded of sand upstream in both the perennial rivers and their ephemeral tributaries. Heath *et al.* (2004) recognise the ongoing impacts of small-scale mining on the sediment loads of our rivers. Whilst South Africa's National Water Act makes provision for control of this activity, it is unfortunate that its regulation is almost entirely absent, largely owing to control of the trade by syndicates and in regions of South Africa with customary governance structures. While the impacts have not yet been quantified and still remain anecdotal or based on first-order hydrological principles in terms of reduction of streamflow regulation process, their effects as a diffuse pollution source are expected to be severe in terms of mobilising fine particulate materials, raising TDS in the main river systems downstream and smothering sensitive aquatic micro-habitats.

Figure 11.3 Sand mining along the Groot Letaba (source: L. Boyd).

11.3 SYNTHESIS

The rivers of the Kruger National Park, as described in this article, are beset with problems of diffuse pollution resulting in a general deterioration of river water quality because of upstream sources of pollution (see Griffin *et al.*, 2014). There is also a clearly distinct variation in the level of understanding of the extent of the problem from river to river. The Crocodile and Olifants river systems have had the most research undertaken into the impacts over the years, and this is probably a reflection of their 'work-horse' nature.

These catchments have both been warranted a management class that accepts a certain form of exploitation with concomitant impact on the environment due to their contribution to the regional and national economies. Interestingly however this research is not only driven by the environmental sector, but also an array of other sectors in which compromised water quality makes an impact: irrigated agriculture, water service authorities and some industry. It must be recognised that in these rivers there is very little opportunity for further water resource development and future development will need to rely on local sources of water. However, it is also important that the ecological water requirements are met through effective and integrated water accounting at basin level and implemented through operational mechanisms such as irrigation and industrial restrictions, and where necessary through conjunctive releases from the various dams. The activities in the upper catchments are of a highly diffuse nature impacting on downstream water users and ultimately international waters shared by the KNP and Great Limpopo Transfrontier Park (GLTP). Therefore water quality management needs to be operationalised together with water quantity.

The principal impacts of diffuse pollution for each of the rivers can be summarised as follows: salts and sedimentation appear to be the most acute problem in the Crocodile River; sedimentation is becoming an issue in the Sabie River system, although the immediate challenge is from point-source effects causing nutrient enrichment; there are a multitude of challenges including sedimentation, increased salinity and metals pollution in the Olifants River system; agricultural residues in the form of nutrient input and raised salinity and turbidity in the Letaba; and in the Luvuvhu River issues include sedimentation and EDC contaminants with the propensity to bio-accumulate in the aquatic ecosystem.

In all cases there is an incomplete understanding of all the areas of concern, especially with respect to metals, pesticides and emerging contaminants.

11.3.1 Policy Status Quo

From a policy perspective all the correct targets are in place for water resource protection, which has evolved using a system to classify the nation's water resources, via an expert informed participatory process, and then set Resource Quality Objective (RQOs) for each catchment. From a KNP water management perspective the mechanisms have now all been determined and gazetted for all the KNP's catchments with the exception of the Luvuvhu River system, using the RQOs, which, although not standardised across catchments, do provide descriptors of quality, quantity, habitat and biotic conditions as a basis from which management actions can be implemented for the sustainable use of all water resources. In other words, a balance is intended between the need to protect and sustain water resources on the one hand, and the need to develop and use them on the other.

Importantly, whilst the policy mechanisms are now largely in place, the national Department of Water and Sanitation highlights the dire need to resolve the diffuse pollution challenges throughout the country, with an ambitious target set to achieve diffuse pollution management strategies inclusive of regulation and land use controls in certain Water Management Areas by 2023. In the Olifants River basin, this has taken a step further with the development of the Water Quality Planning Limits which allow for the 'allocation' of water quality capacity in the river system at manageable local-level units with a focus on the pollution source, as well as attempting to integrate monitoring programmes and encourage catchment collaboration between water users (DWS, 2016b).

11.3.2 Institutional constraints

The challenge though is to manage for these multitude diffuse pollution sources, and whilst there is reasonable historical monitoring data at monthly resolution, it has become ever patchier in recent years, at a time when monitoring should have been more intensive (including the move to near-real time systems) and systemic. Unfortunately, this is hampered by spatial inconsistencies and monitoring programmes have traditionally had a focus on broad system variables. There remains a lack of catchment monitoring in respect of emerging chemicals of concern and endocrine-disrupting chemicals, such as those arising from pesticides. In addition, a broad-spectrum analysis for metals needs to be undertaken throughout the catchments (Griffin *et al.*, 2014).

It is proposed that routine water quality monitoring needs to be expanded to include trace metals, pesticides, organic compounds, microbial contaminants and suspended sediments in order to provide a holistic and systemic understanding of water quality drivers in the nation's river systems (e.g., Ashton & Dabrowski, 2011). Furthermore, this monitoring is troubled by an apparent lack of integration from regional water resources management offices to the national regulator (DWS, 2017).

The national Department of Water and Sanitation has over the past two decades attempted to bring in the waste discharge charge system (WDCS) that would allow the department to penalise a polluter and use the funds received to, for example, remediate a river system or build a water treatment works. This is however still some years away. Another mechanism that has been brought in is the Compliance, Monitoring and Enforcement (CMandE) Directorate within the DWS that will be looking at non-complying water users.

11.3.3 Partnerships and constituency

The national policy is to recognise the value of adaptive management which is rooted in the understanding that there are many different sets of knowledge that must be brought together to address a challenge, including social, political, earth sciences and financial analysis. This is a learning-by-doing approach that is best achieved through partnerships, not only by strengthening those within government but also with civic society organisations, NGOs and private sector role-players. To this end the DWS included the following statement in its recent water quality plan for the Olifants River: '... In addition, the 2017 IWQM Policy is designed to enable government in partnership with civil society and the private sector, to address the issue of water quality across the country. This then underlines the core concept that the management of water quality can no longer be the sole mandate of DWS alone, but rather is a government-wide task, to be implemented under the leadership of the Department of Water and Sanitation, with both the private sector and civil society playing a key role. Within this context, the 2017 IWQM Policy calls for an inclusive approach to managing water quality in the country.' (DWS, 2017)

There are a number of examples where these partnerships have emerged, one example being the revitalised Catchment Management Forums where there is clear dialogue on all matters corresponding to the integrated management of these river systems, including water quality and their diffuse sources.

In the Upper Olifants catchment, a Mine Water Coordinating Body (MWCB) has been established under the Strategic Water Partners Network, a voluntary engagement platform where the private, public and civil society partners can contribute to the development and implementation of effective and innovative solutions to water and sanitation challenges in South Africa. The MWCB comprises mine corporations, the energy regulator, the minerals council, industries and national government. However, while these are well meaning, there has yet to emerge a sustainable business-like arrangement for achieving the mitigation of diffuse pollution impacts, either in the form of a payment for ecosystem services model, or headwater stewardship agreements. The challenge being to identify the appropriate market and entity for these arrangements and having these in the downstream position where the benefits of these arrangements would be felt.

In the meantime, the KNP along with sector partners have formed long established relationships in an attempt to contain and manage these issues, largely through bi-lateral investments such as the recent the RESILM-O (http://award.org.za/index.php/projects/usaid-resilm-o/) program or the iWaGSS (http://www.iwagss.com/wordpress/en/) program. The former being a fairly lengthy program that engendered a community of practice on Integrated Water Resources Management (IWRM) for resilient transboundary outcomes in the Olifants River basin, with a particular focus on securing acceptable flow and quality within the context of a changing climate.

The latter was able to bring specialist technology, one of which is presently in process to assist KNP's neighbouring Water Utility, Lepelle Northern Water, in effective fluvial management of the Phalaborwa Barrage which is beset by significant sedimentation problems (down to 10% storage capacity). This had unfortunately led to the occasional anoxic sediment rich discharges observed downstream into the Olifants River in the KNP (see Figure 11.4). Whilst the KNP is fortunate to have these types of support, again they may not be sustainable until the appropriate pricing mechanisms are established in these catchments. One way to overcome this may well be to use leverage in the form of exemplifying the health risks associated with diffuse contamination on the KNP water resources

Figure 11.4 Example impacts of excessive sediment delivery to the Olifants River at Mica (left, source: J. Venter) and the Phalaborwa Barrage operated by the regional water board, Lepelle Northern Water (right, source: T. Mohlala), both upstream of the KNP, December 2019.

in the water course on which communities around the KNP depend. For instance, the study by Jooste *et al.* (2014) highlighted the risks of sustained consumption of freshwater fish from the Olifants River and the deleterious effect it would have on the regional population. Furthermore, the KNP should expand its mandate into the transboundary space of the GLTP in order to both exemplify these water quality impacts, and use this broadened leverage to negotiate with the River Basin Organisation for international management of water quality improvements upstream and the assimilation benefits that pristine riparian corridors traversing the KNP and GLTP may have for the benefit of the broader users in the landscape, including transboundary neighbours.

11.4 CONCLUSION

Large, protected areas throughout the world, as in Africa, where rivers and wetlands are often key features of a biologically rich landscape, are typically constrained by management actions in upstream catchments. In such areas where the protection of aquatic biodiversity and associated ecosystem services with the terrestrial environment are important then dealing with issues of water quality, and particularly those that arise from diffuse sources, present particular challenges for effective mitigation and management. In the case of the KNP, which is dependent on an effective and well-co-ordinated process to manage its river systems upstream, it is crucial to further forge relationships with key role players in the upstream catchments and coalesce in strong downstream partnerships to leverage and advocate for these changes to ensure an effective and resilient future management of the water resource, not only for the benefit of the park itself, but also in translating that into beneficial ecosystem goods and service provision to the users of the water resource downstream.

REFERENCES

Ashton P. J. (2010). The demise of the Nile crocodile (*Crocodylus niloticus*) as a keystone species for aquatic ecosystem conservation in South Africa: the case of the Olifants River. *Aquatic Conservation: Marine and Freshwater Ecosystems*, **20**, 489–493, https://doi.org/10.1002/aqc.1132

Ashton P. J. and Dabrowski J. M. (2011). An Overview of Surface Water Quality in the Olifants River Catchment. Technical Report KV293-11, Water Research Commission, Pretoria, South Africa.

Ashton P. J., van Zyl F. C. and Heath R. G. (1995). Water quality management in the crocodile river catchment, eastern transvaal, South Africa. *Water Science and Technology*, **32**(5–6), 201–208, https://doi.org/10.1016/0273-1223(95)00664-8

Biggs H. C., Clifford-Holmes J. K., Freitag S., Venter F. J. and Venter J. (2017). Cross-scale governance and ecosystem service delivery: a case narrative from the Olifants River in north-eastern South Africa. *Ecosystem Services*, **28B**, 173–184, https://doi.org/10.1016/j.ecoser.2017.03.008

Boin M. N., Martins P. C. S., da Silva C. A. and Salgado A. A. R. (2019). Pantanal: the Brazilian wetlands. In: The Physical Geography of Brazil, A. A. Rodrigues, S. Leonardo, J. C. Santos and J. C. Paisani (eds), Springer, Cham, Switzerland, pp. 75–91.

Bornman M., Barnhoorn I. E. and Aneck-Hahn N. H. (2009). A Pilot Study on the Occurrence of Endocrine Disruptive Chemicals in A DDT-Sprayed Area. Report KV220-09. Water Research Commission, Pretoria, South Africa.

Bouwman H., Viljoen I. M., Quinn L. P. and Polder A. (2013). Halogenated pollutants in terrestrial and aquatic bird eggs: converging patterns of pollutant profiles, and impacts and risks from high levels. *Environmental Research*, **126**, 240–253, https://doi.org/10.1016/j.envres.2013.06.003

Bouwman H., Booyens P., Govender D., Pienaar D. and Polder A. (2014). Chlorinated, brominated, and fluorinated organic pollutants in Nile crocodile eggs from the Kruger National Park, South Africa. *Ecotoxicology and Environmental Safety*, **104**, 393–402, https://doi.org/10.1016/j.ecoenv.2013.12.005

Breen C., Dent M., Jaganyi J., Madikizela B., Maganbeharie J., Ndlovu A., O'Keeffe J., Rogers K., Uys M. and Venter F. (2000). The Kruger National Park Rivers Research Programme. WRC Report No. TT 130/00, Water Research Commission, Pretoria, South Africa.

Brink K. A. (2009). Effects of DDT on aquatic organisms in the Luvuvhu River. Doctoral dissertation, University of Johannesburg.

Dabrowski J. M. and de Klerk L. P. (2013). An assessment of the impact of different land use activities on water quality in the upper Olifants River catchment. *Water SA*, **39**(2), 231–244

Deksissa J., Ashton P. J. and Vanrolleghem P. A. (2003). Control options for river water quality improvement: a case study of TDS and inorganic nitrogen in the Crocodile River (South Africa). *Water SA*, **29**(2), 209–218, https://doi.org/10.4314/wsa.v29i2.4858

Department of Water Affairs (DWA). (2013a). The determination of water resource classes and associated resource quality objectives in the Inkomati Water Management Area: status quo assessment, IUA delineation and biophysical node identification. Prepared by: IWR Water Resources. Authored by: Mallory S, Louw D, Deacon A, Huggins G, Kotze P, Mackenzie J, Scherman P, Van Jaarsveld P. DWA Report, RDM/WMA05/00/CON/CLA/0213. DWA, Pretoria, South Africa.

Department of Water Affairs (DWA). (2013b). Ecostatus of the Crocodile River Catchment, Inkomati River System. Compiled by: Mpumalanga Tourism and Parks Agency. Authored by: Kleynhans, C.J., Thirion, C., Roux, F., Hoffmann, A., Marais H. and Diedericks, G. DWA, Pretoria, South Africa

Department of Water Affairs and Forestry (DWAF). (2006). Letaba Catchment Reserve Determination Study – Main Report, Directorate: Resource Directed Measures, DWAF, Pretoria, South Africa

Department of Water Affairs and Forestry (DWAF). (2009). Report on Water Quality Assessment of the Inkomati Water Management Area (X Drainage Region) Golder Associates Africa (Pty), Report No. PWMA05/X22/00/1108, DWAF, Pretoria, South Africa

Department of Water and Sanitation (DWS). (2016a). Determination, Review and Implementation of the Reserve in the Olifants/Letaba System: Eco-Classification Report No: RDM/WMA02/00/CON/0116, DWS, Pretoria, South Africa

Department of Water and Sanitation (DWS). (2016b). Development of an Integrated Water Quality Management Plan for the Olifants River System: Water Quality Planning Limits Report, Study Report No. 3. Report No: P WMA 04/B50/00/8916/4, DWS, Pretoria, South Africa

Department of Water and Sanitation (DWS). (2017). Development of an Integrated Water Quality Management Plan for the Olifants River System: Overarching IWQMP for the Olifants River System, Study Report No. 13 Report No: P WMA 04/B50/00/8916/14, DWS, Pretoria, South Africa.

Department of Water and Sanitation (DWS). (2018a). Development of an Integrated Water Quality Management Plan for the Olifants River System: Middle Olifants Sub-catchment Plan, Report: P WMA 04/B50/00/8916/9, DWS, Pretoria, South Africa.

Department of Water and Sanitation (DWS). (2018b). Development of an Integrated Water Quality Management Plan for the Olifants River System: Steelpoort Sub-catchment Plan, Report: P WMA 04/B50/00/8916/11, DWS, Pretoria, South Africa.

Department of Water and Sanitation (DWS). (2018c). Development of an Integrated Water Quality Management Plan for the Olifants River System: Lower Olifants Sub-catchment Plan, Report: P WMA 04/B50/00/8916/10, DWS, Pretoria, South Africa.

Department of Water and Sanitation (DWS). (2018d). National Water and Sanitation Master Plan. Volume 2: Plan to Action v4.2. DWS, Pretoria, South Africa, p. 260.

Du Preez H., Van Der Merwe M. and Van Vuren J. (1997). Bio-accumulation of selected metals in African sharptooth catfish *Clarias gariepinus* from the lower Olifants River, Mpumalanga, South Africa. *Koedoe*, **40**, 77–90, https://doi.org/10.4102/koedoe.v40i1.265

Du Preez M. Govender D. D. and Bouwman H. (2016). Heavy metals in muscle tissue of healthy crocodiles from the Kruger National Park, South Africa. *African Journal of Ecology*, **54**(4), 519–523, https://doi.org/10.1111/aje.12308

Ferreira S. M. and Pienaar D. (2011). Degradation of the crocodile population in the olifants river gorge of Kruger National Park, South Africa, *Aquatic Conservation: Marine and Freshwater Ecosystems*, **21**, 155–164, https://doi.org/10.1002/aqc.1175

Ferreira M., Wepener V. and van Vuren J. H. (2009). Spatial and temporal variation in the macroinvertebrate community structure of the lower elands river, mpumalanga, South Africa. *African Journal of Aquatic Science*, **34**(3), 231–238, https://doi.org/10.2989/AJAS.2009.34.3.4.980

Gerber R., Smit N. J., van Vuren J. H. J., Nakayama S. M. M., Yohannes Y. B., Ikenaka Y., Ishizuka M. and Wepener V. (2015a). Application of a sediment quality Index for the assessment and monitoring of metals and organochlorines in a premier conservation area. *Environmental Science and Pollution Research*, **22**, 19971–19989, https://doi.org/10.1007/s11356-015-5206-z

Gerber R., Wepener V. and Smit N. J. (2015b). Application of multivariate statistics and toxicity indices to evaluate the water quality suitability for fish of three rivers in the Kruger National Park, South Africa. *African Journal of Aquatic Science*, **40**(3), 247–259, https://doi.org/10.2989/16085914.2015.1073139

Gerber R., Smit N. J., van Vuren J. H. J., Nakayama S. M. M., Yohannes Y. B., Ikenaka Y., Ishizuka M. and Wepener V. (2016a). Bioaccumulation and human health risk assessment of DDT and other organochlorine pesticides in an apex aquatic predator from a premier conservation area. *Science of the Total Environment*, **550**, 522–533, https://doi.org/10.1016/j.scitotenv.2016.01.129

Gerber R., Smit N. J., van Vuren J. H. J. and Wepener V. (2016b). Metal concentrations in *Hydrocynus vittatus* (castelnau 1861) populations from a premier conservation area: relationships with environmental concentrations. *Ecotoxicology and Environmental Safety*, **129**, 91–102, https://doi.org/10.1016/j.ecoenv.2016.03.009

Gerber R., Wagenaar G. M., Smith W., Ikenaka Y. and Smit N. J. (2017). Insights into the drivers of histopathological changes and potential as bio-indicator of riverine health of an aquatic apex predator from a premier conservation area: a multiple lines of evidence and multivariate statistics approach. *Ecological Indicators*, **72**, 530–544, https://doi.org/10.1016/j.ecolind.2016.08.048

Griffin N. J., Palmer C. G. and Scherman P. A. (2014). Critical Analysis of Environmental Water Quality in South Africa: Historic and Current Trends. Water Research Commission, Gezina, Pretoria, Report No. 2184/1/14

Harwood A., Johnson S., Richter B., Locke A., Yu X. and Tickner D. (2017). Listen to the River: Lessons From A Global Review of Environmental Flow Success Stories. WWF-UK, Woking, UK

Heath R. G. M. (1999). A Catchment-Based Assessment of the Metal and Pesticide Levels of Fish From the Crocodile River, Mpumalanga. PhD Thesis, University of Johannesburg, South Africa.

Heath R. G. M. and Claassen M. C. (1999). An Overview of the Pesticide and Metal Levels Present in Populations of the Larger Indigenous Fish Species of Selected South African Rivers. Report to the Water Research Commission, Report No. 428/1/99. Water Research Commission, Pretoria, p. 318.

Heath R., Moffet M. and Bannister S. (2004). Water Related Impacts of Small Scale Mining. Water Research Commission Report No. 1150/1/04, ISBN No. 1-77005-119-8, Water Research Commission, Pretoria, South Africa, pp. 100.

Heath R. G. M., Coleman T. and Engelbrecht J. (2010). Water Quality Overview and Literature Review of the Ecology of the Olifants River. WRC report No. TT 452/10, Water Research Commission, Pretoria, South Africa.

Huchzermeyer K. D. A. (2012). Prevalence of pansteatitis in African sharptooth catfish, *Clarias gariepinus* (Burchell), in the Kruger National Park, South Africa. *Journal of the South African Veterinary Association*, **83**, 41–50.

ICMA (2010). The Inkomati Catchment Management Strategy: A First Generations Catchment Management Strategy for the Inkomati Water Management Area. Inkomati Catchment Management Agency, Nelspruit, South Africa.

IUCMA. (2017). Ecostatus of the Sabi-Sand River catchment, Inkomati River system Phase II (2016), Inkomati-Usuthu Catchment Management Agency Report, p. 254.

JIBS. (2001). Joint Incomati Basin Study Report, *Phase 2*. 2001. Consultec Report No: C14-99MRF BKS ACRES Report No: P8491/08, Consultec, Maputo/Johannesburg.

Jooste A., Marr S., Addo-Bediako A. and Luus-Powell W. (2014). Metal bioaccumulation in the fish of the Olifants River, Limpopo province, South Africa, and the associated human health risk: a case study of rednose labeo *Labeo rosae* from two impoundments. *African Journal of Aquatic Science*, **39**, 271–277, https://doi.org/10.2989/16085914.2014.945989

Jooste A., Marr S. M., Addo-Bediako A. and Luus-Powell W. J. (2015). Sharptooth catfish shows its metal: a case study of metal contamination at two impoundments in the Olifants River, Limpopo river system, South Africa. *Ecotoxicology and Eenvironmental Safety*, **112**, 96–104, https://doi.org/10.1016/j.ecoenv.2014.10.033

Josipovic M., Annegarn H. J., Kneen M. A., Pienaar J. J. and Piketh S. J. (2011). Atmospheric dry and wet deposition of sulphur and nitrogen species and assessment of critical loads of acidic deposition exceedance in South Africa. *South African Journal of Science*, **107**(3–4), 1–10.

Joubert S. C. J. (1986). Management and research in relation to ecosystems of the kruger national park. *Koedoe*, **29**(1), 157–163, https://doi.org/10.4102/koedoe.v29i1.528

Kemp M., De Kock K., Wepener V., Roets W., Quinn L. and Wolmarans C. (2014). Influence of selected abiotic factors on aquatic macroinvertebrate assemblages in the Olifants River catchment, Mpumalanga, South Africa. *African Journal of Aquatic Science*, **39**, 141–149, https://doi.org/10.2989/16085914.2014.923371

Kroon F. J., Thorburn P., Schaffelke B. and Whitten S. (2016). Towards protecting the Great Barrier Reef from land-based pollution. *Global Change Biology*, **22**, 1985–2002, https://doi.org/10.1111/gcb.13262

Lorentz S. A., Blight J. and Snyman N. (2013). An Investigation into the effects of atmospheric pollutants on the soil-water-ecosystem continuum in the Eastern Regions of South Africa, Water Research Commission Report, WRC Report No K5/1697/1/11

Marr S. M., Mohlala T. D. and Swemmer A. (2017). The ecological integrity of the lower Olifants River, Limpopo province, South Africa: 2009–2015 part B: tributaries of the Olifants River. *African Journal of Aquatic Science*, **42**(2), 181–190, https://doi.org/10.2989/16085914.2017.1353477

McLoughlin C. A., Deacon A., Sithole H. and Gyedu-Ababio T. (2011). History, rationale, and lessons learned: thresholds of potential concern in Kruger National Park river adaptive management. *Koedoe*, **53**(2), 1–27, https://doi.org/10.4102/koedoe.v53i2.996

Moore C. A., van Veelen M., Ashton P. M. and Walmsley R. D. (1991). Preliminary water quality guidelines for the Kruger National Park rivers. Kruger National Park Rivers Research Programme. (Kruger National Park Rivers Research Programme report; no. 1).

Myburgh J. and Botha A. (2009). Decline in herons along the lower Olifants River – could pansteatitis be a contributing factor. *Vet News*, **3**, 20–23.

Nel J. L., Murray K. M., Maherry A. M., Petersen C. P., Roux D. J., Driver A., Hill L., van Deventer H., Funke N., Swatz E. R., Smith-Adao L. B., Mbona N., Downsborough L. and Nienaber S. (2011). Technical Report for the National Freshwater Ecosystem Priority Areas Project. Water Research Commission Report WRC Report No. 1801/2/11

Novotny V., Olem H. and Olness A. (1994). Water Quality: Prevention, Identification, and Management of Diffuse Pollution. Van Nostrand Reinhold, New York.

O'Keeffe J. and Rogers K. H. (2003). Heterogeneity and management of the lowveld rivers. Chap. 21 In: The Kruger Experience: Ecology and Management of Savanna Heterogeneity, J. T. du Toit, K. H. Rogers and H. C. Biggs (eds.), Island Press, Washington, DC, USA, pp. 447–468.

Okello S. (2019). Improved Hydrological Understanding of A Semi-Arid Subtropical Transboundary Basin Using Multiple Techniques – The Incomati River Basin. CRC Press/Balkema, Leiden, the Netherlands.

Pienaar U. V. (1968). The Freshwater Fishes of the Kruger National Park. National Parks Board of South Africa. Sigma Press Pty Ltd, Pretoria.

Pittock J., Hansen L. J. and Abell R. (2008). Running dry: freshwater biodiversity, protected areas and climate change. *Biodiversity*, **9**, 30–38, https://doi.org/10.1080/14888386.2008.9712905

Pollard S., Du Toit D. and Biggs H. (2011). River management under transformation: the emergence of strategic adaptive management of river systems in the kruger national park. *Koedoe*, **53**(2), 1–14, https://doi.org/10.4102/koedoe.v53i2.1011

Riddell E., Pollard S., Mallory S. and Sawunyama T. (2014). A methodology for historical assessment of compliance with environmental water allocations: lessons from the crocodile (east) river, South Africa. *Hydrological Sciences Journal*, **59**(3–4), 831–843, https://doi.org/10.1080/02626667.2013.853123

Riddell E. S., Govender D., Botha J., Sithole H., Petersen R. M. and Shikwambana P. (2019). Pollution impacts on the aquatic ecosystems of the Kruger National Park, South Africa. *Scientific African*, **6**, e00195, https://doi.org/10.1016/j.sciaf.2019.e00195

Roux D. J., Badenhorst J. E., du Preez H. H. and Steyn G. J. (1994). Note on the occurrence of selected trace metals and organic compounds in water, sediment and biota of the Crocodile River, Eastern Transvaal, South Africa. *Water SA*, **20**(4), 333–340.

Russell I. and Rogers K. (1991). Workshop: Sensitive Fish Species, Skukuza 1991. Kruger National Park Rivers Research Programme.

Soko M. I. and Gyedu-Ababio T. (2015). The spatial and temporal variations of ichythyofauna and water quality in the Crocodile River (East), Mpumalanga, South Africa. *Journal of Water Resource and Protection*, **7**(3), 152–170, https://doi.org/10.4236/jwarp.2015.73013

Stolz M. (2018). Water quality assessment of the Sabie River catchment with respect to adjacent land use. Doctoral dissertation, North-West University, Potchefstroom, South Africa.

USEPA. (1994). Pollution Prevention Directory. EPA/742/B-94/005, USEPA, Washington D.C.

van der Laan M., van Antwerpen R. and Bristow K. L. (2012). River water quality in the northern sugarcane-producing regions of South Africa and implications for irrigation: a scoping study. *Water SA*, **38**(1), 87–96, https://doi.org/10.4314/wsa.v38i1.11

van der Zel D. W. (1977). Analysis of water use in the krokodil river system (ontleding van die watergebruik in die krokodilrivier biosisteem). *South African Forestry Journal*, **103**(1), 8–14, https://doi.org/10.1080/00382167.1977.9629461

Van Vuuren L. (2017). Adapting to survive – Water management and aquatic science in the Kruger National Park. Masters dissertation, North-West University, Potchefstroom, South Africa.

Venter F. J. and Deacon A. R. (1995). Managing rivers for conservation and ecotourism in the Kruger National Park. *Water Science and Technology*, **32**(5–6), 227–233, https://doi.org/10.1016/0273-1223(95)00667-2

Vlok W. and Engelbrecht J. S. (2000). Some aspects of the ecology of the Groot Letaba River in the northern province, South Africa. *African Journal of Aquatic Science*, **25**, 76–83, https://doi.org/10.2989/160859100780177686

Wepener V., Van Vuren J. H. J. and du Preez H. H. (1999). The implementation of an aquatic toxicity index as a water quality monitoring tool in the Olifants River (Kruger National Park). *Koedoe*, **42**, 85–96, https://doi.org/10.4102/koedoe.v42i1.225

Woodborne S., Huchzermeyer K., Govender D., Pienaar D., Hall G., Myburgh J., Deacon A., Venter J. and Lübcker N. (2012). Ecosystem change and the Olifants River crocodile mass mortality events. *Ecosphere*, **3**, 1–17, https://doi.org/10.1890/ES12-00170.1

Wolmarans C., Kemp M., De Kock K., Roets W., Van Rensburg L. and Quinn L. (2014). A semi-quantitative survey of macroinvertebrates at selected sites to evaluate the ecosystem health of the Olifants River. *Water SA*, **40**, 245–254, https://doi.org/10.4314/wsa.v40i2.6

PUBLISHING

doi: 10.2166/9781789061123_187

Chapter 12

Determination of changes in neonicotinoid pesticide concentrations and assessment of the risk posed to aquatic organisms in rivers based on high-frequency sampling

Yuki Kotani[1], Yoshinori Yabuki[2], Arisa Banno[2], Takashi Daifuku[2] and Yukio Komai[3]

[1]Graduate School of Technology, Osaka Institute of Technology, 5-16-1 Omiya, Asahi, Osaka 535-8585, Japan
[2]Research Institute of Environment, Agriculture and Fisheries, Osaka Prefecture, 442 Shakudo, Habikino, Osaka 583-0862, Japan
[3]Department of Environmental Engineering, Osaka Institute of Technology, 5-16-1 Omiya, Asahi, Osaka 535-8585, Japan

Keywords: neonicotinoid pesticides; aquatic ecosystem; risk assessment; SSD; high-frequency sampling

12.1 INTRODUCTION

Crop protection chemicals (CPCs) can protect crops from harmful insects, weeds and diseases, making these chemicals indispensable to stable food production. On the other hand, safe CPCs are those that have the least adverse effect on ecosystems, and such CPCs are needed.

Neonicotinoid pesticides have high selectivity, high persistence against insects, and excellent water solubility (Japan Plant Protection Association, 2016). Since the introduction of imidacroprid, which was registered in 1992, the application amounts of neonicotinoid pesticides have been increasing. Neonicotinoid pesticides have been used as CPCs and for household use. However, their application to crops may result in discharge to the surrounding rivers. There are also concerns about adverse effects on aquatic organisms. Yamamuro *et al.* (2019) demonstrated that the decrease of fishery yields in Lake Shinji, Japan has been affected by neonicotinoide pesticides.

Detailed studies have addressed herbicides in paddy fields and in the water environment (Kawasaki *et al.*, 2008a, 2008b; Sudo & Kawachi, 2006). However, frequent research on neonicotinoid pesticides have not been adequately conducted and the survey periods have often been short and limited. For example, Oyama *et al.* (2019) collected water samples for monitoring neonicotinoid pesticides from several rivers once a week from late May to the end of September, and suggested there was a weak correlation between the concentrations of neonicotinoid pesticides and the timing of pesticides usage in paddy fields. However, the sampling frequency and period were not enough to substantiate their adverse effects on aquatic ecosystems. To evaluate the effects of neonicotinoid pesticides on aquatic ecosystems, it is necessary to perform high-frequency research over a long period including several months before and after rice cultivation. One to three surveys per week would be suitable for determining the seasonal changes in neonicotinoid pesticide concentrations. Since neonicotinoid pesticides are commonly used in both farmland and urban areas, it is also important to investigate the relationship between pesticide concentrations and detailed percentages of land use in river basins upstream from a sampling site identified by GIS.

Editor(s): Puangrat Kajitvichyanukul and Brian D'Arcy

Assessment of the risk posed to aquatic organisms is also very important. In Japan, CPCs are evaluated for human safety and for effects on aquatic organisms according to pesticide registration criteria that are based on the Pesticide Regulation Law.

Pesticide registration criteria are given as a standard to prevent adverse effects to aquatic plants and animals, and these criteria are decided based on toxicity tests on three species of aquatic organisms: the *Cyprinus carpio* fish, the *Daphnia magna* daphnid and the *Pseudokirchneriella subcapitana* alga. However, it is considered difficult to represent the entire ecosystem solely by the effects on these three species, because various species of organisms inhabit aquatic ecosystems. In fact, it has been pointed out that some species are affected at lower concentrations than the pesticide registration criteria. This means that pesticide toxicity differs greatly by species (Nagai, 2017a). The present pesticide registration criteria for aquatic ecosystems are insufficient for assessing aquatic ecological risk when only three species are used (Nagai, 2017a), and to properly assess the effects of pesticides it is better to include more species. In recent years, Species Sensitivity Distribution (SSD) has been developed as a new procedure to assess the risk to the whole ecosystem. To date, there are very few research results (e.g., Oyama *et al.* 2019) with assessment of the risk for aquatic organisms by SSD using the concentration of pesticides in actual rivers.

We conducted high-frequency sampling and long-period investigations to estimate; (1) the concentration and seasonal changes of neonicotinoid pesticide in rivers, (2) the relationship between the concentrations of neonicotinoids and the land use of the river basin, and (3) effects of neonicotinoid pesticides on aquatic ecosystems by an SSD-based risk assessment method.

12.2 METHOD

12.2.1 Sampling and analysis procedure

The locations of the five survey sites are shown in Figure 12.1. St.I and St.IV are on the Ishikawa River, which is a tributary of the Yamato River (a first-class river), in the south-eastern part of Osaka Prefecture, Japan. St.II and St.III are at the ends of two branches of the Ishikawa River, respectively. St.V is located at the middle reaches of the Neya River, which flows into the old Yodo River (a first-class river).

Table 12.1 shows the catchment areas of the survey sites and the breakdown of land use (Ministry of Land, Infrastructure, Transport and Tourism, n.d.) in each catchment area as determined by Arc GIS. Paddy fields account for 41% of the basin upstream of St.II and for 35% of the basin upstream of St.III. Forest accounts for 15% of the basin upstream of St.I and for 5% of the basin upstream of St. IV. Urbanized areas account for 54% of the basin upstream of St.V, which is a greater share than those for the other river basins. The outlet of a sewage treatment plant with a treatment population of 618 900 (Osaka Prefecture, n.d.) was about 0.2 km upstream of St.V.

Water sampling was conducted from April to December 2018, except at St.V, where it started in July 2018. The survey frequency was 1 to 3 times a week from April to September, once a week in October and November, and once a month in December. Surveys were taken a total 45 times for St.I, St,II, St.III and St.IV, and 14 times for St.V.

Water samples were collected in 1 L glass bottles by using a stainless-steel bucket. Water temperature, pH, electrical conductivity (EC), dissolved oxygen (DO) and turbidity were measured *in situ*, and the flow rate was measured at the same time, except at St.V. The water samples were stored refrigerated and were analysed after the pre-treatment shown in Figure 12.2.

Seven neonicotinoid pesticides (acetamiprid, clothianidin, dinotefuran, imidacloprid, nitenpyram, thiamethoxam and thiacloprid) were analysed by LC-MS/MS (LCMS-8050 Shimadzu, Kyoto, Japan) the next day. The results of addition/recovery tests were in the range of 95–107%. The MDL was 0.01 μg/L, and the MQL was 0.02 μg/L.

Assessment by SSD of the risk to aquatic ecosystems was conducted by calculating the single risk per pesticide and the cumulative risk based on all detected pesticides other than acetamiprid. The single risk is called the potentially affected fraction (PAF), and the cumulative risk is called

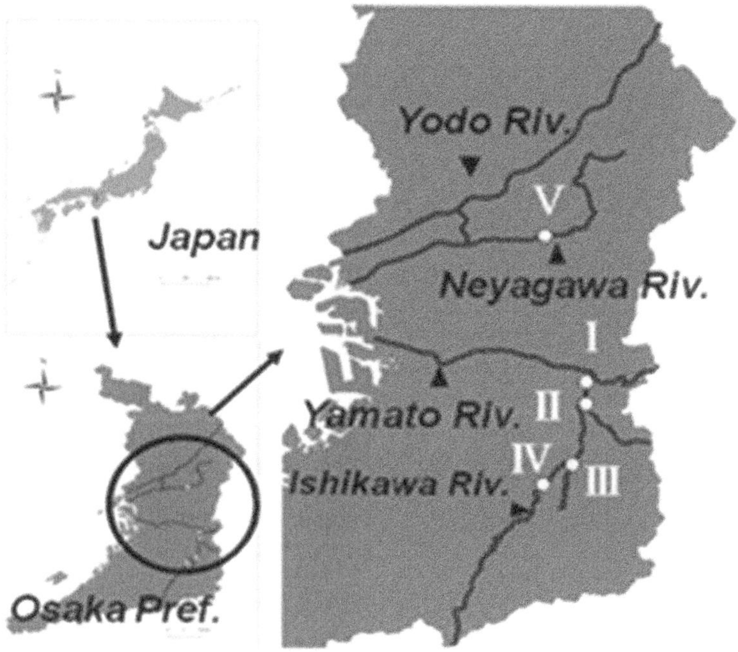

Figure 12.1 Survey sites.

Table 12.1 Land use on survey sites.

	St. I	St. II	St. III	St. IV	St. V
Area (km^2)	216.7	10.3	16.6	103.0	97.5
Farmland (%)	15	41	35	5	6
Urban land (%)	16	19	13	14	54
Forest (%)	64	35	48	77	28

the multi-substance potentially affected fraction (msPAF). A tool (Microsoft Excel spreadsheet) for calculating SSD was downloaded from the website of the National Institute for Agro-Environmental Science (NIAES, n.d.).

12.2.2 Species sensitivity distribution (SSD)

Although it is impossible in practice to conduct toxicity tests for all species in an aquatic ecosystem, it is empirically known that the sensitivity of many species to chemicals conforms to a lognormal distribution that can be expressed as a cumulative probability distribution (Nagai, 2017b).

The affected species can be divided into four levels: (1) not detected (percentage of affected species is less than 0.1%); (2) low (0.1–5%); (3) medium (5–50%); and (4) high (greater than 50%). The fifth percentile of a distribution (called the 5% hazardous concentration or HC5) has been used in the United States, the European Union, and Australia to derive threshold concentrations that will not affect most species in a community. Because the hazardous concentration calculated by SSD is normally lower than the concentration calculated by other tests for various pesticides, HC5 evaluation would be more useful than mesocosm and other assessment methods in terms of addressing the aquatic environment on the

Take 250 mL of sample with a measuring cylinder and transfer it to a medium bottle

↓←Add 50 μL of surrogate solution (200 μg / L)

Pass water through the column (Pharma FF) at a flow rate of 10 mL / min for adsorption of the pesticides in the sample

↓

Centrifuge for 5 minutes to dewater

↓

Dry for 30 minutes with nitrogen gas

↓

Elute 8 mL of acetone at a flow rate of 1 mL / min

↓

Concentrate to 1 mL or less with nitrogen gas

↓

Fill concentration tube (10 mL) up to 2 mL with methanol

↓

Re-concentrate to 1 mL or less with nitrogen gas

↓

Fill concentration tube (10 mL) to 1 mL with methanol

Figure 12.2 Sample pre-treatment methods.

safe side (Nagai, 2017b). SSD is used only to quantify the risk, and by doing so it expresses the current level of risk and the results after risk-reduction measures are implemented. In other words, lowering the risk value of the current SSD leads to efficient and steady mitigation of impacts on aquatic life.

12.3 RESULTS AND DISCUSSION

12.3.1 Characteristics of concentrations and frequencies of neonicotinoid pesticides

All the neonicotinoid pesticides other than nitenpyram and thiacloprid were detected at all survey sites. No data on dinotefuran or imidacloprid were obtained on 24 August 2018. Basic statistics on the pesticide concentrations at each survey site and on the detection frequency (detected samples as a share of all samples for each pesticide, expressed as a percentage) are shown in Table 12.2. Dinotefuran was detected in more than 90% of the surveys at all survey sites, except at St. IV. The detection frequency of imidacloprid was 22–38%. Three other pesticides showed different trends at each survey site. Dinotefuran was found to have the highest maximum concentration of any pesticide at all five survey sites, and imidacropridwas found to have higher maximum concentrations than those of the other three pesticides, except at St.V. The maximum allowable concentrations specified in the pesticide registration criteria for acetamiprid, clothianidin, dinotefuran, imidacloprid, and thiamethoxam are 2.5, 2.8, 12, 1.9, and 3.5 μg/L, respectively. Maximum values for five of the neonicotinoid pesticide concentrations did not exceed the standards.

The changes in the concentrations of dinotefuran and imidacroprid at each survey site from April to December 2018 are shown in Figure 12.3. The peaks of dinotefuran were found from late August to early September at St.I, St.II, St.III, and St.IV. Since rainfall (about 50 mm/day) was observed on September 1 (Japan Meteorological Agency, n.d.), the river water collected on September 3 might have been diluted and two peaks of concentration were detected at St.III. If the neonicotinoid pesticide concentrations had been affected by precipitation, then the concentrations would have been lower because of dilution by the precipitation. However, a single concentration peak was detected at St.I and St.II in early September after precipitation. Therefore, dilution by rainfall might not be a strong factor affecting the detection of single or multiple concentration peaks.

Table 12.2 Basic statistics of pesticide concentration at each survey point.

		Detection Frequency (%)	Number of Samples	Max	Min	Mean	Geometric Mean	Median
St.I	Acetamiprid	0	44	0.00	0.00	–	–	–
	Clothianidin	2		0.02	0.02	0.02	0.02	0.02
	Dinotefuran	100		0.74	0.02	0.12	0.07	0.06
	Imidacloprid	39		0.35	0.03	0.11	0.08	0.08
	Thiamethoxam	2		0.03	0.03	0.03	0.03	0.03
St.II	Acetamiprid	25		0.34	0.02	0.07	0.04	0.03
	Clothianidin	18		0.06	0.02	0.04	0.03	0.03
	Dinotefuran	100		0.72	0.04	0.12	0.09	0.07
	Imidacloprid	36		0.28	0.04	0.11	0.08	0.08
	Thiamethoxam	0		0.00	0.00	–	–	–
St.III	Acetamiprid	7		0.04	0.02	0.03	0.03	0.03
	Clothianidin	7		0.06	0.03	0.04	0.04	0.04
	Dinotefuran	100		1.29	0.04	0.21	0.12	0.09
	Imidacloprid	39		0.33	0.02	0.15	0.11	0.17
	Thiamethoxam	5		0.03	0.02	0.02	0.02	0.02
St.IV	Acetamiprid	0		0.00	0.00	–	–	–
	Clothianidin	0		0.00	0.00	–	–	–
	Dinotefuran	41		0.21	0.02	0.08	0.05	0.06
	Imidacloprid	23		0.10	0.03	0.07	0.06	0.08
	Thiamethoxam	0		0.00	0.00	–	–	–
St.V	Acetamiprid	0	14	0.00	0.00	–	–	–
	Clothianidin	7		0.03	0.03	0.03	0.03	0.03
	Dinotefuran	100		0.11	0.02	0.05	0.04	0.05
	Imidacloprid	0		0.00	0.00	–	–	–
	Thiamethoxam	7		0.02	0.02	0.02	0.02	0.02
								(μg/L)

There were differences in the concentrations of pesticides among the survey sites. The peak concentrations were lower at St. IV and St.V than at St.I, St.II and St.III.

12.3.2 Relation between neonicotinoid pesticides detected and land use near the survey sites

The difference in the detection of concentration peaks of neonicotinoid pesticides among the survey sites explained in the preceding section could be attributed to the following. The land around the survey sites was mainly used as farmland and residential land; however, farmland as a share of all land use and residential land as a share of all land various use among the sites. The basins of St. II and St. III had a higher share of farmland than the other basins had. The shares of land used for agriculture in the basins of St.II and St.III were 41% and 35%, respectively. In contrast, farmland accounted for only 15% the land use in the basin upstream of St.I. In this study, neonicotinoid pesticides were detected at survey sites for which farmland accounted for more than 15% of the land use in the basin.

Although, neonicotinoid pesticides are widely used insecticides on farmland and residential land, the concentration peaks of the pesticides obviously differed depending on the share of farmland and urban

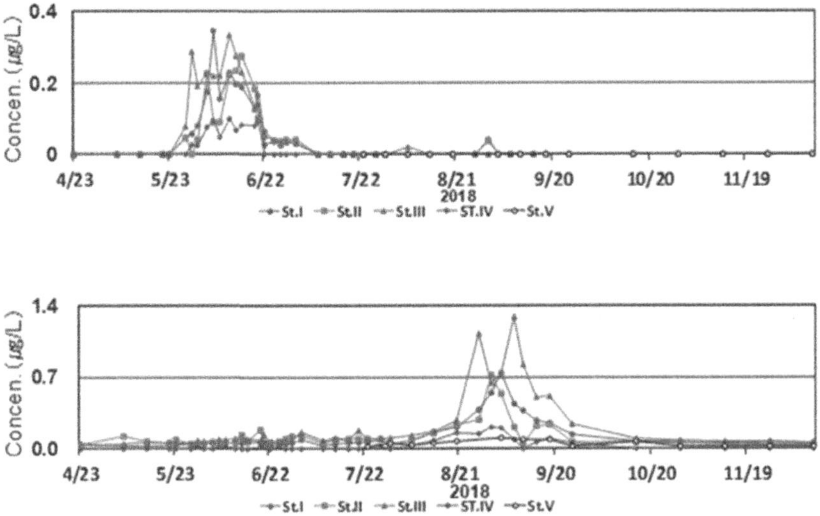

Figure 12.3 Changes in concentrations of (upper graph) imidacloprid; (lower graph) dinotefuran.

area in the river basin. Notable results were obtained for imidacloprid and dinotefuran at St.I, St.II, and St.III. However, the maximum concentrations were slightly lower at St.I than at St.II and St.III, and the concentration peak of imidacloprid was observed in early June and that of dinotefuran was recognized in early September. Rice planting takes place in early June and the flowering season comes in early September. Imidacloprid is commonly applied in rice cultivation as an agent against planthoppers in nursery box treatment. Dinotefuran is often used to protect young brown rice from stinkbugs, which cause spots on rice that reduce its commercial value. The peaks of imidacloprid and dinotefuran were observed during pesticide usages in paddy fields. Since both imidacloprid and dinotefuran are water soluble (Table 12.3), they would be discharged from rice fields into surrounding rivers.

This could explain the significant detection of the pesticides at St.II and St.III, which are mostly surrounded by farmland. Moreover, the field water and paddy soil half-lives (DT50) of imidacloprid are 2.0 days and 12.5 days, respectively (Phong *et al.*, 2009), and the paddy water and paddy soil half-lives (DT50) of dinotefuran are 5.4 days and 12.0 days, respectively (Yokoyama *et al.*, 2015), which could indicate that the pesticides flowed downstream before degrading. In these river basins, rice transplants were carried out approximately from 20 May to 10 June 2018. This also suggest that the concentration peaks of the pesticides at St.I are not attributable only to pesticide effluent from nearby farmland but are also affected by rivers in the basin areas upstream of St.II and St.III.

These results were similar to those obtained by Oyama *et al.* (2019) in which the high concentrations of imidacloprid and dinotefuran were observed in early July and from late August to early September, respectively. The concentration peaks of St.I and St.II and St.III located in the tributaries of St.I were

Table 12.3 Properties of imidacloprid and dinotefuran.

	Water Solubility (µg/L)	Field Water Half-life (day)	Paddy Soil Half-life (day)
Imidacloprid	4.8×10^5	2.0[1]	12.5[1]
Dinotefuran	4.0×10^7	5.4[2]	12.0[2]

[1]Phong *et al.* (2009); [2]Yokoyama *et al.* (2015).

observed. These results would suggest that the peaks of pesticides in river water derived from leaks and runoffs shortly after usage in paddy fields.

12.3.3 Risk assessment by SSD

As acetamiprid was not listed in the SSD database, basic statistics on the risk assessment by SSD of pesticides other than acetamiprid detected at each survey site are shown in Table 12.4. The values in the table show the percentage of species affected by pesticides.

None of the single risks of pesticide exceed HC5 (5%). However, imidacloprid is seen to be close to HC5, with three sites exceeding 3% (St.I, St.II, and St.III). The pesticide concentrations at St.I and the risk assessment by SSD are shown in Figure 12.4 as representative of the survey results.

Thus, even though the concentration of imidacloprid was lower than the concentration of dinotefuran, the risk value of imidacloprid exceeded that of dinotefuran. By using the SSD method to qualify the risk, we were able to understand the current risk not as a concentration but as a risk value. Although none of the single pesticide-specific risks at any site exceeded HC5, the cumulative risk posed by dinotefuran, imidacloprid, clothianidin and thiamethoxam temporarily exceeded HC5 at St.I, St.II and St.III. Oyama *et al.* (2019) also observed a similar tendency at St.I and St.III in June 2017.

Table 12.4 Risk assessment by SSD.

		Max	Min	Mean	Geometric Mean	Median
St.I	Clothianidin	0.3	0.0	0.0	–	0.0
	Dinotefuran	2.0	0.2	0.5	0.4	0.4
	Imidacloprid	3.7	0.0	0.5	–	0.0
	Thiamethoxam	0.1	0.0	0.0	–	0.0
	msPAF	8.2	0.0	1.6	–	0.3
St.II	Clothianidin	0.9	0.0	0.1	–	0.0
	Dinotefuran	2.0	0.3	0.6	0.5	0.4
	Imidacloprid	3.1	0.0	0.5	–	0.0
	Thiamethoxam	0.0	0.0	0.0	–	0.0
	msPAF	7.3	0.0	1.6	–	0.2
St.III	Clothianidin	0.9	0.0	0.0	–	0.0
	Dinotefuran	2.8	0.3	0.8	0.6	0.5
	Imidacloprid	3.5	0.0	0.6	–	0.0
	Thiamethoxam	0.1	0.0	0.0	–	0.0
	msPAF	8.0	0.0	2.0	–	0.6
St.IV	Clothianidin	0.0	0.0	0.0	–	0.0
	Dinotefuran	0.9	0.0	0.2	–	0.0
	Imidacloprid	1.2	0.0	0.2	–	0.0
	Thiamethoxam	0.0	0.0	0.0	–	0.0
	msPAF	3.9	0.0	0.7	–	0.0
St.V	Clothianidin	0.4	0.0	0.0	–	0.0
	Dinotefuran	0.6	0.2	0.3	0.3	0.3
	Imidacloprid	0.0	0.0	0.0	–	0.0
	Thiamethoxam	0.0	0.0	0.0	–	0.0
	msPAF	0.9	0.0	0.1	–	0.1
						(%)

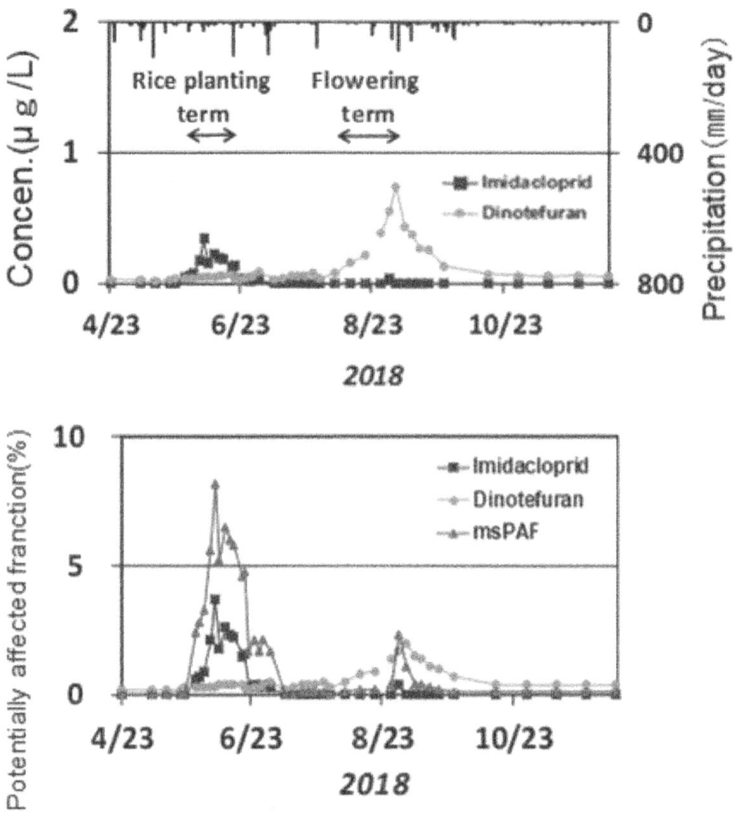

Figure 12.4 (a) High-frequency sampling results; (b) risk assessment with SSD (St.I).

The results of the msPAF by SSD suggested that neonicotinoid pesticides may have some influence on aquatic organisms, even though the concentrations of neonicotinoid pesticides were lower than the standard values. Therefore it is necessary to continue monitoring for water quality and ecosystem disturbance and to evaluate the risk by the assessment method based on SSD. On the other hand, the peak of SSD was also observed at St.V, where the river runs through urban areas and farmland. The effluent from some sewage treatment plants in urban area should be considered in addition to that from farmland. To prevent harm to aquatic organisms, it is important to perform continuous monitoring and assessment that can quantify the environmental risks of pesticides.

12.4 CONCLUSIONS

The concentrations of dinotefuran and imidacloprid were clearly observed to peak immediately after the period when these pesticides were used on farmland in the basins upstream of the survey sites. Relatively high concentrations of neonicotinoid pesticides were observed at sites where the land at the upper basin was mainly used as farmland. Their concentrations were lower than the maximum allowable concentrations specified in the pesticide registration criteria. The concentration of dinotefuran was much higher than that of imidacloprid. However, the risk to the aquatic environment posed by dinotefuran was lower than that posed by imidacloprid when the risk posed to aquatic organisms was assessed by SSD. The cumulative risk of all detected neonicotinoids other than acetamiprid was calculated as higher than HC5. The risk assessment by SSD suggested that neonicotinoid pesticides might have temporary effects on several species in rivers. These results suggested that visualizing the risk posed by pesticides on a numerical scale using the SSD method would more reliably assess the risk posed to aquatic organisms than estimating the risks by the concentrations.

REFERENCES

Japan Meteorological Agency (n.d.). Past weather data search. http://www.data.jma.go.jp/obd/stats/etrn/index.php (accessed 1 July 2019)

Japan Plant Protection Association (2016). Noyaku-handbook 2016 (Pesticides Handbook). Japan Plant Protection Association, Tokyo. (In Japanese).

Kawasaki E., Okubo T. and Sudo M. (2008a). Herbicide discharge from paddy fields located around lake biwa basin and development of simple model for estimation of herbicide. *Journal of Japan Society on Water Environment*, **31**, 375–382, https://doi.org/10.2965/jswe.31.375

Kawasaki E., Sudo M., Miki T. and Shibahara F. (2008b). Application of simple model to estimation of herbicide losses from seven paddy fields located in Lake Biwa basin. *Journal of Japan Society on Water Environment*, **31**, 677–683, https://doi.org/10.2965/jswe.31.677

Nagai T. (2017a). Studies on ecological risk assessment of pesticide using species sensitivity distribution. *Journal of Pesticide Science*, **42**, 124–131, https://doi.org/10.1584/jpestics.J17-03

Nagai T. (2017b). Species sensitivity distribution as the basis of analysis from laboratory to realistic environment. *Japanese Journal of Pesticide Science*, **42**, 133–137 (In Japanese), https://doi.org/10.1584/jpestics.W17-23

Ministry of Land, Infrastructure, Transport and Tourism (MLIT) (n.d.). National Land Numerical Information. MLIT, Tokyo, http://nlftp.mlit.go.jp/ksj/ (accessed 1 July 2019).

National Institute for Agro-Environmental Sciences (NIAES) (n.d.). Species Susceptibility Distribution Analysis for Pesticide Ecological Risk Assessment. NIAEWS, Tokyo http://www.naro.affrc.go.jp/archive/niaes/techdoc/press/160325/ (accessed 15 July 2019).

Osaka Prefecture (n.d.). Neyagawa basin sewer. http://www.pref.osaka.lg.jp/gesui_jigyo/ryuiki/neyagawa.html (accessed 15 July 2019).

Oyama K., Yabuki Y. and Banno A. (2019). Investigation of seasonal changes and ecological risk assessments of in rivers in Osaka, Japan. *Journal of Japan Society on Water Environment*, **42**, 277–284, https://doi.org/10.2965/jswe.42.227

Phong T. K., Nhung D. T. T., Motobayashi T., Thuyet D. Q. and Watanabe H. (2009). Fate and transport of nursery-box-applied tricyclazole and imidacloprid in paddy fields. *Water Air Soil Pollution*, **202**, 3–12, https://doi.org/10.1007/s11270-008-9953-z

Sudo M. and Kawachi T. (2006). Herbicide discharge from paddy fields located around Lake Biwa basin and factors affecting herbicide losses. *Journal of Japan Society on Water Environment*, **29**, 715–721, https://doi.org/10.2965/jswe.29.715

Yamamuro M., Komuro T., Kamiya H., Kato T., Hasegawa H. and Kameda Y. (2019). Neonicotinoids disrupt aquatic food webs and decrease fishery yields. *Science*, **366**, 620–623, https://doi.org/10.1126/science.aax3442

Yokoyama S., Ito M., Nagasawa S., Morohashi M., Ono M., Todate Y., Kose T. and Kawata K. (2015). Runoff and degradation of aerially applied dinotefuran in paddy fields and river. *Bulletin of Environmental Contamination and Toxicology*, **94**, 796–800, https://doi.org/10.1007/s00128-015-1554-0

doi: 10.2166/9781789061123_197

Chapter 13

Addressing adverse water quality in Strathclyde Loch, Scotland

Mark D. Wyer[1], David Kay[1], Calum McPhail[2], Chris Kay[1], Carol Francis[3], Cheryl Davies[3], John Watkins[3] and Julie Bolland[2]

[1]Centre for Research into Environment and Health, Department of Geography and Earth Sciences, Aberystwyth University, Aberystwyth, Ceredigion SY23 3DB
[2]Scottish Environment Protection Agency, Angus Smith Building, 6 Parklands Avenue, Eurocentral, Holytown, North Lanarkshire, Scotland, ML1 4WQ
[3]CREH Analytical, Hoyland House, 50 Back Lane, Horsforth, Leeds, West Yorkshire, LS18 4RS

Keywords: Faecal indicators; phage tracer; water quality; water monitoring

13.1 INTRODUCTION

Strathclyde Loch is a constructed recreational water, used for water sports such as competitive rowing. However, activities also include triathlon events that involve total immersion. It impounds riverine waters draining from the industrial centres and historical industrial towns within the South Calder Water catchment in Lanarkshire. This riverine system provides the main input to Strathclyde Loch (Figure 13.1). The 80 ha artificial lake, located in Strathclyde Country Park, was chosen as the site for triathlon events in the 2014 Glasgow Commonwealth Games. However, concerns over water quality in Strathclyde Loch were raised following an earlier norovirus outbreak amongst participants at an open water swimming event held at the loch in June 2012 (Logan, 2013). In this outbreak, which followed exceptionally heavy rainfall, 85% (i.e., 60 of 71 participants) reported symptoms of gastroenteritis (GI) (diarrhoea and vomiting) and five out of six participants tested proved to have norovirus present in their faeces. Of six kayakers, who did not swim and were acting as stewards for the event, three became ill with GI symptoms whilst five attendees at the event who did not come into contact with the water did not present with any symptoms. It was strongly suspected that the principal source of pathogens, and associated faecal indicator organisms (FIOs), in the loch originated from the South Calder Water catchment. The river receives treated sewage effluent inputs from three sewage treatment works (STWs) (sites 301, 302 and 303 in Figure 13.1), serving an estimated human population equivalent of 33 000, plus untreated sewage discharges from combined sewer overflows (CSOs), that discharge following heavy rainfall, which occurred prior to the competitive event. Strathclyde Loch drains to the River Clyde via sites 101, 102, 103 and 104 (Figure 13.1).

The very high norovirus attack rate, suggested in the Public Health Report for Lanarkshire (2012–13) by the Director of Public Health, caused great concern to the management agencies involved with the proposed Commonwealth Games, to be held there in 2014, and this study was established to provide empirical evidence on the risk to health, information on pollution pathways, source tracking

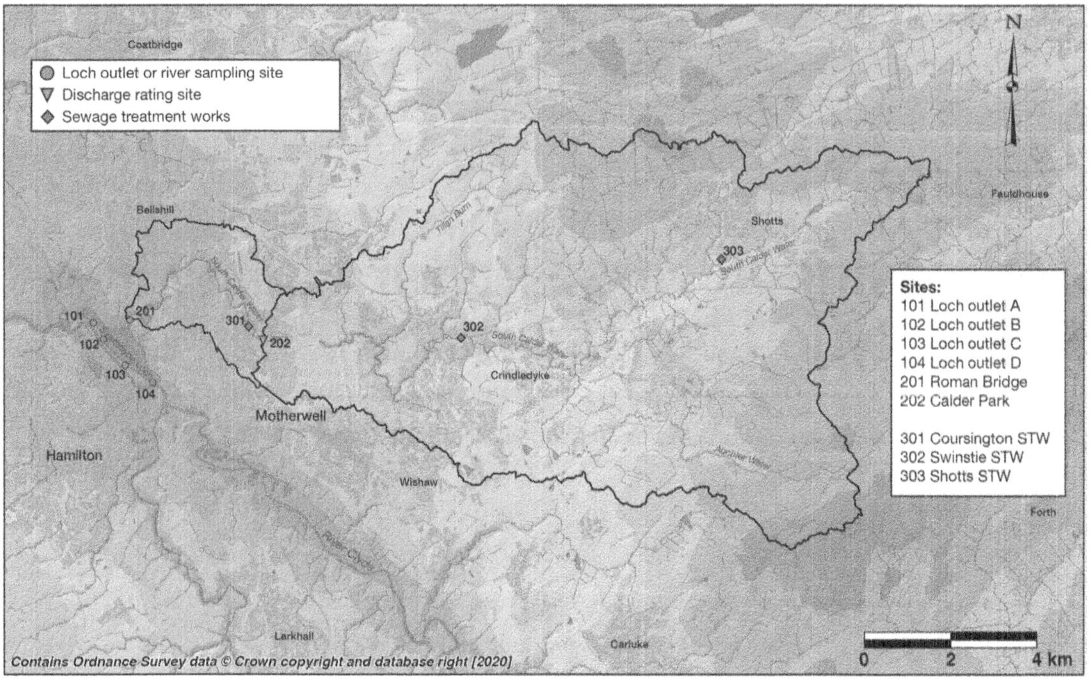

Figure 13.1 Location of Strathclyde Loch outlet sampling sites (101–104), South Calder Water catchment outlet river sampling site (201), South Calder Water discharge gauging site (202), catchment boundaries (black line), and sewage treatment works sites (STWs) (301, 302, 303).

and, ideally, how it could be mitigated. Two investigations were carried out: (i) estimation of the FIO loading for Strathclyde Loch and (ii) testing of a physical curtain barrier designed to isolate an open swimming area from transmission and contamination by viral pathogens, such as norovirus.

The microbial loading study focused on FIOs: *Escherichia coli* (*E. coli*) and intestinal enterococci (IE), which are commonly measured as compliance parameters in many water quality standard systems across the world (e.g., the European Bathing Water Directive (CEU, 2006) and the World Health Organisation Guidelines for Safe Recreational Water Environments (WHO, 2003, 2009)). The direct measurement of specific pathogens, such as norovirus was not regarded as viable based on two considerations: (i) that presence of human pathogens in surface waters is highly episodic and driven by rates of pathogen shedding in the human population, and (ii) work conducted for the EU to inform viral parameter choice for the regulation of bathing waters suggested that the available enumeration techniques (i.e., rtQPCR) produce relatively imprecise measures of the virus. In relation to point (i) above, when the contributing population is not shedding pathogenic viruses the flux for upstream catchments into the loch is likely to be very sparse or absent. In respect of point (ii), European studies (Wyn-Jones *et al.*, 2011) have shown that measures of norovirus contamination did not correlate well with FIOs, such as *E. coli*, which is known to index health risk in recreational freshwaters (Wiedenmann *et al.*, 2006).

Testing the physical barrier system involved the release and detection of a bacteriophage tracer organism, *Serratia marcescens* (*S. marcescens*) PJN1 phage. This organism is a harmless virus, originally isolated from sea water, and should theoretically behave like other virus particles (e.g., pathogenic viruses) in the Strathclyde Loch environment.

13.2 MATERIALS AND METHODS

13.2.1 Sampling programmes

A catchment-scale microbial flux study was designed to define the FIO loading from the South Calder Water Catchment to Strathclyde Loch and from the loch to the River Clyde. Figure 13.1 shows the loch and river water quality sampling sites defined for the full study. These included the South Calder Water river inlet to the loch at Roman Bridge in Strathclyde Country Park (site 201; Figure 13.1). This defines the quality of inflow to the loch and it is reported here. Likewise, the FIO export from the loch is defined by water quality measured at sites 101–104 inclusive (Figure 13.1).

The sampling period covered 38 days between 00:00 BST on 12 August 2013 and 23:45 BST on 18 September 2013. The sampling regime comprised of routine dry weather sample collection supplemented by opportunistic response to rainfall-induced hydrograph events. At site 201 samples were collected by lowering a sampling can into the water using a rope. To minimize cross-contamination between sites, the container was dried and then cleaned with laboratory disinfectant wipes (Azo wipes) prior to each sample and then filled and emptied three times before decanting the water into a pre-sterilized container.

The phage tracer study sites comprised a tracer release point, near the inlet of South Calder Water to the Loch, and nine sampling sites around the shore of the loch (Figure 13.2). These included the outlets from the loch to the River Clyde (sites 101–104 inclusive, Figures 13.1 and 13.2) and paired sites adjacent to the curtain barrier (Sites 105A/B and 106A/B, Figure 13.2). An hourly sampling programme ran for 73 hours between 09:00 GMT on 11 November 2013 to 11:00 GMT on 14 November

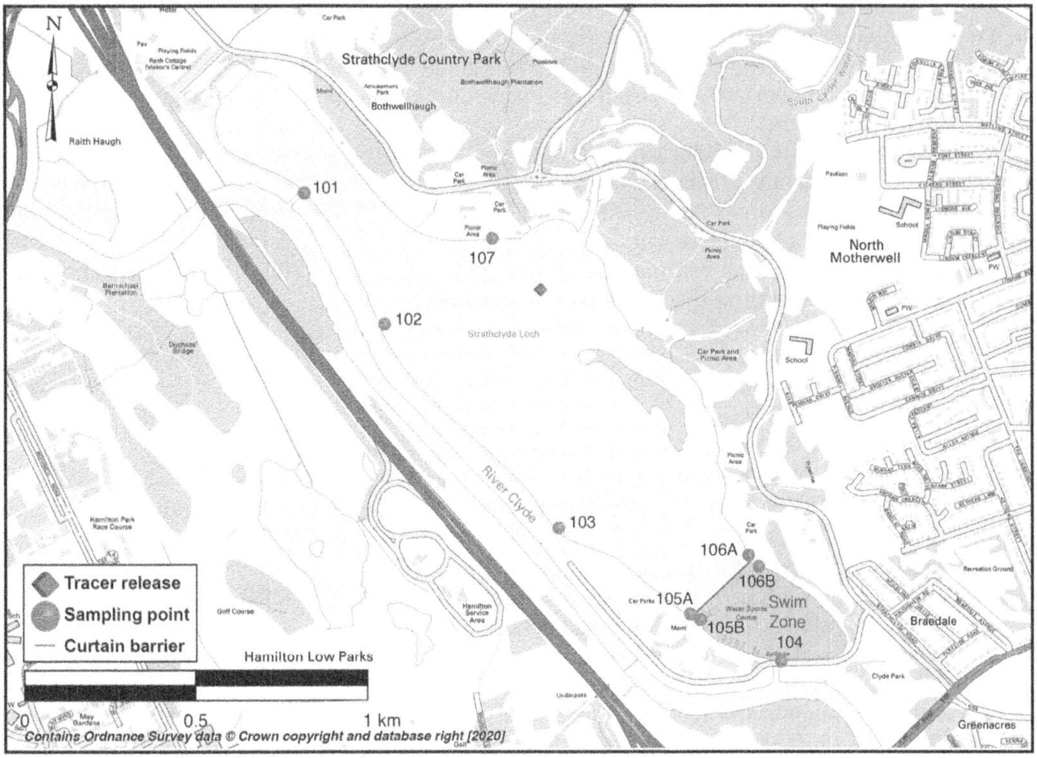

Figure 13.2 *Serratia marcescens* bacteriophage tracer release and sampling points and the location of curtain barrier and swim zone in Strachclyde Loch.

2013. Thus, the sampling programme commenced prior to the *S. marcescens* phage release at 10:17 GMT on 11 November 2013.

Samples were collected aseptically in pre-sterilized containers using a pole with a laboratory clamp attached. This allowed sampling from the shore without personnel entering the water. Samples were placed inside insulated cool boxes on melting ice. This ensured that samples remained in dark, cool conditions, prior to and during transport to the laboratory. Dedicated sets of sampling equipment (i.e., poles, clamps and cool boxes) were assigned to sites within the swim zone (i.e., Sites 104, 105B and 106B, Figure 13.2) and the remaining sites. Each sampler in the two-person sampling team was dedicated to a specific group of sites (i.e., inside and outside the swim zone) for the duration of each sampling shift. Equipment was routinely cleaned using Azo wipes after collection of each sample and disposable gloves were routinely changed on an hourly basis. This regime was adopted to minimize any cross contamination from the process of sample collection, due to the detection sensitivity of the phage tracer used.

13.2.2 Hydrometric monitoring and catchment analysis

A sentinel discharge gauging site for South Calder Water was established at Calder Park (site 202, Figure 13.1). This included a pressure transducer system (Eijkelkamp Diver), logging river level (m) at 0.25 hour intervals and a 1 m staff gauge board. A programme of open channel discharge measurements (Q, m^3/s), across a range of river levels, was undertaken using standard procedures (Environment Agency, 2003; Herschy, 1985). Velocities, at 0.6 of the depth from the water surface, were measured using electromagnetic velocity meters (Sensa RC2) and the average of three measurements was recorded at regular intervals across the channel profile. The resulting discharge measurements were related to staff gauge readings to generate a discharge rating curve for the station. Pressure transducer level data from site 202 were related to regular staff gauge readings to produce a continuous river level record for the study period. The discharge rating curve was applied to this record to produce a continuous discharge record.

Topographic catchment boundaries, and corresponding areas, were generated from a digital terrain model (DTM), using Ordnance survey data, in a geographical information system (GIS) using the Geographic Resources Analysis Support System (GRASS 6.4) software hydrologic module watershed analysis tools (Neteler & Mitasova, 2008). Discharge (Q_o) at site 201 was estimated using the topographic catchment area estimate for the gauging station (site 202; A_g) and the sampling site (A_o) to adjust the gauging station discharge (Q_g) accordingly:

$$Q_o = Q_g \times \left(\frac{A_o}{A_g} \right) \tag{13.1}$$

13.2.3 Laboratory preparation and analysis

Analysis of samples for *E. coli* and IE followed standard procedures based on membrane filtration as specified in relevant European Directives (e.g., CEU, 2006). Most catchment samples were analysed at the Scottish Environment Protection Agency (SEPA) Laboratory in East Kilbride, with backup analysis at the CREH Analytical laboratory in Horsforth, Leeds. The CREH Analytical laboratory analysed all samples related to the swim zone barrier study.

E. coli were enumerated using equivalent chromogenic media as outlined in SCA (2009, 2011). The SEPA laboratory used Tryptone Bile X-glucoronide agar (TBX) whilst the CREH Analytical laboratory employed membrane lactose glucoronide agar (MLGA, Oxoid/Glycosynth). Membranes were incubated for 4 h at 30°C, followed by 14 h at 44°C (±0.5°C). All blue colonies were counted as *E. coli*.

Enterococci were isolated using membrane enterococcus agar (MEA, Oxoid) by incubation for 4 h at 37°C, followed by 44 h at 44°C (±0.5°C) (SCA, 2012). All maroon colonies were counted as presumptive enterococci. Membranes were then transferred to kanamycin aesculin azide agar (KAAA, Oxoid) and incubated for 6 h at 44°C (±0.5°C). All colonies that developed black halos were counted as confirmed IE.

The lower limit of detection for the analyses was variable, depending on the serial dilutions applied to individual analyses from each site. The minimum lower limit of detection was 100 cfu/100 ml at the SEPA laboratory and 9 cfu/100 ml at the CREH Analytical laboratory. Appropriate positive and negative controls were included with each batch of samples.

A suspension of the *S. marcescens* (PJN1 isolated from sea water, host bacterium: *S. marcescens* SM24 (NCIMB 10644)) was prepared by growing the host bacterium in high nutrient medium (brain heart infusion (Oxoid), casamino acid (Difco) broth) (SCA, 2000) in a batch fermenter (L H Fermentation 500 series) and monitoring growth rate by the level of carbon dioxide in the effluent gas (ADC Gas Analysis, Hoddesdon). At an appropriate stage in the growth cycle, the fermenter was infected with the bacteriophage, followed by subsequent lysis. The lysate was cleaned by centrifugation (3600 × g for 0.5 h at 4°C) and the supernatant shaken with 10% polyethylene glycol 6000. The resultant precipitate was concentrated by centrifugation as above, mixed with an equal volume of glycerol, and titrated using the agar overlay method outlined below to determine the tracer concentration. Concentrates were stored below −20°C until required.

Bacteriophage enumeration followed the double agar overlay method (Adams, 1959; Havelaar & Hogeboom, 1984; SCA, 2000). Host bacterium broth cultures were grown in the same high nutrient medium used for fermentation, incubated at 37°C overnight. 0.1 ml of host bacterium and 1 ml of sample were pipetted into 4 ml of molten semi-solid overlay (45°C) and plated onto blood agar base in 90 mm triple-vent Petri dishes. Overlays were allowed to set and the plates dried by partly removing the lids and allowing the plates to stand for 15 minutes. The lids were then replaced, the plates inverted and incubated (37°C for 15 ± 3 h). Following incubation, areas of lysis (plaques) were counted and reported as plaque forming units (pfu)/ml.

13.2.4 Statistical analysis

Statistical analyses were performed using the SPSS statistical software package (SPSS, 2010). The parametricity of distributions was assessed using Shapiro-Wilk (S–W) and Kolmogorov-Smirnov (K–S) normality tests and the Skewness statistic. Where distributions were positively skewed, a $\log_{10}$ data transformation was applied to improve parametricity. Here, any FIO results that were below the lower limit of detection were assigned the detection limit value. General descriptive statistics included the mean, standard deviation (SD), range and the 95% confidence interval (CI) for the mean. The independent sample Student's t-test was used to compare means between two groups (e.g., two sites or flow conditions). The outcome of the corresponding Levene test for homogeneity of variances was used to determine the appropriate type of t-test; based on either (i) separate or (ii) pooled variance estimates. Single sample Student's t-tests were used to compare a sample mean with a particular value.

Robust analysis of variance (ANOVA) was employed to examine the significance of differences between more than two mean values. Here, the significance of the ANOVA was judged on (i) the Levene test for homogeneity of group variances, and (ii) whether the numbers of observations (n) in groups (k) can be considered equal. Where variances can be considered homogenous and n values are equal, the significance (p) of the F statistic was used. Where n values were equal but variances not homogenous, then the Brown–Forsyth statistic p value was used. Finally, when n values were unequal and variances not homogeneous, the significance of the Welch statistic was employed. The Levene test also drives the selection of an appropriate *post-hoc* test to explore the significance of multiple paired comparisons between means. Where variances were homogenous the Tukey test was selected, whilst the Tamhane test was employed when variances were not homogenous. Additionally, Dunnett's test was used for comparisons of means with one specific group mean.

The overall statistical significance of all tests was evaluated at the 95% confidence level.

13.2.5 Discharge component and sample categorization

Previous studies (e.g., Kay *et al.* 2005, 2008; Wyer *et al.* 1998) have shown that FIO concentrations in rivers and streams tend to be significantly elevated during hydrograph response to rainfall. The discharge sequence for the gauged catchment was, thus, split into two components: (i) base flow, dry weather discharge ($_{Qb}$) and wet weather, high flow event discharge (Q_h). This was achieved by detailed inspection of individual hydrograph events. This split was then used to categorize the samples collected from river sampling sites into two corresponding groups. Stream level records from the pressure transducers at individual sites and site observations were also used in the sample categorization.

13.2.6 Faecal indicator organism load and budget calculations

The total loading of FIOs (L_t, organisms) from each site (i) during the study period was calculated as:

$$L_{it} = L_{ib} + L_{ih} \tag{13.2}$$

where:

L_{ib} is the dry weather FIO load (organisms), and
L_{ih} is the wet weather event FIO load (organisms)

computed as:

$$L_{ib} = V_{ib} \times C_{ib} \tag{13.3}$$

$$L_{ih} = V_{ih} \times C_{ih} \tag{13.4}$$

where:

V_{ib} is the dry weather discharge volume at the site (m^3),
C_{ib} is the dry weather FIO concentration at the site (organisms/m^3),
V_{ih} is the wet weather event volume discharged at the site (m^3), and
C_{ih} is the wet weather event FIO concentration at the site (organisms/m^3).

The FIO concentrations, C, used in the calculations are typically geometric mean (GM) values derived from statistical analysis of the FIO concentration data. The totals are used to examine the proportion (%) of the FIO load at each site under the two discharge conditions.

13.3 RESULTS AND DISCUSSION

13.3.1 Hydrometric monitoring and discharge estimates

Figure 13.3a shows the results of discharge gauging and the resulting discharge rating curve derived for South Calder Water at Calder Park (site 202; Figure 13.1). The best rating curve was a power function and exhibited a high level of explained variance (adjusted r^2: 99%). Figure 13.3d shows the discharge time series for the study period, generated by applying the rating to the staff gauge corrected pressure transducer record. Only 3% of observations exceeded the upper threshold of the rating (2 m^3/s) during the study period. Table 13.1 summarises the DTM derived catchment area estimates for Calder Park and Roman Bridge.

The total discharge estimate for South Calder Water at Calder Park was 2.48×10^6 m^3 (Table 13.1) of which approximately 71% was associated with dry weather conditions and 29% associated with hydrograph response (Figure 13.3b) in relation to rainfall at Shotts STW (Figure 13.3c). The hydrograph response discharge component occupied 159 h (17.43%) of the study period (Figure 13.3d). During the dry weather periods of the study the discharge pattern displayed a marked daily 'saw tooth' pattern (Figure 13.3d). This probably relates to the upstream input from the catchment STWs, the closest of which is Swinstie STW (site 304; Figure 13.1).

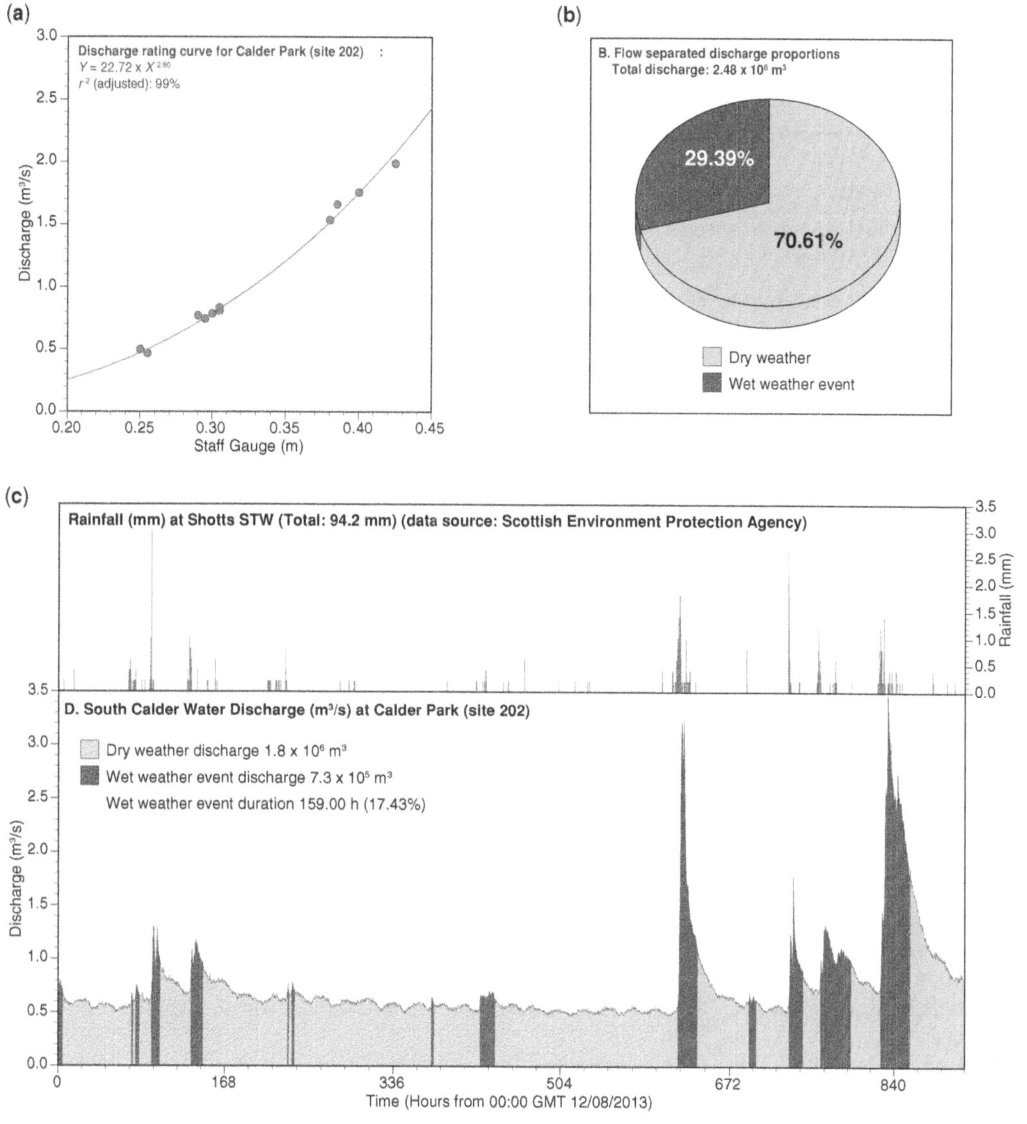

Figure 13.3 Discharge rating curve for South Calder Water at Calder Park, rainfall (mm) at Shotts Sewage Treatment Works (STW) and the flow separated discharge (m^3/s) for the river between 00:00 BST on 12/08/2013 and 23:45 BST on 18/09/2013 (see Figure 13.1 for site details).

Catchment area scaled discharge estimates for the input to Strathclyde Loch at Roman Bridge are given in Table 13.1. The same scaling was used to generate a temporal discharge sequence for the input, based on the time series for Calder Park shown in Figure 13.3b.

13.3.2 Faecal indicator organism concentrations

A total of 29 samples were collected at Roman Bridge during the study period. Of these, 16 (55%) were categorized as rainfall response hydrograph event samples. *E. coli* results for two event samples were

Table 13.1 Catchment areas, area factors and discharge estimates for river sampling sites in the South Calder Water catchment between 00:00 BST on 12/08/2013 and 23:45 BST on 18/09/2013.

Site	Name	Area (km^2)	Area Factor[a]	Q_b (m^3)[b]	Q_h (m^3)[c]	Q_t (m^3)[d]
201	Roman Bridge	95.91	1.0897	1.91×10^6	7.95×10^5	2.70×10^6
202[e]	Calder Park	88.02	1.0000	1.75×10^6	7.29×10^5	2.48×10^6

[a] A_o/A_g, where A_g is the gauged catchment area (site 203) and A_o is the corresponding sample site catchment area.
[b] Dry weather discharge estimate: $Q_{bg} \times$ Area Factor, where Q_{bg} is the dry weather discharge estimate for the gauged catchment (site 203).
[c] Wet weather hydrograph event response discharge estimate: $Q_{hg} \times$ Area Factor, where Q_{hg} is the wet weather response discharge estimate for the gauged catchment (site 203).
[d] Total: $Q_b + Q_h$.
[e] Gauged catchment (see Figure 13.2).

not reported (i.e., $n = 14$). Normality tests showed that distributions of FIO concentrations at Roman Bridge were $\log_{10}$ normal (S-W $p > 0.05$). Further statistical analysis of the riverine input FIO data was, thus, performed on $\log_{10}$ transformed values.

Statistical summaries of the FIO concentration data for Roman Bridge are given in Table 13.2. The mean $\log_{10}$ concentrations of both FIOs showed statistically significant elevations during hydrograph event response (Student's t-test $p < 0.05$). The elevation in *E. coli* concentration ranged from 2500 cfu/100 ml during dry weather to 12 000 cfu/100 ml under high flow event, conditions. The corresponding elevation for IE was from 440 cfu/100 ml to 8100 cfu/100 ml.

Comparison of these GM concentrations with typical values published for UK catchments (Kay *et al.*, 2008) suggests that the observed values are not atypical. However, the corresponding catchment classifications were inconsistent, with GM *E. coli* concentrations typical of agricultural catchments with >75% improved pasture during dry weather conditions and >75% rough grazing under event conditions. The results for IE probably reflect the catchment type more closely, fitting with GM IE concentrations typical of a semi-urban catchment under both flow conditions.

Corresponding sets of 29 samples were collected from the four outlets from Strathclyde Loch to the River Clyde (Sites 101–104, Figure 13.1). Analytical results showed a high proportion of censored values below the limit of detection, affecting 36.2% of *E. coli* results and 62.1% of IE results. To allow use of the data all values recorded as <100 cfu/100 ml were assigned a maximum possible concentration of 99 cfu/100 ml as a 'worst case'. Twenty IE results were reported with lower detection limits (16 results as <9 cfu/100 ml and 4 results as <91 cfu/100 ml) and were assigned the corresponding detection limit values.

Based on all available valid results from the loch outlets (74 *E. coli* results and 44 IE results), normality tests showed statistically significant departures from normality for both raw and $\log_{10}$ transformed FIO concentrations (S–W $p < 0.05$). However, the $\log_{10}$ transformation did reduce skewness to <1 for both FIOs. Further analyses were thus based on $\log_{10}$ concentrations.

Table 13.2 summarizes FIO concentrations at the four loch outlets. With the exception of *E. coli* at site 103, Student's t-tests showed no significant differences between mean $\log_{10}$ FIO concentrations in the dry weather and wet weather response groups (Student's t-test $p > 0.05$). In the case of *E. coli* at site 103, the mean $\log_{10}$ value was significantly lower in the wet weather group (Student's t-test $p < 0.05$) and thus did not indicate a significant deterioration in water quality under wet weather event conditions.

ANOVA was used to compare the 'worst case' mean $\log_{10}$ FIO values for the loch outlets with the corresponding mean $\log_{10}$ value for the South Calder Water input to the loch (i.e., site 201, Roman Bridge). The comparisons, using all data and data split by the dry weather and wet weather event groupings, consistently demonstrated that the mean $\log_{10}$ FIO concentrations at the loch outlets each

Table 13.2 Summary of faecal indicator organism concentrations (colony-forming units (cfu)/100 ml at outlet sampling sites from Strathclyde Loch and the South Calder Water input to the loch at Roman Bridge between 00:00 BST on 12/08/2013 and 23:45 BST on 18/09/2013.

Site[a]	GM[b]	95% CI[c] Lower	95% CI[c] Upper	Min.	Max.	n[d]
A. *Escherichia coli* – dry weather						
101 Loch outlet A	1.3×10^2	9.9×10^1	1.8×10^2	9.9×10^1	4.0×10^2	13
102 Loch outlet B	2.2×10^2	1.3×10^2	3.9×10^2	9.9×10^1	1.9×10^3	13
103 Loch outlet C	5.1×10^2	3.1×10^2	8.2×10^2	9.9×10^1	2.8×10^3	13
104 Loch outlet D	1.3×10^2	1.0×10^2	1.7×10^2	9.9×10^1	3.0×10^2	13
All Loch outlet samples	2.1×10^2	1.7×10^2	2.7×10^2	9.9×10^1	2.8×10^3	52
201 Roman Bridge	2.5×10^3	1.4×10^3	4.5×10^3	5.0×10^2	3.6×10^4	13
B. *Escherichia coli* – wet weather hydrograph response						
101 Loch outlet A	1.0×10^2	9.5×10^1	1.1×10^2	9.0×10^1	1.8×10^2	16
102 Loch outlet B	1.4×10^2	9.5×10^1	1.9×10^2	9.9×10^1	1.0×10^3	16
103 Loch outlet C	1.7×10^2	1.0×10^2	2.9×10^2	4.0×10^1	1.0×10^3	16
104 Loch outlet D	1.3×10^2	1.1×10^2	1.6×10^2	9.9×10^1	2.8×10^2	16
All Loch outlet samples	1.3×10^2	1.1×10^2	1.6×10^2	9.9×10^1	1.0×10^3	64
201 Roman Bridge	1.2×10^4	3.8×10^3	4.0×10^4	2.0×10^2	1.8×10^5	14
C. Intestinal enterococci – dry weather						
101 Loch outlet A	9.9×10^1	9.9×10^1	9.9×10^1	9.9×10^1	1.0×10^2	13
102 Loch outlet B	9.9×10^1	9.9×10^1	9.9×10^1	9.9×10^1	1.0×10^2	13
103 Loch outlet C	1.1×10^2	9.5×10^1	1.3×10^2	9.9×10^1	2.0×10^2	13
104 Loch outlet D	9.9×10^1	9.9×10^1	9.9×10^1	9.9×10^1	1.0×10^2	13
All Loch outlet samples	1.0×10^2	9.8×10^1	1.1×10^2	9.9×10^1	9.9×10^1	52
201 Roman Bridge	4.4×10^2	2.5×10^2	7.9×10^2	1.0×10^2	3.1×10^3	13
D. Intestinal enterococci – wet weather hydrograph response						
101 Loch outlet A	7.9×10^1	4.9×10^1	1.3×10^2	9.0×10^0	2.0×10^2	16
102 Loch outlet B	6.1×10^1	3.6×10^1	1.0×10^2	9.0×10^0	3.0×10^2	16
103 Loch outlet C	1.2×10^2	7.3×10^1	1.9×10^2	9.0×10^0	5.5×10^2	16
104 Loch outlet D	9.3×10^1	6.6×10^1	1.5×10^2	1.0×10^1	3.5×10^2	16
All Loch outlet samples	8.5×10^1	6.8×10^1	1.1×10^2	9.0×10^0	5.5×10^2	64
201 Roman Bridge	8.1×10^3	3.8×10^3	1.7×10^4	5.0×10^2	6.0×10^4	16

[a]See Figure 13.1 for details.
[b]Geometric mean.
[c]95% confidence interval for the GM.
[d]Number of observations.

showed a statistically significant lower mean than site 201 (Dunnett's $p < 0.05$). Given that the outlet means would actually be lower than those used, due to the censored values, the differences between the outlets and the South Calder Water input would be even greater. These results demonstrate that the loch has an attenuating effect on FIO concentrations between the river inlet and the outlets, through dispersion, sedimentation and die-off processes in the loch.

13.3.3 Discharge and faecal indicator organism budgets

The total *E. coli* load estimate from South Calder Water to Strathclyde Loch during the study period amounted to 1.44×10^{14} organisms. Just over two-thirds (67.22%) of this input was associated with wet weather event conditions that accounted for 29.39% of the total discharge and associated with 17.43% of the sampling period. The total IE input was estimated at 7.26×10^{13} organisms, with 88.39% associated with high flow event conditions, following rainfall. This reflects the large, statistically significant, increases in FIO concentrations during hydrograph event response. This illustrates the importance of rainfall event driven mobilisation of FIOs from the contributing catchment to Strathclyde Loch. Such conditions were implicated in June 2012 norovirus outbreak.

This is further illustrated in Figure 13.4, which shows temporal load estimates for FIOs to Strathclyde Loch through the study period and the pulses of elevated FIO flux associated with short term hydrograph response. Typical dry weather *E. coli* flux ranged between 1.25×10^7 and 4.83×10^7 organisms/s (mean: $1/74 \times 10^7$ organisms/s). The peak load estimate for *E. coli* was 4.62×10^8 organisms/s, 1.4 $\log_{10}$ orders higher than during dry weather. The dry weather IE flux ranged from 2.22×10^6 to 8.60×10^6 organisms/s (mean: 3.11×10^6 organisms/s). The estimated peak IE flux during the study period was 3.06×10^8 organism/s, 2.0 $\log_{10}$ orders of magnitude higher than the dry weather (background) IE flux and more extreme than the *E. coli* elevation. These results highlight the importance of elevated FIO loading to Strathclyde Loch associated with short term rainfall response hydrograph events, which can increase FIO flux from the river by between 1.4 and 2.0 $\log_{10}$ orders of magnitude (i.e., a 25–100 fold increase) compared to the dry weather FIO flux. If a viral pathogen, such as norovirus, was present in the community it is likely that it would be rapidly mobilized in the catchment from infected sewage sources under high flow event response episodes when the discharge of untreated sewage effluent, through the network CSOs, would also be greatest.

FIO load estimates from Strathclyde Loch to the River Clyde were constructed assuming that the overall discharge to the loch (i.e., discharge at site 201) and the outlet discharge were balanced through the study period. The GM *E. coli* and IE concentrations from all outlet samples were used in the calculations (Table 13.1). The results suggest an overall FIO load attenuation through the loch of 96%. The attenuation estimates were lower during dry weather (*E. coli*: 91.51%; IE: 76.91%). The corresponding wet weather event FIO reductions were 99%. This suggests that, overall, the loch provides a hostile environment for FIO survival, favouring bacterial die-off and sedimentation processes. However, these estimates do not consider any temporal variations in loading evident in Figure 13.4. For example, at peak loading 99% attenuation would still leave an estimated residual flux of 1.66×10^{10} *E. coli*/h and 1.12×10^{10} IE/h from the loch to the River Clyde.

The evidence of the 2012 norovirus outbreak suggests that any similar attenuation of the pathogenic virus input from South Calder Water through the loch was not great enough to prevent a related disease outbreak following rainfall response event conditions. This is also likely to be the case for other viral pathogens (e.g., adenovirus) and bacterial pathogens present in the catchment and corresponding community. Thus, the realistic solution to exclude pathogens from the open swimming area in the loch, defined for the 2014 Commonwealth games triathlon event, was to design and construct a physical curtain barrier. Exclusion of microbial pathogens from the swim zone would reduce the health risk to participants from water contact exposure to pathogens such as norovirus. The barrier comprised a combination of floating pontoon and flexible PVC liner barrier systems. At the time of the study only the latter had been installed, with two flexible barriers separated by a stretch of open water.

13.3.4 Swim zone curtain barrier efficacy study using a bacteriophage tracer

The results of pre-study sampling at the sites shown in Figure 13.2 on 05 November 2013 confirmed the absence of the *S. marcescens* phage. With the exception of the sample collected from site 101, the phage was also absent in samples collected from the sites prior to the tracer release at 10:17 GMT on

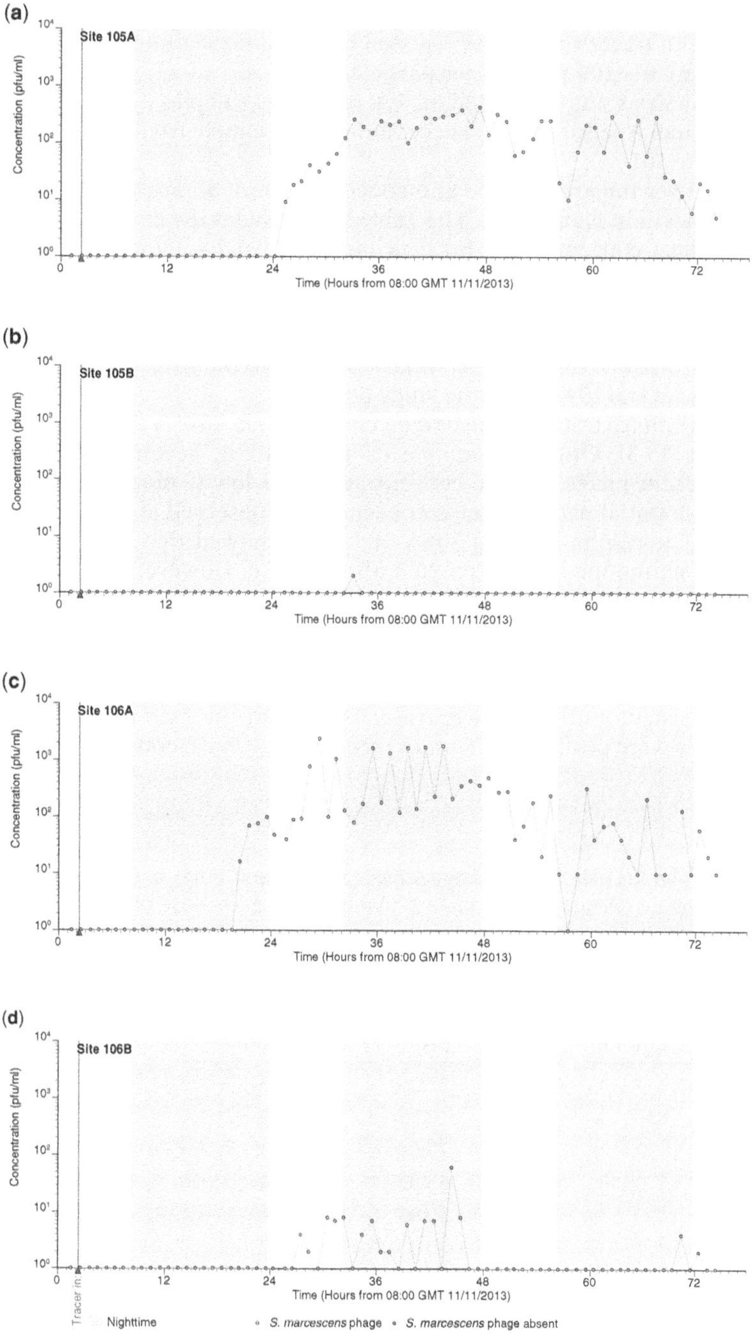

Figure 13.4 Concentrations of *Serratia marcescens* phage (plaque forming units per ml (pfu/ml)) at pairs of sites outside the barrier swim zone (sites 105A and 106A) and inside the barrier swim zone (Sites 105B and 106B) following *S. marcescens* phage release to the Strathclyde Loch at the South Calder Water inlet (see Figure 13.2 for site details).

11 November 2013. The concentration in the positive sample was 1 pfu/ml, which is the lower limit of detection. The litre of tracer suspension released into Strathclyde Loch close to the South Calder Water input contained 1.0×10^{16} phage virus particles.

Successful phage analysis was completed on 660 of the 666 samples (i.e., 99.1%) collected through the 73 h sampling period. Analysis was unsuccessful for one sample from each of sites 103, 106A and 107 and three from site 101.

Table 13.3 summarises the arrival time and concentration of *S. marcescens* phage at each of the nine sampling sites shown in Figure 13.2. The Table also includes the corresponding concentrations. Time series plots of phage concentrations for sites immediately adjacent to the barrier curtain outside and inside the swim zone are shown in Figure 13.4. The short arrival times (<15 h) at sites 101, 102 and 107 suggest a north-westward movement of tracer away from the swim zone, with the sluice at site 102 providing a rapid exit for contaminated water to the River Clyde (arrival time <9 h). Management of the water level in the loch was set for preferential discharge to the River Clyde via sites 101, 102 and 103, with no discharge at site 104 during the study period.

Transmission of the phage tracer to the barrier curtain at sites 105A and sites 106A took between 18 h and 23 h (Table 13.3). Only one sample from site 105B, beyond the curtain barrier, was positive for *S. marcescens* phage and the concentration was low (2 pfu/100 ml, Figure 13.5b). This corresponded with the initial peak in phage concentration observed at site 105A outside the swim area (Figure 13.5a). The results for sites 106A and 106B showed that the phage was transmitted across the barrier, reaching the swim area 25 h after release. However, the peak concentration at site 106B inside the swim zone (61/pfu/ml) was much lower than outside the barrier (2360 pfu/ml) (Figure 13.5c and 13.5d). Observations at the time of sampling suggested issues with sealing of the barrier at the landfall of the structure at the lock shore, intermittent closure of the two curtain walls and wind-driven wave action allowing water from outside the barrier to enter the swim zone. These observations are consistent with the phage tracer results. At the extremity of the swim zone (site 104) only two samples were positive for *S. marcescens* phage, with concentrations at the detection limit of 1 pfu/ml.

Table 13.3 Summary of *Serratia marcescens* phage arrival times (hours), times to peak concentration (hours) and corresponding concentrations (plaque forming units (pfu)/ml) following a release of the phage close to the South Calder Water input to Strathclyde Loch.

Site[b]	Arrival		Initial Peak Concentration		Maximum Concentration[a]	
	Time[c] (h)	Concentration (pfu/ml)	Time[c] (h)	Concentration (pfu/ml)	Time[c] (h)	Concentration (pfu/ml)
101	13.72	7	23.78	1870		
102	8.87	79	12.78	11 000		
103	15.87	7	23.00	40	41.85	392
104[d]	44.08	1	63.03	1		
105A	23.08	9	30.93	259	41.85	430
105B[d]	30.93	2	30.93	2		
106A	18.17	16	27.18	2360		
106B[d]	25.13	4	42.13	61		
107	6.43	190	7.38	800		

[a]If different from initial peak.
[b]See Figure 13.2 for details.
[c]Time from tracer release at 10:17 GMT 11/11/2013.
[d]Sites within the swim zone area beyond the curtain barrier.

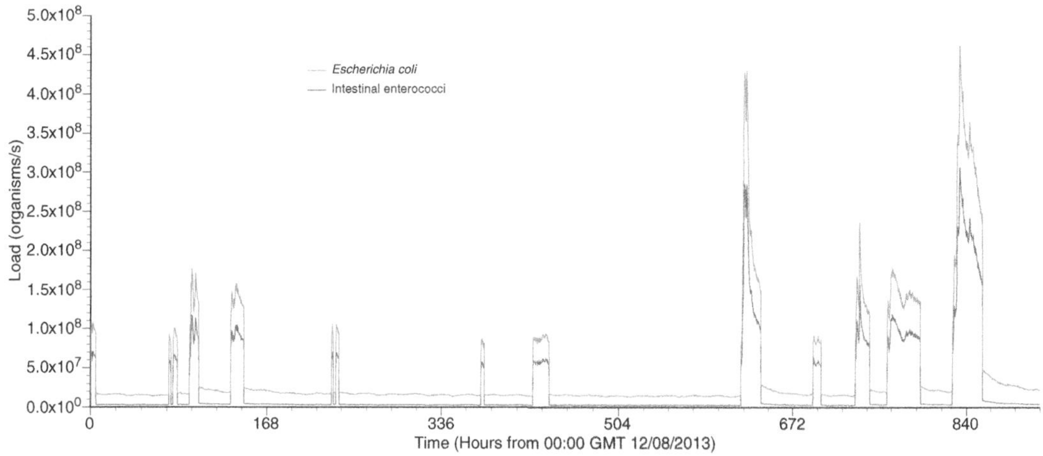

Figure 13.5 Estimates of temporal flux of faecal indicator organisms (organisms/s) from South Calder Water (site 201, Roman Bridge) to Strathclyde Loch between 00:00 BST on 12/08/2013 and 23:45 BST on 18/09/2013.

Overall, the bacteriophage tracer study suggested that the curtain barrier was largely successful at excluding virus particles from the swim zone area. Management of the barrier at the north shore would be likely to improve virus exclusion further. However, conditions during the study were not extreme (6.0 mm rain recorded at Shotts STW (site 303, Figure 13.1)) and more rapid transmission of virus through the loch might be expected under event conditions. Under such conditions the barrier would require careful, detailed, management to prevent contamination of the swim area.

Corresponding FIO concentrations on either side of the curtain barrier during the study period showed lower GM concentrations in the swim zone than outside the barrier (Figure 13.6). ANOVA showed that these differences were statistically significant ($p < 0.05$). This analysis also showed that GM FIO concentrations were not significantly different between the two sites outside the swim zone (sites 105A and 106A). In addition, the GM IE concentrations inside the swim zone (sites 105B and 106B) were also not significantly different from each other. The differences in concentration between the pairs of sites on either side of the curtain barrier were large enough to affect the compliance outcome when compared to the European Bathing Water Directive standards (CEU, 2006), with the swim zone FIO concentrations demonstrating an 'Excellent' water quality classification for both FIOs. In contrast the corresponding outcome for the samples taken at 106A outside the swim zone was 'Poor' for both FIOs, whilst site 105A showed a 'Sufficient' classification for *E. coli* and 'Good' classification for IE. This suggests that the curtain barrier also had a beneficial effect on FIO concentrations, acting to exclude FIOs in the wider loch from the swim zone area.

13.4 SUMMARY AND CONCLUSIONS

Discharge in South Calder water was dominated by dry weather flow (70.6%). In contrast, FIO load estimates for the river input to Strathclyde Loch was dominated by hydrograph response event conditions, which accounted for 67.22% of the *E. coli* load estimate and 89.39% of the IE load estimate. The FIO loading to Strathclyde Loch was thus dominated by hydrograph response conditions, which occupied a relatively short part of the study period (17.43). Using the event duration during the study period, *E. coli* flux estimates for the South Calder Water input to Strathclyde Loch increased by almost one $\log_{10}$ order of magnitude from 6.28×10^{10} organisms/h during dry weather, to 6.10×10^{11} organism/h under hydrograph response conditions. The IE flux estimates at this site showed an even

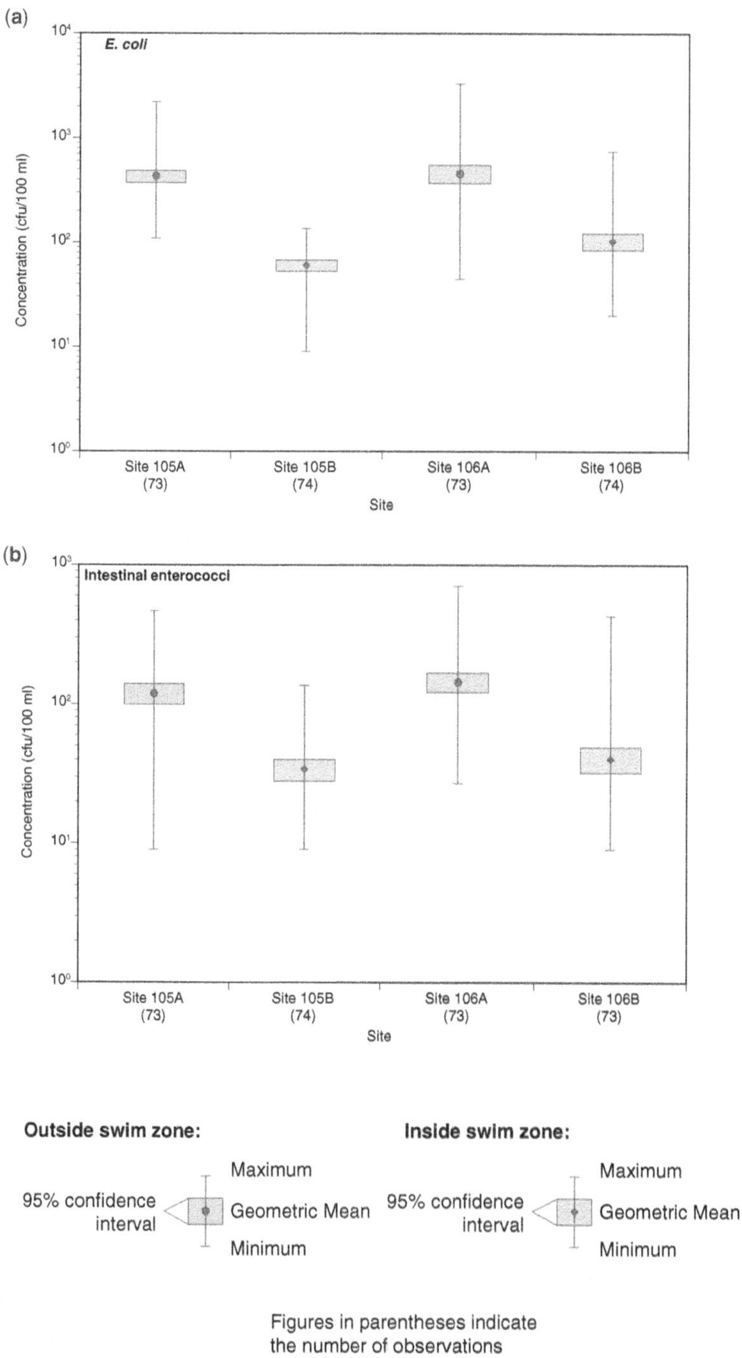

Figure 13.6 Statistical summaries of *Escherichia coli* and intestinal enterococci (colony forming units per 100 ml (cfu/100 ml) in water samples from Strathclyde Loch between 08:00 GMT on 11/11/2013 and 11:00 GMT on 14/11/2013 (see Figure 13.2 for site details).

greater increase, by 1.56 $\log_{10}$ orders of magnitude, from 1.12×10^{10} organisms/h during dry weather to 4.04×10^{11} organisms/h during hydrograph response. Instantaneous FIO flux estimates for the input to the loch suggested that the peak *E. coli* loading during the study period was 25 × higher than the typical dry weather flux, whilst the peak IE flux to the loch was 100 times greater. This suggests that associated pathogen loading might also demonstrate a similar response, with maximum loading to the loch following hydrograph response.

Assuming a balance of discharge entering and leaving the loch, load estimates for the output to the River Clyde suggested 91.51% attenuation of the *E. coli* input to the loch during dry weather 76.91% for IE. Under wet weather response conditions, the attenuation for both FIOs was higher at 99%. This suggests that the loch provides a natural 'self-cleaning' environment, favouring conditions conducive to bacterial dispersion, die-off and sedimentation. However, at peak loading even 99% attenuation would result in a high residual FIO flux (1.66×10^{10} *E. coli*/h and 1.12×10^{10} IE/h). This appears likely to be large enough to increase any pathogen loading to an unacceptable level, increasing the risk of illness to exposed individuals such as triathletes swimming in the loch.

The main remediation approach applied in this instance was a flexible curtain barrier to prevent water exchange between a swim zone at the south western end of the loch, near the water sports centre, and the main body of the impoundment directly receiving riverine inputs. The release of a harmless *S. marcescens* bacteriophage virus to the loch, followed by a 72 h sampling programme, demonstrated that the curtain barrier was effective at excluding the phage from the swim zone. Evidence of slight contamination, particularly on the north shore, suggested that the barrier would require careful management to reduce contamination further. However, the complete barrier system, including the pontoon barrier, had not been installed at the time of the experiment. Results of FIO monitoring during the tracer study suggested that the curtain barrier also acted to reduce FIO levels in the swim zone, improving the water quality classification to 'Excellent'. The exclusion of phage virus and FIO reduction afforded by the curtain barrier was considered a successful strategy to reduce health risks from microbial contamination. As a result, the Commonwealth Games triathlon events scheduled to take place in Strathclyde Loch were therefore allowed to proceed in July 2014. The events attracted wide international media coverage and were regarded as being very successful. All of the strict water quality standards required by the International Triathlon Union were met on the two event days.

ACKNOWLEDGEMENTS

This research project could not have been completed without the assistance of many colleagues at the Scottish Environment Protection Agency (SEPA), Scottish Water, North Lanarkshire Council and Aberystwyth University. From Aberystwyth University we are indebted to: Dan Bennett, Tom Chibnall, Ifan Edwards, Paul Fingleton, Beth Homes, Dan Jones, Noreen Kay and Joe Pearce who undertook the instrumentation and water sampling programme; Dave Whitehead who provided logistical support for sample transport; Paula Hopkins who provided logistical support with vehicle hire and accommodation and Aiysha Ali, Emmie Boulde and Duncan Ewin who provided expert technical support for the microbiological analyses. From SEPA, Morag MacLachlan and Dominic Habron provided logistical support with catchment data whilst John McKechnie, Amanda Reilly and Laurie Flannigan provided environmental data and intelligence on the barrier engineering works. We are also indebted to Brian Devlin (North Lanarkshire Council – Strathclyde Loch Watersports Centre) and his team who provided access and support for completion of the phage tracer releases and sampling programme. In addition, John Caldwell (Scottish Water) provided operational data for sewage treatment works and storm overflows and provided logistical support for access to Scottish water infrastructure sites.

This work was funded by SEPA and the Scottish Government and completed with the guidance and support of Ian Speirs from the Scottish Government.

REFERENCES

Adams M. H. (1959). Bacteriophages. Inter-science, New York, p. 592.

Council of the European Union (CEU). (2006). Directive 2006/7/EC of the European Parliament of 15 February 2006 concerning the management of bathing water quality and repealing Directive 76/160/EEC. Official Journal of the European Union No L 64 4.3.2006, p. 37–51.

Environment Agency. (2003). Draft Hydrometric Manual. Chapter 4: Instantaneous Flow Measurement. Environment Agency, Bristol.

Havelaar A. H. and Hogeboom W. M. (1984). A method for enumeration of male-specific bacteriophages in sewage. *Journal of Applied Bacteriology*, **56**, 439–447, https://doi.org/10.1111/j.1365-2672.1984.tb01372.x

Herschy R. W. (1985). Stream Flow Measurement. Elsevier Applied Sciences Publishers, London and New York, p. 547.

Kay D., Wyer M., Crowther J., Stapleton C. M., Bradford M., McDonald A., Greaves J., Francis C. and Watkins J. (2005). Predicting faecal indicator fluxes using digital land use data in the UK's sentinel water framework directive catchment: The ribble study. *Water Research*, **39**, 3967–3981, https://doi.org/10.1016/j.watres.2005.07.006

Kay D., Crowther J., Stapleton C. M., Wyer M. D., Anthony S., Bradford M., Edwards A., Fewtrell L., Francis C., Hopkins M., Kay C., McDonald A. T., Watkins J. and Wilkinson J. (2008). Faecal indicator organism concentrations and catchment export coefficients in the UK. *Water Research*, **42**, 2649–2661, https://doi.org/10.1016/j.watres.2008.01.017

Logan J. (2013). The strathclyde loch norovirus outbreak. Section 2.6 in chapter 2. In: Public Health 2012/13. The Annual Report of the Director of Public Health, H. S. Kohli (ed.), National Health Service, Lanarkshire, pp. 21–22.

Neteler M. and Mitasova H. (2008). Open Source GIS – A GRASS GIS Approach. 3rd edn. Springer, Berlin, p. 406.

SPSS. (2010). IBM SPSS Statistics 19 Brief Guide. IBM Company, New York, USA, p. 161.

Standing Committee of Analysts (SCA). (2000). The Microbiology of Recreational and Environmental Waters. Methods for the Examination of Waters and Associated Materials. Environment Agency, Bristol.

Standing Committee of analysts (SCA). (2009). The Microbiology of Drinking Water (2009) – Part 4 – Methods for the Isolation and Enumeration of Boliform and *Escherichia coli* (Including *E. coli* O157:H7). Methods for the examination of waters and associated materials. Environment Agency, Bristol, UK, p. 101.

Standing Committee of analysts (SCA). (2011). The microbiology of recreational and environmental waters (2011) – part 3 – methods for the isolation and enumeration of coliform and *Escherichia coli* (including *E. coli* O157:H7). Methods for the Examination of Waters and Associated Materials. Environment Agency, p. 111.

Standing Committee of analysts (SCA). (2012). The microbiology of drinking water (2012) – part 5 – methods for the isolation and enumeration of enterococci. Methods for the Examination of Waters and Associated Materials. Environment Agency, p. 31.

Wiedenmann A., Krüger P., López-Pila J. M., Szewzyk R. and Botzenhart K. (2006). A randomized controlled trial assessing infectious disease risks from bathing in fresh recreational waters in relation to the concentration of *Escherichia coli*, intestinal enterococci, *Clostridium perfringens*, and somatic coliphages. *Environmental Health Perspectives*, **114**(2), 228–236, https://doi.org/10.1289/ehp.8115

World Health Organization (WHO). (2003). Guidelines for Safe Recreational Water Environments. Volume 1: Coastal and Fresh Waters. World Health Organization, Geneva, Switzerland, p. 219.

World Health Organization (WHO). (2009). Addendum to the WHO Guidelines for Safe Recreational Waters. Volume 1: Coastal and Fresh Waters. List of Agreed Updates. World Health Organization, Geneva, Switzerland, pp.1–33.

Wyer M. D., Kay D., Crowther J., Whittle J., Spence A., Huen V., Wilson C., Carbo P. and Newsome J. (1998). Faecal-indicator budgets for recreational coastal waters: a catchment approach. *Journal of the Chartered Institution of Water and Environmental Management*, **12**(6), 414–424, https://doi.org/10.1111/j.1747-6593.1998.tb00210.x

Wyn-Jones A. P., Carducci A., Cook N., D'Agostino M., Divizia M., Fleischer J., Gantzer C., Gawler A., Girones R., Höller C., de Roda Husman A., Kozyra I., López-Pila J., Muscillo M., São José Nascimento M., Papageorgiou G., Rutjes S., Sellwood J., Szewzyk R., Wyer M. D. and Kay D. (2011). Surveillance of adenoviruses and noroviruses in European recreational waters. *Water Research*, **45**, 1025–1038, https://doi.org/10.1016/j.watres.2010.10.015

doi: 10.2166/9781789061123_213

Chapter 14

Socio-economic impacts of pollution and restoration: Loch Leven case study, UK

B. J. D'Arcy[1], L. Kuhfuss[2], L. May[3] and R. Audsley[4]

[1]Independent Consultant, Kinross, UK
[2]James Hutton Institute, Aberdeen, UK
[3]UK Centre for Ecology and Hydrology, UK
[4]SRUC, Auchincruive, UK

Keywords: Algal blooms; environmental regulation; Loch Leven restoration; socio-economic issues; tourism; water pollution

14.1 INTRODUCTION

14.1.1 Economic costs of pollution

The costs to a polluter of major industrial incidents or failure to address chronic pollution can run into millions of US dollars (Harrison & D'Arcy, 2018). That reality may be a reason for a belief that environment comes only at a major cost to business, and that society has to make a stark choice between economic development (with pollution) or a clean healthy environment. Although that view is long gone in most modern industries (e.g., D'Arcy, 1991; D'Arcy *et al.*, 1999) it is a persistent dogma in the popular press and with many politicians. A sustainable industrial and business sector (including agriculture) requires an international level playing field for integrating environmental and socio-economic considerations for business (Cable, 2018). Waste minimisation initiatives highlight how businesses can save money by preventing pollution, and the development of the ecosystem services approach offers measures of value from protecting the environment. Some ecosystem services, however, are difficult to quantify in simple economic terms. Environmental regulation is part of the business environment and has an important role in achieving restoration goals. The well documented success of lake restoration at Loch Leven, Scotland, where diffuse pollution has been the largest source of nutrients to the loch over the past twenty-five years, provided an opportunity to investigate the socio-economic value for a local community of the restoration actions taken, and the interplay of drivers and counter-pressures.

14.1.2 Loch Leven

Loch Leven is the largest lowland lake in Scotland, with a surface area of 13.3 km^2, a mean depth of 3.9 m, and maximum depth 25.5 m (May & Spears, 2012, after Smith, 1974). The natural catchment area is 145 km^2 (ECN 2020) and the loch has a flushing period of 140–180 days (Bailey-Watts &

Editor(s): Puangrat Kajitvichyanukul and Brian D'Arcy

Table 14.1 Costs incurred by algal bloom episode in June 1992 (from LLAMAG, 1993).

Sector	Estimated Cost to Local Businesses	Notes
Kinross shops, hotels and bed & breakfast businesses	£673,000	A conservative estimate
Angling on the Loch	£110,000	Loss of revenue from boat hire
Paper mills downstream of the loch (on River Leven)	£110,000	Increased filtration costs for river intakes, one mill stopped production for a day.

Kirika, 1999). Land use in the catchment is dominated by agriculture: arable crops (38.6%); improved pasture (31.5%); upland moor (11.6%), coniferous woodland (3.8%), heathland (3.5%), rough grazing (3.5%), suburban/rural development (2.2%) and the rest (deciduous woodland, bog, bare ground, inland water) (5.3%) (Loch Leven ECN Data Centre, 2019).

At the turn of the 20th Century, Loch Leven was a relatively unpolluted ecosystem with diverse flora and fauna (Dudley *et al.*, 2012; May & Spears, 2012). Deterioration over subsequent decades involved polluting inputs from: a textile mill, town sewage discharges, farm pollution, septic tanks, and pollution incidents (Bailey-Watts *et al.*, 1987; D'Arcy *et al.*, 2006a). A consequent shift to an algal-dominated lake community occurred with an associated loss of macrophytes (Dudley *et al.*, 2012). One major bloom (13 June 1992) achieved extensive publicity, with local economic impacts reported as totalling almost £1M (LLAMAG, 1993). The economic impact also included costs incurred by paper mills abstracting water downstream (Table 14.1).

The problems at Loch Leven were widely reported in 1992 in the national press in Scotland (e.g., *The Scotsman* newspaper and in UK angling magazines), as well as locally. Subsequently, economic decline included the permanent closure of an angling shop in Kinross, and the demise of a previously popular parking area for mobile homes and caravans on the loch shore in Kinross. That was closed and the land sold for housing, thereby losing forever a significant income stream from passing visitors. Further closures of pubs, hotels and shops followed in that part of Kinross. The Loch Leven fishery had been in decline for years, and anglers believed it was affected by pollution (Winfield *et al.*, 2012). In the crisis year of 1992, an alliance of anglers, birdwatchers and local people stimulated local campaigns and actions. The latter included the introduction of green infrastructure landscape measures for farms, forestry and urban runoff, as well as improvements to major point source effluents. An initial account of restoration progress was published in 2006 (D'Arcy *et al.*, 2006b), and ecological surveys in 2008 indicated great improvements (Dudley *et al.*, 2012; Gunn *et al.*, 2012). Figure 14.1 shows the trend for recovery, as indicated by the maximum depth at which rooted macrophytes occur (water clarity permits growth at depth).

A review of the restoration strategy, the measures taken and prospects for sustained good quality, is given in D'Arcy *et al.* (2019) (and paper in press).

14.1.3 Research opportunity

The cost to the local economy of the major cyanobacterial bloom in summer 1992 noted above, has been followed, after a significant delay, by documented economic improvements. There was also a delay before the benefits of actions to address point source and diffuse inputs became evident in the loch. In 2008 and subsequently, a walking and cycling trail around Loch Leven began to be established by the TRACKS partnership (The Rural Access Committee for Kinross-Shire). An evaluation of that initiative showed significant benefits to the local economy (Donnell, 2015). Expenditure by visitors at businesses along the Loch Leven Heritage Trail was estimated by asking 239 visitors about their spending at those businesses. The economic value to the local economy was estimated at about £2M per year (Donnell, 2015).

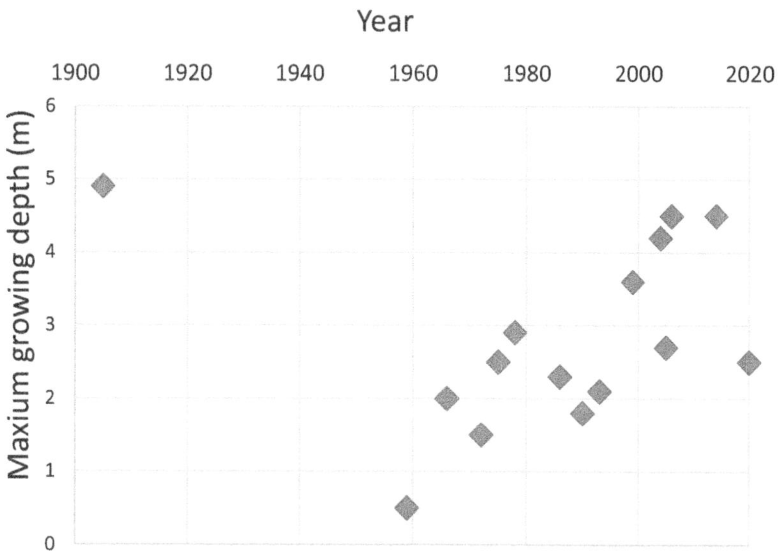

Figure 14.1 Maximum depth at which rooted plants occur in Loch Leven; a measure of recovery.

The aim of this research was to investigate any links between the effective action to reduce pollution impacts, and the locally significant economic growth once the restoration was approaching completion (Figure 14.1). In addition, the opportunity was taken to evaluate how businesses (including farmers), viewed the potential or actual impacts on decision making of environmental regulation and other factors, and the willingness to continue with pollution prevention measures.

14.2 METHODS

14.2.1 Identification of businesses in relation to Loch Leven improvements

The investigation of economic benefits associated with the creation of the Loch Leven Heritage Trail in 2014–15 (Donnell, 2015), identified seven businesses where visitors to the Trail spent money. From local knowledge, an additional five businesses were added for this study. Information was also provided by contacts with the Kinross and District Rotary Club and local councillors. Each of the businesses was dependent on tourism, to some degree, deriving income from visitor (local or otherwise) spending in shops, cafes or restaurants, loch view holiday accommodation, or activities on the loch (angling, wildlife watching or visiting historic sites).

14.2.2 Semi-structured interviews with local businesses

Semi-structured interviews are a data collection technique which can provide quantitative information from closed questions, but also valuable insights and more nuanced views in response to open questions (Thomas, 2003). This technique has been used to investigate perceptions and influencing factors for the farming sector in a neighbouring part of Scotland, alongside the Loch Leven catchment (MacGregor & Warren, 2016). Whilst the time taken to conduct and process the information obtained from interviews precludes a large number of samples, the value of the information obtained is enhanced by the opportunities to explore complex issues and seek validation of answers by attention to non-verbal communication. Face-to-face communication can also encourage a positive degree of participation (Gorden, 1975).

Table 14.2 Closed questions responses from semi-structured interviews with 10 local businesses, Autumn 2019.

Factors Influencing Your Decision to Invest in the Business at Loch Leven	Unimportant			Important		1 + 2 Scores	4 + 5 Scores
Relative importance scores	1	2	3	4	5		
General influences on your business							
Local (Kinross-shire) environment?	1	0	0	2	7	1	9
General economic situation in UK	1	0	3	5	1	1	6
General business outlook for your own business sector	0	0	2	6	2	0	8
Potential positive factors for decision making							
Market readiness locally	1	0	1	1	7	1	8
Loch Leven environmental quality	1	1	2	0	6	2	6
Availability of finance for development	2	0	0	5	3	2	8
Potential negative factors for decision making							
Diffuse pollution & SUDS regulations	2	4	2	1	1	6	2
Local economy	0	1	3	4	2	1	6
Pollution of Loch Leven	1	1	3	3	2	2	5

Key: Numbers of responses in each value category from (1) Unimportant to (5) Important.

Only a dozen businesses were relevant to the primary research question here (Section 14.2.1), so the semi-structured interview was an ideal technique. Interviews on that basis were undertaken with proprietors of existing or new water-related or trail-related businesses (identified as above). The focus was the relation, if any, of local environmental issues (including regulatory requirements) to business investment decisions. Three categories of closed questions were used, with follow-up discussion points for open questions:

- general influences on your business
- potential positive factors for decision making
- potential negative factors for decision making.

Details of the closed questions, together with responses, are indicated in Table 14.2, and the open questions in Table 14.3. The interviews sought to discover if measures to manage diffuse pollution had included sustainable urban drainage systems (SUDS) and rural best management practices (BMPs), as well as ask about the economic drivers for business decisions taken by the selected businesses.

14.2.3 Focus groups

Focus groups are a cost-effective technique in terms of time and demands on the sector being investigated. Although, in contrast to extensive questionnaires, they cannot produce quantitative data, the inferences gained, are often comparable or better. That is because of the freedom from constraints inherent in questions constructed by the researcher; a focus group allows participants to voice opinions that can be counter to preconceptions, to add insights not anticipated in a draft question, and to discuss amongst themselves, drawing out consensus views or deeper insights. For focus groups to be effective, a maximum number of eight participants is optimal; larger groups tend to fragment making conversations and comments difficult to capture. Smaller groups limit the sample of views presented and may be dominated by one or two individuals. A weakness of a focus group (or voluntary questionnaire) is that less motivated representatives of a study sector may not be willing to participate.

Table 14.3 Summary of responses to open questions in semi-structured interviews with local businesses around Loch Leven.

'Factors Influencing Your Decision to Invest in the Business at Loch Leven'	Example answers
What factors influenced your decision to develop/expand?	The Heritage Trail, availability of premises, finances, market perception
Were you aware of the algal blooms history at Loch Leven?	'Yes' (but no recent bad press when investing). Or, not considered because of priority independent business considerations
How was the quality of the loch when you decided to invest?	'No idea' was a typical response
Do you think investment and effort to restore the loch will benefit/has benefitted your business?	'Yes' was a typical response
What measures did you put in place to meet SEPA's requirement for: diffuse pollution control?	Responses ranged from not applicable (coffee business in town) to provision of SUDS on car parks, and rural BMPs on farms*

*Rural BMPs (Best Management Practices) as set out in USEPA (1993) and advocated by SEPA from 1996 (D'Arcy and Frost, 2001).

Focus groups were organised to gain information from two sectors:

(a) local businesses (all sectors and categories if possible)
(b) farmers and their advisors.

Group (a) comprised members of Kinross Rotary Club. They provided an opportunity because their membership was predominantly older people, with memories of events spanning the period under investigation, and they were from a wider variety of local businesses than those selected for the individual interviews. The discussion points for the Business Focus Group centred around perceptions of past and present water quality in the Loch Leven catchment, the reasons behind changes in water quality, including actions by them, the role of pollution control regulation, and future actions needed. The same questions were put to Group (b), the farmers focus group, with additional questions about on-farm measures taken to reduce diffuse pollution.

14.2.4 Review of planning applications

The local planning authority for developments in the Loch Leven catcgment is Perth and Kinross Council (PKC). New and expanding businesses are required to seek formal planning approval. A review of their submissions via the local council's Planning Portal (Perth & Kinross Council, 2020) could provide incontrovertible evidence of any awareness of the step-change major improvmeents prior to their application, if that was a factor in their decision. Such a written submission, if it included a reference to the improved quality of the loch in their business development proposals would be more reliable than recollections in interviews or focus groups some years later.

14.2.5 Assessment of the impact of environmental regulations

The statements given in semi-structured interviews or in focus groups, included refererence to measures taken to reduce diffuse pollution such as sustainable urban drainage systems (SUDS) for the tourist-related businesses, and rural best management practices (BMPs) for the farmers, including comments about the effectiveness of regulatory requirements for such measures. To validate that information, two methods were used:

(a) the prevalence of SUDS – a legal requirement in Scotland – was assessed by a survey of developments in the town and villages of the catchment built since the legislative requirements came into force;

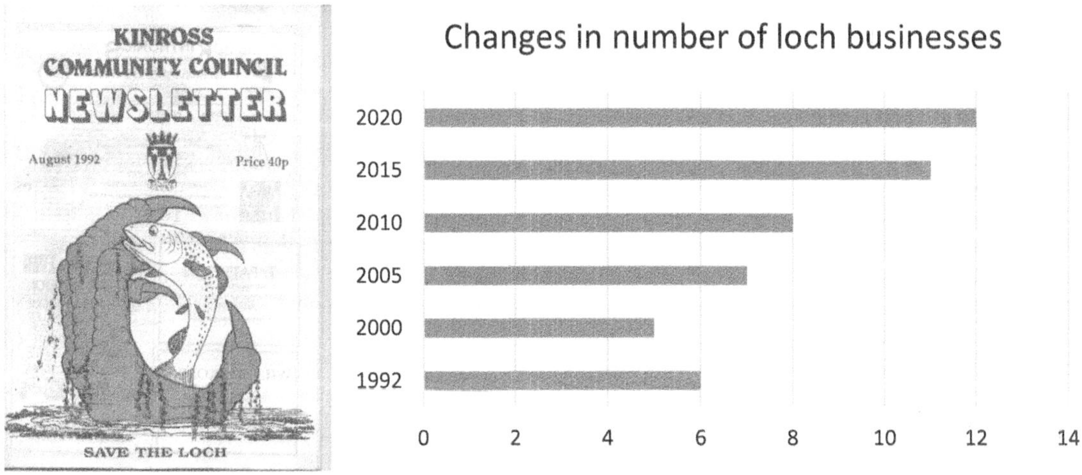

Figure 14.2 The high profile locally (Kinross) of the eutrophication problems in Loch Leven in 1992 and the post-recovery increase in new or expanded informal leisure businesses around the loch.

(b) the uptake of rural BMPs information provided in semi-stuctured interviews and the focus groups, was explored by further dialogue with some additional farmers and farm advisors, and spot checks at three farms which had originally been leaders of the local buffer strips initiative.

14.3 RESULTS

14.3.1 Growth in numbers of tourist businesses around Loch Leven

Post restoration improvements, the new, re-opened or expanding businesses comprised tourist-related (locals and visitors) enterprises with links to the loch: cafes, restaurants, shops, loch view accommodation, and walking trails (Figure 14.2). One relevant business had gone (angling shop) after 1992, but other existing businesses subsequently continued, re-opened or expanded. The number of loch-related businesses has doubled since the worst period after the 1992 bloom and negative publicity.

14.3.2 Semi-structured interviews with local businesses (2019–20)

Twelve businesses around the loch likely to be influenced by visitors to the Loch Leven Heritage Trail were interviewed in 2019–20. The questionnaire used was entitled 'Factors influencing your decision to invest in the business at Loch Leven'. The businesses were: TRACKS (a partnership who created the Heritage Trail), the Kinross Estate Company (angling business), local cafes, restaurants, and shops, including farm shops and chalets. Only one business was unwilling to complete a survey form, but nonetheless offered useful insights into customers' responses to pollution of the loch: 'when there are algal blooms people … ask why, who is to blame?'. Another business also supplied useful information but was unable to complete the data collection form in time for the paper. Only one of the new businesses was uninfluenced by the loch or pollution – their café was peripheral to the sales of coffee roasted there. The other businesses all had an interest commercially in the loch and an aversion to bad publicity about it. The average values for responses to the closed questions are given in Figure 14.3, and individual responses in Table 14.2.

The local environment for a business was explained as referring to everything from the prevalence of litter in the streets, vandalism, dereliction and so on to being in a well maintained, affluent and attractive locality; a high score indicated an aversion to the poorer character and the importance of the alternatives.

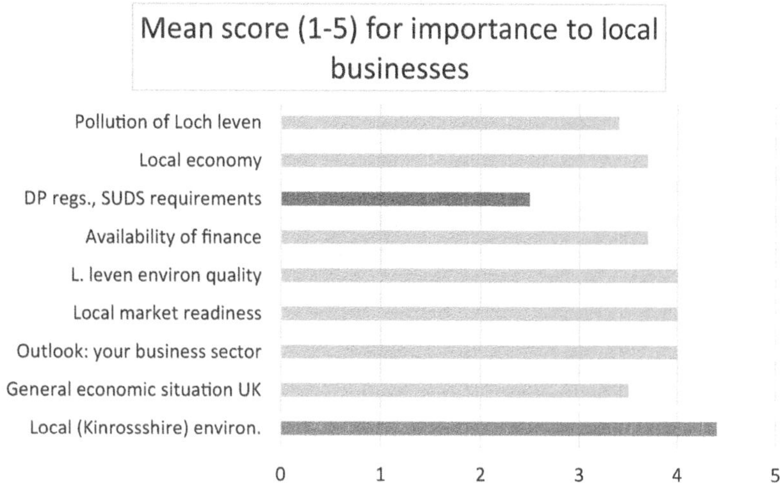

Figure 14.3 The relative importance of the issues put to local businesses (zero indicates unimportant and five means very important; coloured bars indicate lowest and highest scores). DP: diffuse pollution, SUDS: sustainable urban drainage systems.

The least important issue for the local businesses was diffuse pollution regulations and their associated requirements. The latter included sustainable drainage systems for car parks and driveways, and rural BMPs for farmers. That result was surprising and might be attributable to two causes:

(a) the businesses are comfortable with the light touch regulatory regime in place in Scotland for managing diffuse pollution;
(b) there may have been insufficient effort by the environmental regulator to engage with all sectors to explain and promote that light touch regulatory regime, and in recent years an absence of enforcement actions.

In the Loch Leven catchment, local farmers especially, have had a long involvement with development and roll-out of measures to address diffuse pollution, so for them at least, explanation (a) is most likely.

Table 14.2 shows the numbers of responses to each of the value choices put to them and summarises the positive and negative responses to highlight the strength of particular views. Local environment was most important for investment decisions, together with economic considerations such as business sector buoyancy, market readiness locally and availability of finance. Most businesses reported good water quality in Loch Leven had some importance, but mainly they wanted no bad publicity for the area, which might reduce tourist/visitor numbers.

Of the measures taken to protect and improve water quality in the loch several businesses had created carparks with drainage treated by gravel features, grass edges, or a settlement pond. For the farm-based businesses, rural BMPs included physical measures such as buffer strips, and economic practices such as nutrient budgets and pesticide targeting, yet most reported those requirements as being unimportant as factors in their decisions to invest in their businesses.

For the individual businesses, discussions around the open questions followed a consistent pattern for only two of the questions: 'How was the quality of the loch when you decided to invest in the business?' (almost no awareness of it) and 'Do you think investment and effort to restore the loch will/has benefit or has benefitted your business?' (yes). One business commented 'totally unaware of the improvements in the loch' yet that business overlooks the loch and the location is important ('overlooks

Loch Leven' is specified on notices advertising the site). Most of the interviewed businesses expressed concerns that if the loch is polluted it attracts adverse publicity. Most businesses were aware of the history of algal blooms in the loch. No-one, however, was aware of the step-change improvements (see Figure 14.1 and Dudley *et al.*, 2012) that had been achieved in the loch (Table 14.3).

14.3.3 Focus groups

14.3.3.1 Business focus group

The most relevant comments from the business focus group (which included some farmers who were not at the subsequent farmers focus group) are grouped under headings directly related to the primary objectives of this study:

Loch Leven, tourism and the local economy

So great was the reputation of the loch and its fishery many years ago, that 'Loch Leven trout were exported around the world'. For tourism, 'Loch Leven is the only asset and attraction' for Kinross and district. Bad publicity from algal blooms has damaged the reputation of the loch; two hotels for example have closed since 1992. Day visits however, 'have increased following the creation of the Loch Leven Heritage Trail' between 2008–2014, but overnight stays are needed for full economic recovery. Various suggestions were made for water contact sports on the loch (a National nature reserve and Ramsar wetland) 'But toxic blooms would prevent that kind of tourism realising its potential'. No-one mentioned improvements in water quality achieved in the loch. 'There have always been algal blooms – angling records from 1800s'.

Business perceptions on causes of pollution and effectiveness of remedial actions

Participants were well aware of the history of pollution '1970s it was the mill, but then housing increased and changes in farming practices too'. Concerns were expressed about 'Combined sewers, industrial effluent discharges and sewage effluents'. But it was suggested that the 'Biggest factor was raising local awareness of the need to protect the loch'. Several people expressed concerns about increasing numbers of houses in the district: 'Too much housing – where is the legislation?' and 'New housing – doesn't seem like enough legislation'. Actions taken to manage pollution risks were identified as 'Using P-free detergents'; 'Upgraded sewage treatment works' and a farmer noted 'Livestock exclusion fencing on farms to protect the stream'.

Future prospects

Several participants identified increasing housing developments in the catchment as a concern for the future: 'Limit the number of houses' ... 'Loch has deteriorated as number of houses has increased'. Others remarked that 'In agriculture the encouragement/requirement to install silt-traps and so on in 1990–2000s has lapsed' ... 'Changed pressure from government, and reduced budgets'. For managing the loch and pollution control it was suggested that a Friends of Loch Leven group be set up, since loch interests were too fragmented currently.

In summary, at this focus group there was strong enthusiasm for protecting the loch and efforts to maintain or improve good water quality. But during the presentation afterwards to the participants, it was clear that there was no awareness of the step-change improvements achieved. There was clearly concern for the local economy and community if any deteriorations. Algal blooms were recognised as a cause of bad publicity with associated risks for business and the local economy. Awareness of the relative importance of the various causes of pollution and which actions had been most effective was limited.

14.3.3.2 Farmer focus group

The focus group meeting was divided into two parallel sessions: (a) farmers and (b) farm advisors, to allow optimum participation by all (15 participants in total). Comments are from both groups are combined here, together with perceptions from a wrap-up plenary session.

Recollections of water quality and actions taken to address pollution

Farmers' perceptions of historic water quality were similar to the previous focus group, referring to use of the shore, swimming, and also algal blooms. The deterioration was recalled in terms of blooms, but also perceived as an impact on fishing; associated bad publicity was recognised as a significant issue for the local economy. One farmer pointed out that 'Impacts are on water users, not farmers.' Nevertheless, farmers had taken up recommended changes ranging from nutrient budgets (including organic sources as well as commercial inorganic fertiliser) and applying soil erosion measures, to moving the intensive land-use back from the streams by providing buffer strips, some of which also played a role in stabilising steep banks on arable fields. Drivers included government payments and associated requirements, the presence of the Scottish Environment Protection Agency (SEPA) and farmers' desire to help, as Loch Leven locals. Constructive dialogue, rather than top-down advice, was seen as an important factor in achieving the actions taken by farmers (a view consistent with Blackstock *et al.* (2010), who write about 'knowledge exchange' being the preferred approach rather than a mindset of 'knowledge transfer'. Farmers had few concerns about environmental regulation, except in relation to restrictions on their freedom to dig sediment out of ditches and streams.

Perceptions about improvements in loch water quality

Farmers, like other business sectors locally, had little awareness of the extent of recovery from pollution. Questions they asked in the focus group included: 'Has P reduced? 'Have measurements shown improvements?' and 'For whose benefit are we making it clean?'. The farmers and advisors also noted that 'While water quality is good no-one notices. But if bad, it would be an issue and detract from the area'.

Future prospects

Advances in technology for closer targeting of fertilisers and pesticides should continue to help reduce losses to the water environment. Motivating factors for farmers included rising costs of agrochemicals 'No-one wastes any'. The farmers stated that the measures they had taken were generally cost neutral; savings were made with nutrient budgets, and 'Buffer strips are usually a marginal cost issue; getting crops off that margin is not worth it.' There were additional benefits from having buffer zones which would ensure they remained as farm features (e.g., peace of mind that chemicals applied will be on the field, not in the stream, compliance with regulatory and government requirements too). In-field measures might increase: 'Min-till, direct drilling – helps reduce soil erosion. And easy in spring to … sow next crop.' On hill-farms livestock predominated, with a need for fenced-off watercourses and safe access to watering points. Farmers also highlighted the importance for flood risk management of retaining vegetation and trees, citing recent flooding problems downstream, and sediment erosion when harvesting plantations. Negative drivers were possible changes in funding, and also public access along buffer strips, creating litter and dog fouling problems for farmers.

In summary: local farmers have been responsive to the need for an even-handed (across all contributing sectors) effort to reduce P inputs to the loch since the mid-1990s (D'Arcy and Frost, 2001; SAC, 1994). That willingness to 'play their part' in the catchment to protect the loch was still evident in the November 2019 focus group session.

14.3.4 The details submitted with planning applications

This proved to be a disappointing line of investigation, due to an inconsistent presence/absence of documents on the planning archive. Full details of documents submitted to the planning authority were supplied by TRACKS, for the Loch Leven Heritage Trail. Whilst not referring to the recovery (unaware of it – see section 14.3.2, Table 14.3), their documents only referred to the lack of negative impact which the development would present for Loch Leven. Of the remaining eleven local businesses who opened, re-opened or expanded their businesses at or after the recovery of the loch, in the dialogue for interviews (section 14.3.2 above) apparently none had noted the improvements achieved in the loch in their supporting statements about their plans.

14.3.5 The impact of regulatory changes

14.3.5.1 Urban diffuse pollution

Legislation for the management of diffuse pollution was enacted in Scotland in 2006 by the Water Environment and Water Services Act of 2003 (commonly referred to as the 'WEWS Act'), which implemented the EU Water Framework Directive in Scotland. Daughter regulations ('the CAR regulations') of the WEWS Act, The Water Environment (Controlled Activities) (Scotland) Regulations (2005) included an explicit requirement for new developments, and redevelopments or extensions, to use sustainable urban drainage systems, SUDS (D'Arcy & Hemingway, 2018; SEPA, 2020). Perhaps surprisingly, as indicated in Figure 14.3 and Table 14.2, all the businesses interviewed reported minimal or no issues with those regulations. To see whether such a view was atypical, a survey of the town of Kinross and the local villages of the district was undertaken to assess wider compliance in the district with that statutory requirement for SUDS. The survey found that at in every development since the regulations came into force in 2006, feature/s intended to comply with SUDS requirements had been provided: 32 examples were found, suggesting generally good compliance with regulatory requirements. The quality of some of the SUDS features was not good, however, although some others were exemplary.

14.3.5.2 Rural diffuse pollution and measures for agriculture

Consultations with government advisors reported that by 2020 the width of landscape features such as buffer strips in much of the catchment, had reduced to only 2 m, following the introduction of a statutory minimum width specified in the CAR regulations. Exceptions included the farmers in the Pow/Greens Burn tributary stream who had been the original leaders in the buffer strips partnership of the mid-1990s. Steep banks on highly erodible soil remained stabilised as grass buffer zones in 2020, 25 years after the initial action (Figure 14.4).

14.4 DISCUSSION

14.4.1 Economic impacts of pollution

The adverse impact of the algal blooms on the local economy in the mid 1990s had a strong influence on local businesses and was a major motivation for the creation of the TRACKS partnership which created a Trail around the loch. The area has still not fully recovered from the decline in businesses twenty five years ago. All of the new businesses interviewed would be concerned about pollution and poor publicity for the area. The people who set up the TRACKS enterprise to create the trail right around the loch, which has been the primary driver for the economic revival (Donnell, 2015), conceived of the idea in 1993, just after the economic impact of the algal bloom and national adverse publicity. The formal plan was submitted in 2005, coincidentally when the recovery of the loch was well established as a result of the greatly reduced P input over the antecedent decade. The research here suggests that a visible deterioration could threaten the economic revival.

14.4.2 The economic importance of tourism and informal recreation

The economic up-turn for the local community around Loch Leven has been driven by the spending by locals and by visitors to the area. As noted in the Rotary Club Business Focus group (Section 3.3 above) Loch Leven is the principal economic asset of the area. The importance of informal recreation and well-being (including regular use by local people) is evident from comments offered in the interviews and focus groups reported here. They were compatible with positive views on the importance of natural environment, clear water, and other parameters identified in the findings of the earlier study by Donnell (2015), who reported that 85% of respondents strongly agreed that the trail around Loch Leven benefited their physical health, and 82% the same for mental health. That study also indicated the importance of good ecological quality and wild places when respondents were asked which aspects

Figure 14.4 Partnership between agricultural advisors and environmental regulator in 1995, Loch Leven; (left) the 1995 Buffer Strips initiative, and (right) the riparian field slopes stabilised in grass buffer zones.

of the natural environment of Loch Leven were most important to them. Responses showed that over four fifths of visitors valued Unspoilt and Wild (88%); Peace and Tranquillity (87%); Surrounding hills (82%); and Woodland (81%). More than two thirds valued nature reserve status, clear water and wetlands (Donnell, 2015). Concern about any possible deteriorations and associated adverse publicity was registered in all the lines of investigation reported here, and is especially important for economic success.

14.4.3 Diffuse pollution and motivation to address it

It is vital to sustain interest, and belief, in the need to persist with pollution prevention activities. For diffuse sources that involves influencing a great many individual sources, more than can be reached by one to one regulation (Campbell *et al.*, 2004; D'Arcy & Hemingway, 2018; D'Arcy *et al.*, 2006a). A continuing sense of community and acceptance of a need to sustain efforts, was characteristic of the farmers' responses in the focus groups and in interviews with farmer-owned businesses. But it should be noted that this research was undertaken with a sector steeped in diffuse pollution and associated remedial mesasures for 25 years, sustained by active joint partnership initiatives by agricultural advisory organisations and regulators. The long period of positive engagement may be a factor in positive views of local farmers in interviews and focus groups in 2019. For example a buffer strips initiative in the 1990s was a flagship way of introducing a range of techniques for mitigating diffuse pollution risks; it was led by local farmers. The photograph of the partnership booklet published in 1994, is shown in Figure 14.4 alongside a grass stabilised bank in a field of very erodable sandy soil – established at the start of the partnership, and still there when photographed on 1 September 2020. Similar awareness and long-term commitment may not be so prevalent elsewhere.

River-borne P loadings have been the greatest proportion of the external load into Loch Leven throughtout the recent decades. P associated with agricultural sources, including soil erosion, is probably the largest proportion of that, and benefits are externalities for farmers (Frost and McTernan 1998). Since the measures achieved to control industrial sources in the mid-1980s, and treated sewage effluents in mid 1990s (D'Arcy *et al.*, 2006b, 2019), the collective input from the agricultural inputs is likely to be the largest contributing sector. Nonetheless, the largest individual input is still probably the sewage effluent from Kinross, and therefore of continuing interest to the famers and others asked to address their own individual inputs. The town and villages in the catchment are on combined sewers, but storm sewage discharges have never been quantified in the surveys of input loads reported to date. Concern about storm sewage discharges was raised at all of the focus groups, and has also been a concern of the local Kinross-shire Civic Trust (2020) and local councillors (e.g., Robertson, 2020). Farmers and others want a fair, even-handed approach to pollution control. Regulation was not a negative issue, but its importance was recognised (setting targets for behaviours and evident in comments such as 'why is that permitted?') and it must be seen to be fair.

14.4.4 Good news is not news

The consistent evidence from focus groups and from business interviews that people were unaware of the remarkable improvements achieved, was surprising, especially given the national outcry in 1990s about the loch. One reason might be the more nuanced impact on the angling at Loch Leven, where a tangential interest (numbers of trout caught by anglers) was not necessarily the same as judgement focused directly on the ecological parameters for step-change improvements observed in water quality. An undoubted decline in the brown trout fishery was reported in a recent review but was contrasted with the improvement in the current excellent condition (Winfield *et al.*, 2012). More recently, the fishery business has diversified and found economic resilience accordingly, for example the pike fishery as well as brown trout (Loch Leven Fisheries, 2020). For others surveyed for this paper (and most people surveyed did not refer to angling) the reason for lack of awareness of the radical improvements must be inadequate dissemination of the evidence of recovery. Although press releases associated with publication of a set of papers on the loch (May & Spears, 2012) were published in the national press, they were an ephemeral impact. The local monthly Newsletter in Kinross provides an easy outlet for reaching the local population with positive as well as other news stories and should be better used in future (Kinross-shire Newsletter, 2019; and the example from 1992 in Figure 14.2). Wider dissemination is possible now using social media (e.g., Loch Leven Fisheries, 2020); the environment agencies and local businesses and others could be far more proactive in use of those opportunities to reach sectors and the general public in future. The press release is now a largely out-of-date technique, especially for good news messages.

14.5 CONCLUSIONS

The socio-economic importance of pollution and abatement actions is evident from visitor and business surveys reported here, and in other surveys (Donnell, 2015). Natural or wild quality is highly valued, and benefits to physical and mental health are perceived (Donnell, 2015). The reported overall economic impact of the developments post clean-up exceeds the cost of the earlier pollution crisis in 1992 and should be an annual feature of the local economy; approximately £2M p.a. (Donnell, 2015).

Tourism businesses (used by visitors and locals) comprised the expanded enterprises benefitting from the proximity to the loch and from the Loch Leven Heritage Trail around it. But this study suggested they could be in jeopardy from deterioration in water quality, especially visual issues such as algal blooms.

Awareness of pollution abatement actions is usually only evident in relation to problems and negative press publicity. 'No news is good news' seems to have been an assumption; no disincentive to invest.

It may be that since most ecological improvements, (e.g., macrophyte species diversity and depth of growth in the loch at several metres) are less visible to the public than for example algal blooms or sewage pollution, they are easily overlooked, creating additional requirements for dissemination of the good news at a sustained local level.

Nationally, for the press 'Good news is no news'. Social media however, could provide a powerful new opportunity for good news stories to reach a wider audience. Otherwise creeping deterioration is likely to occur until such time as bad press resumes with consequent calls for action.

Continuing action requires continuing evidence so monitoring must be maintained, with informative feedback to sectors and public.

Environmental regulation did not feature as an important economic issue for developing businesses in this study. For farmers, it was only identified as a problem in relation to maintenance of ditches and streams. When in alignment with the economic environment of businesses, and when supported by a positive community engagement effort by the environmental regulator, the diffuse pollution control regime in Scotland can be effective.

The price of light touch regulation is community engagement; it requires budget and effort by the regulators (D'Arcy & Hemingway, 2018; D'Arcy *et al.*, 2006a). That is particularly vital for diffuse pollution, owing to the large number of often individually minor, but collectively significant sources. It is demonstrably not difficult to achieve but does require a very different approach to the historic practices for managing point-source effluent discharges (Campbell *et al.*, 2004; D'Arcy *et al.*, 2006b).

ACKNOWLEDGEMENTS

George Lawrie, Andy Vinten, Vanessa Burns, Scotinform, Alan Hendry, Jamie Montgomery, participating local businesses, Kinross and District Rotary Club, local community councils, participating farmers and advisors. Some of the data were collected under award number NE/R016429/1, as part of the NERC funded UK-SCaPE programme.

REFERENCES

Bailey-Watts A. E., Sargent R., Kirika A. and Smith M. (1987). Loch Leven Phosphorus Loading. Report to Department of Agriculture and Fisheries for Scotland, Nature Conservancy Council, Scottish Development Department and Tayside Regional Council. Institute of Terrestrial Ecology, Edinburgh, UK, 46pp.

Bailey-Watts A. E. and Kirika A. (1999). Poor water quality in loch leven (Scotland) in 1995 in spite of reduced phosphorus loadings since 1985: the influences of catchment management and inter-annual weather variation. *Hydrobiologia*, **403**, 135–151, https://doi.org/10.1023/A:1003758713050

Blackstock K. L., Ingram J., Burton R., Brown K. M. and Slee B. (2010). Understanding and influencing behaviour change by farmers to improve water quality. *Science of the Total Environment*, **408**, 5631–5638, https://doi.org/10.1016/j.scitotenv.2009.04.029

Cable V. (2018). Foreword. In: Wealth Creation Without Pollution: Designing for Industry, Ecobusiness Parks and Industrial Estates, B. J. D'Arcy, L.-H. Kim and M. Maniquiz-Redillas (eds.), IWA Publishing, London, pp. xxv–xxvi. ISBN: 9781780408330.

Campbell N. S., D'Arcy B. J., Frost C. A., Sansom A. and Novotny V. (2004). Economic instruments, regulation and education. In: Diffuse Pollution – an Introduction to the Problems and Solutions, IWA Publishing, London, Chap. 7, pp. 198–230. ISBN 1-900222-53-1.

D'Arcy B. and Frost A. (2001). The Role of Best Management Practices in Alleviating Water Quality Problems Associated with Diffuse Pollution. *The Science of The Total Environment*, **265**, 359–367.

D'Arcy B. J. (1991). Legislation and the control of dye-house pollution. *Journal of the Society of Dyers and Colourists*, **107**, 387–191, https://doi.org/10.1111/j.1478-4408.1991.tb01282.x

D'Arcy B. J. and Hemingway A. (2018). The regulatory regime for bringing SUDS into routine use for industrial estates and business parks in Scotland, UK. In: Wealth Creation Without Pollution: Designing for Industry, Ecobusiness Parks and Industrial Estates, B. J. D'Arcy, L.-H. Kim and M. Maniquiz-Redillas (eds.), IWA Publishing, London, pp. 205–218. ISBN: 9781780408330.

D'Arcy B. J., Todd R. B. and Wither A. W. (1999). Industrial effluent control and waste minimisation: case studies by UK regulators. *Water Science and Technology*, **39**(10), 281–287, https://doi.org/10.2166/wst.1999.0668

D'Arcy B. J., Schmulian K. and Wade R. (2006a). Regulatory options for the management of rural diffuse pollution. In: Managing Rural Diffuse Pollution, Proceedings of SAC-SEPA Binennial Conference, Edinburgh, April 2006, L. Gairns, K. Crighton and B. Jeffrey (eds.), SAC, Edinburgh, pp. 192–200.

D'Arcy B. J., May L., Long J., Fozzard I. R., Greig S. and Brachet A. (2006b) The restoration of loch leven, Scotland. *Water Science Technology*, **53**(10), 183–191, https://doi.org/10.2166/wst.2006.311

D'Arcy B. J., Lindsay J., Clunie A., Mendey S., Hamilton A. and May L. (2019). Restoration of Loch Leven: sustaining ecosystem services. Presentation at the 8th International Conference of the Environmental Engineering Association of Thailand, 23–24 May 2019, Bangkok.

Donnell N. (2015). Loch Leven Heritage Trail, Visitor Survey 2014–15. SCOTINFORM Report for TRACKS, August 2015, Leith Edinburgh. Available from: www.scotinform.co.uk

Dudley B., Gunn I. D. M., Carvalho L., Proctor I., O'Hare M. T., Murphy K. J. and Milligan A. (2012). Changes in aquatic macrophyte communities in loch leven: evidence of recovery from eutrophication? *Hydrobiologia*, **681**, 49–57, https://doi.org/10.1007/s10750-011-0924-9

ECN. (2020). Site information: Loch Leven. UK Environmental Change Network, available online from: data.ecn.ac.uk/sites/ecnsites.asp?site=L06 (accessed 20 July 2020)

Frost A. and McTernan J. (1998). Cost benefit analysis of proposed erosion control at loch leven. In: Diffuse Pollution and Agriculture II, A. M. Petchey, B. J. D'Arcy and C. A. Frost (eds.), Scottish Agricultural College, Aberdeen, pp. 134–144, ISBN 1854826654.

Gorden R. L. (1975). Interviewing: Strategy, Techniques, and Tactics. Dorsey Press, Homewood, IL, USA.

Gunn I. D. M., O'Hare M. T., Maitland P. S. and May L. (2012). Long-term trends in loch leven invertebrate communities. *Hydrobiologia*, **681**, 59–72, https://doi.org/10.1007/s10750-011-0926-7

Harrison P. T. C. and D'Arcy B. J. (2018). Accidents and pollution: industry impacts. In: Wealth Creation Without Pollution: Designing for Industry, Ecobusiness Parks and Industrial Estates, B. J. D'Arcy, L.-H. Kim and M. Maniquiz-Redillas (eds.), IWA Publishing, London, pp. 25–36. ISBN: 9781780408330.

Kinross-shire Civic Trust. (2019). Loch Leven Blues… and greens. Kinross Community Council Newsletter, Kinross-shire Civic Trust October 2019, pp. 18–19, Kinross. http://kinrossnewsletter.org/archive/2019/10_october/october2019.pdf

Kinross-shire Civic Trust. (2020). Kinross Community Council Newsletter, Kinross-shire Civic Trust, September 2020, p. 69. Available from: http://www.kinrossnewsletter.org/

Loch Leven – ECN Data Centre. (2019). data.ecn.ac.uk/sites/ecnsites.asp?site=L06

Loch Leven Fisheries. (2020). Accessed 9th September 2020, www.fishlochleven.co.uk

LLAMAG. (1993). Loch Leven: the Report of the Loch Leven Area Management Advisory Group, Forth River Purification Board, Edinburgh (available subsequently from Scottish Environment Protection Agency, Stirling, Scotland).

Macgregor C. J. and Warren C. R. (2016). Evaluating the impacts of nitrate vulnerable zones on the environment and farmers' practices: a scottish case study. *Scottish Geographical Journal*, **132**, 1–20, https://doi.org/10.1080/14702541.2015.1034760

May L. and Spears B. M. (2012). Managing ecosystem services at loch leven, Scotland, UK: actions, impacts and unintended consequences. *Hydrobiologia*, **681**, 117–130, https://doi.org/10.1007/s10750-011-0931-x

Perth and Kinross Council. (2020). Planning Portal, https://www.pkc.gov.uk/article/15174/Community-Council-Planning-Portal (accessed 8 September 2020)

Robertson W. (2020). Loch Leven Water Quality, *Kinross Community Council Newsletter*, Kinross-shire Civic Trust, April 2020, p. 45. Available from: http://kinrossnewsletter.org/

SAC. (1994). Buffer Strips. Scottish Agriculture College (SAC), Forth River Purification Board and Farming and Wildlife Advisory Group (FWAG), SAC, Edinburgh, UK.

SEPA. (2020). RM08. Available at: http://www.sepa.org.uk (accessed 8 September 2020).

Smith I. R. (1974). The structure and physical environment of loch leven, Scotland. *Proceedings of the Royal Society of Edinburgh*, **B74**, 81–100.

Thomas D. R. (2003). A General Inductive Approach for Qualitative Data Analysis [Online]. University of Aukland, New Zealand. Available at: http://www.fmhs.aukland.ac.nz/soph/centres/hrmas/_docs/Inductive2003.pdf (accessed 27 August 2020)

USEPA. (1993). Guidance Specifying Management Measures for Sources of Nonpoint Pollution in Coastal Waters, USEPA, Washington, DC.

Winfield I. J., Adams C. E., Armstrong J. D., Gardiner R., Kirika A., Montgomery J., Spears B. M., Stewart D. C., Thorpe J. E. and Wilson W. (2012). Changes in the fish community of loch leven: untangling anthropogenic pressures. *Hydrobiologia*, **681**, 73–84, https://doi.org/10.1007/s10750-011-0925-8

doi: 10.2166/9781789061123_227

Chapter 15

Diffuse pollution and contamination of pesticides in natural water and water supply: case study of Northern Thailand

Puangrat Kajitvichyanukul[1], *Suthatip Sinyoung*[2] *and Ruediger Anlauf*[3]

[1]*Department of Environmental Engineering, Faculty of Engineering, Chiang Mai University, Chiang Mai 50200, Thailand*
[2]*Environmental Assessment and Technology for Hazardous Waste Management Research Center, Department of Civil and Environmental Engineering, Faculty of Engineering, Prince of Songkla University, Hat Yai, Songkhla 90112, Thailand*
[3]*Osnabrueck University of Applied Sciences, Faculty of Agricultural Sciences and Landscape Architecture, Osnabrueck 49009, Germany*

Keywords: Pesticides; risk assessment; water contamination; water supply

15.1 INTRODUCTION

From the past to present, agriculture has played a role as the most important sectors of Thailand economy that has generated food and living incomes for Thai people. Pesticides are widely used to ensure high crop productions and supply the constantly expanding food demand to the country and for the export market. Thailand has relied heavily on pesticide use, and the pesticide consumption has increased significantly over time. In recent decades, Thai agriculture has experienced significant transformations in changing type of crops from rice to the economic crops. The shift from rice to various high-value crops such as maize, baby corn, and other field crops such as vegetables and fruit is recently reported (Chalermphol *et al.*, 2014; Kajitvichyanukul *et al.*, 2012). Use of chemicals such as pesticides, growth hormones, and fertilizers, along with modern technology and modern cultivation practices is the common feature of commercial vegetable production in recent periods.

Change in the agriculture practices is, even more, dynamic in northern Thailand compared to other regions of the country (Schreinemachers *et al.*, 2011). A few decades ago, the landscape of this area could be characterized as the cultivation of irrigated paddy in valleys. Currently, the highland agriculture in this region is shifted to be the cultivation of maize or rubber trees on the slopes (Kajitvichyanukul *et al.*, 2012). The deforestation becomes the major problem due to the intrusion of illegal highland farming. The intensive production of maize and rubber trees was becoming the dominant type of land use, especially in those areas where infrastructure and markets are both well-developed. Among the provinces in northern region, Nan has the largest area covered with the highland farming among the provinces in the country. Also, this forestry area is the origin of Nan river that is one of the most important tributaries of the Chao Phraya River, the biggest river in Thailand. During the crop transformation in this Nan watershed, massive amounts of herbicides are applied to the highland farming to control the weed and enhance crop yield. Pesticides such as atrazine, glyphosate, and paraquat, are restricted in several countries but not in Thailand. These chemicals are widely used in this region, and they are ranked among the top-ten imported pesticides in the year 2013 (Tawatsin *et al.*, 2015). Also, the average quantity of active ingredients per hectare of

Editor(s): Puangrat Kajitvichyanukul and Brian D'Arcy

farmland has gone up from 1.2 kg/ha in 1997 to 3.7 kg/ha in 2010 with the average increase of 11% per year (Praneetvatakul *et al.*, 2011).

This highly intensive use of pesticides leads to the associated environmental and health risks to the people living in the agriculture and neighborhood areas. A large number of pesticides have been detected in stream sediment and soil (Kajitvichyanukul *et al.*, 2012). Pesticide exposure through potable water has become a major concern. Because farmers have only a limited knowledge and awareness about the proper use of pesticides (Praneetvatakul *et al.*, 2013), the contamination of pesticides in the environment, especially in water, can become the threatening problem for the people. In the Nan watershed, drinking water sources include groundwater and surface water from Nan River and tributaries. Water treatment units are available in many villages located on the central plain of the catchment. However, the direct use of groundwater and river water for drinking water and household water is generally found in many remote and highland areas by the farmer families. The herbicides and insecticides detected in the drinking water were those expected from agricultural landuse dominated by maize and rubber production and included atrazine, paraquat, glyphosate, and chlorpyrifos. In this present study, we aimed to assess the environmental and health risk of these pesticide residues in surface and drinking water of 8 sub-river basins in Nan watershed. The communities obtaining natural water from catchments that are predominantly crop lands have a high tendency of being exposed to relatively high concentrations of pesticides in drinking and household water. We demonstrated the contamination of pesticides in the environment, especially in water, and the removal of pesticides by water treatment plants. The reduction of environmental and health risks through treatment plants is a goal of this work as this is a part of the implementation of Integrated Pest Management (IPM) (Räsänen *et al.*, 2015).

15.2 STUDY SITES

The upper Nan watershed lies in the middle reaches of the Nan river. It contains many rivers and tributaries. The total area drained by the Nan and its tributaries is 57 947 km^2, and the area of upper Nan watershed is 11 472 km^2. Approximately 500 000 people (2018) reside in this area and more than 80% of its people are farmers. Agriculture plays a significant role in the livelihood of the local populations. Conservation laws protect large areas of the province (89.56%); hence, agriculture and forest activities in this area are limited. In this work, the study sites were grouped and divided in to 8 areas following the sub-river basins as described in Table 15.1 and shown in Figure 15.1.

Water, soil, and sediment samples from eight sub-river basins associated with 65 surface waters, 15 groundwaters, 50 soils, 35 sediments, and 21 community water treatment plants in the Nan watershed were taken for pesticide analysis. The surface water samples were taken from the stream and rivers directly.

Table 15.1 Typical information of sub-river basins identified for pesticide analysis.

Sub-river Basin	Major River	Catchment Area (km^2)
A – Huay Nam Yao River	Nam Yao river, Nam Yod river, and Numrin river	863.54
B – Upper Part of Mae Nam Nan River	Nan river, Num Kan river, Num Pua river	2224.77
C – Second Part of Mae Nam Nan River	Nan river and Namyao river	2046.46
D – Nam Wa River	Nam Wa river	2203.64
E – Nam Sa River	Nam Sa river	753.62
F – Nam Samun River	Nam sanian river, Nam Samun river	583.55
G – Nam Hang River	Nam Hang river	1045.03
H – Third Part of Mae Nam Nan River	Nan river	3376.98

Figure 15.1 Study zones of eight sub-river basins for sampling of soil, sediment, surface water, and ground water.

15.2.1 General framework

This paper evaluats the impact of pesticides on human health and ecosystems on the basis of a risk assessment of pesticides in water. The life cycle of pesticides was integrated to this study to follow the appearance of pesticides in water. The pesticide utilization by farmers and the pesticide loading to the maize and rubber fields in upper Nan watershed was the first step in analysis. Monitoring of the pesticide residue in soil, sediment, and natural water were conducted. The evaluation of the pesticide removal by each unit within water treatment plants was determined. Finally, the transfer of pesticides from water to human was assessed in term of risk analysis. The mitigation approaches were proposed to reduce the health risk of these pesticides.

15.2.2 Water treatment

Water treatment in the communities can be divided into two groups, surface water treatment, and groundwater treatment. Surface water treatment includes coagulation–flocculation with alum dosing, lime softening (optional), and filtration, and finally chlorination before distribution to the community. The groundwater treatment typically comprises of an aeration unit to remove iron and manganese, settling tank, activated carbon adsorption (optional), and chlorination as a final step of the treatment. Three communities had used membrane filtration in their treatment process for drinking water. Many communities used water directly from river and groundwater without any treatment method.

15.2.3 Water sampling and analysis

The water samples for pesticide analyzes were collected and stored in two separate 1 L amber glass bottles for herbicides (paraquat, glyphosate, and atrazine), and chlorpyrifos, the organophosphorus insecticide. The acid herbicide samples were preserved with 2 mL of concentrated, pesticide-grade sulfuric acid, and all samples were maintained at 4°C in the dark until analysis.

Herbicides and organophosphorus insecticide water samples (1 L) were extracted with dichloromethane. The acid herbicide water samples (1 L) were first acidified to pH 2 with 50% sulfuric acid and then extracted with dichloromethane. The pesticide extracts were concentrated (~5 mL) using Kuderna–Danish (KD) evaporation and quantitatively transferred to a test tube; isooctane (2 mL) was added, and then the sample was evaporated to approximately 1.0 mL using a gentle stream of nitrogen gas. The chlorpyrifos extracts were transferred to silica gel (deactivated with 10% water) cleanup columns and eluted with 10% acetone in hexane. The herbicide extracts were transferred to Florisil cleanup columns and eluted with 2% methanol in dichloromethane. Eluates from both cleanup columns were concentrated to 1 mL volume before gas chromatographic analysis.

Chlorpyrifos and atrazine in water sample were extracted by gas chromatography. A model 7890 gas chromatograph (GC) nitrogen phosphorous detector (Agilent Technologies, Wilmington, DE, USA) operated in selected ion monitoring mode with a DB-5 column (30 m × 0.25 mm i.d.; 0.25 μm film thickness; Agilent Technologies, Palo Alto, CA, USA) with helium as the moderating gas. The concentration of paraquat and glyphosate was determined by LC/MS/MS (Agilent 6470B, triple quadrupole). The sample extraction of paraquat was followed the EPA 549.2 method for the analysis of paraquat and diquat with the reversed phase/ion-pair extraction C8 SPE cartridges followed by ion-pair liquid chromatography.

Soil and sediment samples were extracted following USEPA method 3550; sonication extraction for low concentrations of organics and pesticides (US EPA, 1989). Approximately, 4 g of soil or sediment was ground and mixed with anhydrous $MgSO_4$ (1:1 wt/wt for sediment, 5:1 wt/wt for soil). The mixtures were transferred to extraction tubes and extracted using a 10 ml mixture of dichloromethane, acetone, ethyl acetate, and cyclohexane (2:1:1:1) and sonicating using ultrasonic homogenizer (Zhejiang, China) for two minutes. The soil or sediment samples were centrifuged for 5 min to separate the extract from the mixture. The obtain extracts were evaporated under a nitrogen stream, redissolved in hexane before analysis using either GC or LC/MS/MS.

All pesticide concentrations were quantified against five-point calibration curves from the targeted analyte mixture (Sigma-Aldrich Canada Ltd., Oakville, Ontario, Canada), and the two most abundant isotopic ions of each parent ion were monitored. An analysis of variance (ANOVA) was used to test the significance of the results, and $p < 0.05$ was statistically significant.

15.3 PESTICIDE CONTAMINATION

15.3.1 Pesticide loadings and environmental contamination in soil, sediment, surface water and groundwater

In crop production, the expense of pesticides used in controlling pests and weeds was roughly estimated as about half of the total expenditure for agriculture activities. All investigated pesticides are imported products. Glyphosate and paraquat were the most intensively used herbicides with 27 and 13 million kg in Thailand in 2013, respectively (Tawatsin *et al.*, 2015). Atrazine is also used heavily in Thailand with over a million kilograms used per year. These herbicides are known as the hazard chemicals, and some of them are banned in other countries. Paraquat is very toxic to human skin and mucous membranes, and it is lethal if ingested. It is also suspected that paraquat exposure increases the risk of Parkinson's disease (Dinis-Oliveira *et al.*, 2006). Glyphosate is one of the world's most widely produced herbicide by volume. Glyphosate has been linked to tumors in mice and rats and DNA damage to human cells from exposure to glyphosate (Guyton *et al.*, 2015). Atrazine also induces mammary tumors in aging female rats. However, epidemiological studies provide no support for a causal relationship between atrazine exposure and breast cancer (Simpkins *et al.*, 2011). Although several studies (Guyton *et al.*, 2015; Simpkins *et al.*, 2011) have shown that people who work with herbicides especially glyphosate and atrazine, seem to be at increased risk of cancer, many reviews stated that there is limited evidence for a link to cancer in humans for both chemicals (Guyton *et al.*, 2015; Simpkins *et al.*, 2011). These herbicides are banned in many countries in Europe. Atrazine was banned in Europe in 2003 due to concerns about its ubiquity as a water pollutant. Paraquat has been banned or use disallowed, in 32 countries (including the countries of the European Union).

Also, due to the World Health Organization's recent declaration that the glyphosate chemicals are 'likely carcinogenic', many countries such as Brazil, Argentina, Germany, the Netherlands, and Sri Lanka, are on their way to banning the use of this chemical. Chlorpyrifos is the primary insecticide imported to Thailand in quatities of more than a million kilogram per year. Such organophosphates can inhibit the activity of the neurotransmitter acetylcholinesterase in humans and thus affect both the central nervous system and the peripheral nervous system, impacting on the muscles, glands and smooth muscles that make the body's organs function (Alavanja *et al.*, 2004). These four chemicals are allowed to be used in Thailand. The loadings of these chemicals in maize and rubber production in the Nan watershed at the study sites are shown in Table 15.2.

Table 15.3 lists the concentrations of glyphosate, paraquat, atrazine, and chlorpyrifos in soil and sediment. The limit of detection (LD) reached for all pesticides in soil and sediment samples was 1.00×10^{-2} ppb and the limit of quantification (LQ) was 5.00×10^{-2} ppb. The concentration range of pesticides in soil and sediment depends on the type of pesticide, the characteristics of soil (such as organic matter, moisture content, and cation exchange capacity), and the physical and chemical properties of the pesticide (such as partition coefficient, and soil adsorption ability).

Two key properties of pesticides that control their accumulation in soil and sediment are hydrophobicity and persistence. Both glyphosate and paraquat were detected in high concentrations in soil and sediment, due to their environmental persistence and their extensive use in the past for

Table 15.2 Average loadings of pesticides in maize and rubber production in the upper Nan watershed.

Pesticides	Type of Pesticide	Pesticide Loadings/Area (kg/ha)
Glyphosate-isopropylammonium	Herbicide	1.9
Paraquat dichloride	Herbicide	1.5
Atrazine	Herbicide	1.1
Chlorpyrifos	Insecticide	3.1

Table 15.3 Concentrations of investigated pesticides in soil and sediment in each sub-river basin.

Sub-river Basins	Concentration (ppb or μg/kg)			
	Glyphosate	Paraquat	Atrazine	Chlorpyrifos
Soil				
Sub-river basin A	4.75 E + 02–1.01 E + 03	3.99 E + 01–1.18 E + 02	3.81E–01–1.55 E + 00	1.38 E + 00–2.16 E + 01
Sub-river basin B	1.84 E + 02–1.13 E + 03	9.11 E + 00–1.84 E + 02	1.10 E–01–2.25 E + 00	2.21 E–01–1.23 E + 02
Sub-river basin C	1.45 E + 02–2.56 E + 03	2.02 E + 01–1.94 E + 02	2.30 E–01–1.19 E + 02	1.01 E–01–1.43 E + 01
Sub-river basin D	1.65 E + 02–3.31 E + 03	1.62 E + 01–2.91 E + 02	9.01 E–02–1.54 E + 00	5.82 E–01–1.70 E + 01
Sub-river basin E	1.71 E + 02–2.37 E + 03	9.15 E + 00–1.70 E + 02	5.10 E–02–3.43 E + 01	<LD – 1.14 E + 01
Sub-river basin F	1.92 E + 02–3.99 E + 02	6.75 E + 00–2.70 E + 01	8.01 E–02–1.85 E + 02	<LD – 4.05 E + 01
Sub-river basin G	2.53 E + 02–2.18 E + 03	1.21 E + 01–1.52 E + 02	5.22 E–01 –1.76 E + 00	5.98 E + 00–6.65 E + 00
Sub-river basin H	2.02 E + 02–2.36 E + 03	9.45 E + 00–1.25 E + 03	8.01 E–02–8.58 E–01	<LD – 7.70 E + 00
Sediment				
Sub-river basin A	1.68 E + 02–3.56 E + 02	7.95 E + 00–1.55 E + 01	4.61E–01–7.32 E + 00	3.03 E + 01–7.36 E + 01
Sub-river basin B	1.32 E + 02–2.76 E + 03	1.46 E + 01–1.02 E + 02	5.80 E–01–3.04 E + 00	1.21 E + 01–4.60 E + 00
Sub-river basin C	1.92 E + 02–2.35 E + 03	1.05 E + 01–1.17 E + 02	9.00 E–02–1.41 E + 01	<LD–1.03 E + 01
Sub-river basin D	1.07 E + 02–3.91 E + 03	2.35 E + 01–1.47 E + 02	6.61 E–01–2.37 E + 00	2.93 E + 00–2.02 E + 01
Sub-river basin E	1.88 E + 02–1.52 E + 03	1.05 E + 01–1.50 E + 02	1.10 E–01–5.96 E + 01	1.11 E–01–7.74 E + 01
Sub-river basin F	2.40 E + 02–9.17 E + 02	2.27 E + 01–5.38 E + 01	1.61E + 00–6.96 E + 01	2.21 E–01–1.38 E + 01
Sub-river basin G	2.41 E + 02–2.01 E + 03	2.41 E + 02–2.01 E + 03	4.32 E + 00 –1.17 E + 02	4.32 E + 00–1.17 E + 02
Sub-river basin H	1.59 E + 02–1.73 E + 03	1.59 E + 02–1.73 E + 03	4.01 E–01–3.08 E + 01	4.01 E–01–3.08 E + 01

LD: limit of detection.

30 years (Table 15.3). For glyphosate, the soil adsorption coefficient ($K_d = 61$ g/cm^3) and a very low octanol/water coefficient ($K_{ow} = 0.00033$) suggest that glyphosate has low mobility and only a slight tendency to leach in the soil. Glyphosate is classified as very slightly mobile in soil (Linders *et al.*, 1994) and it is inactivated by soil adsorption; it has low leaching potential and very low volatility (Franz *et al.*, 1997). Paraquat also binds strongly to soil particles and tends to remain firmly bound for a long time in an inactive state, although it can also desorb again and become biologically active (Watts, 2011). The half-life of paraquat in the soil can be up to 20 years.

Spatial distributions of four pesticides in soil and sediment of the upper Nan watershed are shown in Figures 15.2 and 15.3, respectively. The concentration of pesticides detected during the study was almost in the same range for all eight sub-river basins ($p > 0.05$, $F = 0.23$). Glyphosate and paraquat in soil and sediment were highly concentrated in some areas, such as zone D – Nam Wa river, G – Nam Hang river, and H – the third part of Nam Nan river. The difference in pesticide concentration comes

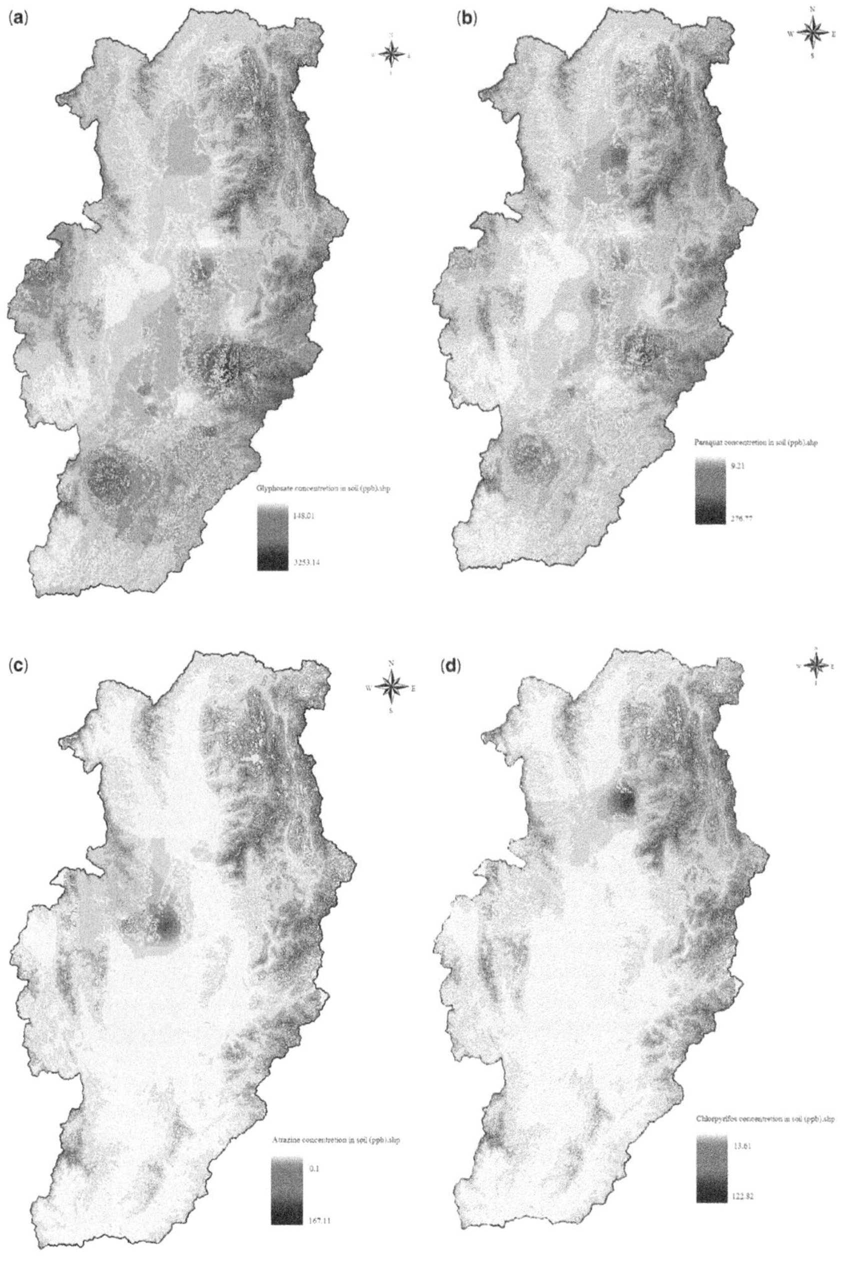

Figure 15.2 Soil contamination of four pesticides in upper Nan watershed.

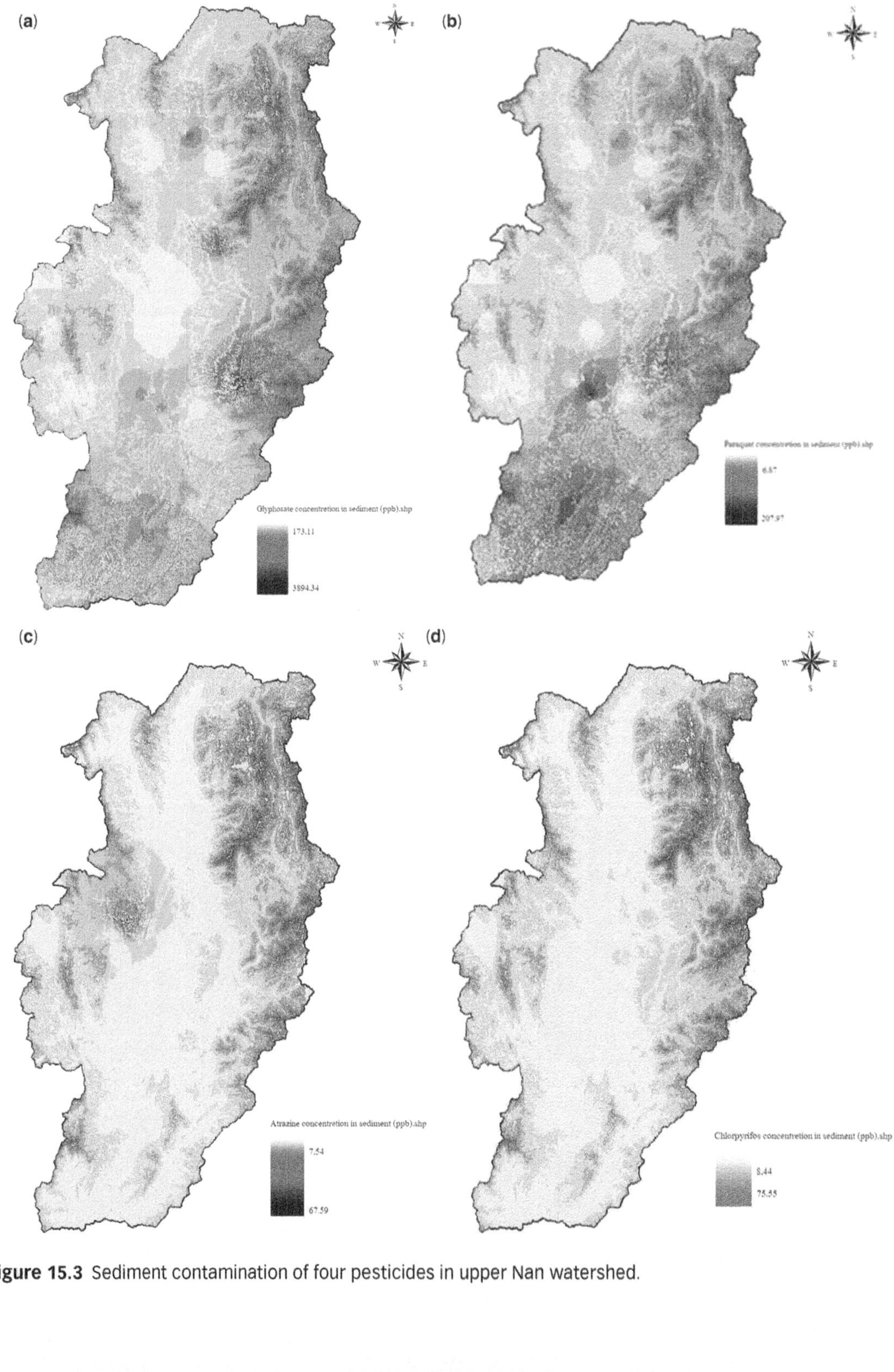

Figure 15.3 Sediment contamination of four pesticides in upper Nan watershed.

from the vast agriculture area and the high volume of pesticide loadings used in highland farming. Moreover, zone H is downstream of Nan river in this watershed.

Table 15.4 lists the concentrations of those pesticides which appeared in surface water and groundwater. The limit of detection (LD) reached for all pesticides in water and was 1.00×10^{-1} ppb, and the limit of quantification (LQ) was 5.00×10^{-1} ppb. The occurrence of pesticide in a stream and sediment depends on the type of pesticide, the sources of that pesticide in the sub-river basins, the

Table 15.4 Concentrations of investigated pesticides in surface water and groundwater in each sub-river basin (LD: limit of detection, NA: not available).

Sub-river Basins	Concentration (ppb or μg/L)			
	Glyphosate	**Paraquat**	**Atrazine**	**Chlorpyrifos**
Surface water				
Sub-river basin A	3.68 E + 00-2.33 E + 01	2.40 E−01-1.07 E + 00	7.25 E + 00-6.21 E + 02	1.50 E−01-1.96 E + 02
Sub-river basin B	2.95 E + 00-2.96 E + 01	3.80 E−01-2.54 E + 00	<LD - 9.74 E + 01	<LD - 6.78 E + 00
Sub-river basin C	4.59 E + 00-4.75 E + 01	2.40 E−01-8.10 E−01	2.30 E−01-2.75 E + 02	<LD - 1.36 E + 01
Sub-river basin D	4.57 E + 00-5.79 E + 01	2.92 E−01-6.51 E−01	2.66 E + 00-2.27 E + 02	1.23 E−01-3.64 E + 01
Sub-river basin E	6.11 E + 00-1.49 E + 01	<LD - 5.50 E−01	1.75 E + 00-1.04 E + 01	<LD - 6.84 E−01
Sub-river basin F	4.24 E + 00-4.31 E + 01	2.20 E−01- 5.50 E−01	2.33E + 00-5.45 E + 01	<LD -1.19 E + 00
Sub-river basin G	5.01 E + 00-1.74 E + 01	2.22 E−01-4.98 E + 00	2.32 E + 00-7.53 E + 01	<LD - 4.18 E + 01
Sub-river basin H	3.71 E + 00-8.09 E + 01	2.31 E−01-2.96 E + 00	2.31 E + 00-8.32 E + 02	1.01 E−01-6.22 E + 00
Groundwater				
Sub-river basin A	NA	NA	NA	NA
Sub-river basin B	7.13 E + 00-7.22 E + 00	4.31 E−01-1.03 E + 00	1.73 E + 00- 1.80 E + 00	<LD - 1.4 E−01
Sub-river basin C	4.42 E + 00-3.05 E + 01	<LD - 1.93 E + 01	4.77 E + 00-2.08 E + 01	<LD - 2.1 E−01
Sub-river basin D	3.86 E + 00-8.55 E + 00	3.61 E−01-6.21 E−01	2.33 E + 00-26.05 E + 00	<LD - 3.6 E−01
Sub-river basin E	5.64 E + 00-9.70 E + 00	3.82 E−01-4.11 E−01	3.57 E + 00- 5.12 E + 00	<LD - 4.1 E−01
Sub-river basin F	3.86 E + 00-3.11 E + 01	2.91 E−01-6.21 E−01	2.33 E + 00-1.61E + 02	2.19 E + 00-4.24 E + 00
Sub-river basin G	5.64 E + 00-9.79 E + 00	3.49 E−01-4.10 E−01	1.11 E + 00-2.05 E + 02	<LD - 3.6 E + 00
Sub-river basin H	5.51 E−01-2.41 E + 01	<LD - 2.12 E + 00	1.21 E + 00- 2.61 E + 01	<LD

characteristics of the stream (such as water flow), and the physical and chemical properties of the pesticide (such as water solubility). Surface runoff can transport pesticides to surface water bodies including rivers, lakes, and streams. Most pesticides that are highly detected in natural water have high water solubility and shorter soil half-life. From Table 15.4, all four pesticides were found in both surface water and groundwater. In many areas (52 out of 85 sites), atrazine concentration was even greater than that permitted by the US. EPA maximum contamination levels (MCL) and Thailand standard as of 3 ppb or μg/L. They are not as likely to accumulate in the soil and sediment but will easily infiltrate into groundwater. Conversely, paraquat was detected in the low concentration range in both surface water and groundwater. The concentration of this chemical exceeds MCL value (3 ppb) at only 2 out of 80 study sites. Glyphosate was detected in both surface water and groundwater with the concentration lower than MCL value (0.7 ppm). Owing to chemical properties that rapidly and tightly bind to soil particles, both paraquat and glyphosate have a high tendency to be immobile in the soil and hardly leach to the water. However, these chemicals were still detected in the water. The concentrations of chlorpyrifos varied in both surface water and groundwater from different sub-river basins. However, as the reference dose (RfD) of this chemical is relatively small (3.5×10^{-4} mg/kg-day), the detection of chlorpyrifos in water can still pose an adverse effect to human health.

15.3.2 Pesticide residues in treated waters

Water supply and drinking water treatment processes in upper Nan watershed can be divided into two groups, surface water treatment, and groundwater treatment units. Water sampling from 11 surface water treatment plants and 9 groundwater treatment plants were collected to analyze the residual pesticide from each unit. Results of the pesticide residues after treatment of the natural waters for all 8 sub-river basins are shown in Table 15.5. It is clearly seen that the performance of water treatment unit in the removal of the four chemicals from the natural water was relatively low. The percent reduction (based on individual herbicides detected in both natural water supply and drinking water samples) was highly variable from one treatment facility to another. In comparison, water treatment plants can remove chlorpyrifos but hardly remove atrazine from the influent.

Table 15.5 Concentrations of investigated pesticides in water supply in each sub-river basin.

Sub-river Basins	**Concentration (ppb or μg/L)**			
	Glyphosate	**Paraquat**	**Atrazine**	**Chlorpyrifos**
Sub-river basin A	8.13 E + 00	1.94 E + 00	4.62 E + 00	<LD
Sub-river basin B	6.17 E + 00–1.34 E + 01	2.20 E−01–4.46 E−01	2.11 E + 00– 4.94 E + 00	<LD - 2.50 E−01
Sub-river basin C	4.74 E + 00–8.26 E + 00	4.40 E−01–4.59 E + 00	1.41 E + 00–6.20 E + 01	<LD - 1.50 E−01
Sub-river basin D	3.15 E + 00–1.06 E + 01	4.80 E−01–2.19 E + 00	4.53 E + 00–2.07 E + 01	<LD - 2.40 E−01
Sub-river basin E	5.15 E + 00–6.22 E + 00	3.00 E−01–6.70 E−01	3.79 E + 00–4.17 E + 01	<LD - 3.90 E−01
Sub-river basin F	6.22 E + 00	3.1 E−01	4.37 E + 00	<LD
Sub-river basin G	8.81 E + 00–2.18 E + 01	2.50 E−01–7.80 E−01	1.81 E + 00 –1.07 E + 01	<LD - 1.70 E−01
Sub-river basin H	3.09 E + 00–5.25 E + 01	3.30 E−01–4.67 E + 00	4.71 E + 00–7.75 E + 01	<LD - 1.95 E + 01

LD: limit of detection, NA: not available.

Table 15.6 Performance of water treatment unit in removal of investigated pesticides.

Water Treatment Unit	Removal Percentage (%)			
	Glyphosate	Paraquat	Atrazine	Chlorpyrifos
Groundwater treatment unit				
Aeration	0	0	0	0
Activated carbon	15.5–61.7	35.7–94.5	16.7–90.9	66.9–86.0
Chemical oxidation – MnO_2	0	0	0	18.2–31.0
Chlorination – NaOCl	3.5–24.6	3.4–18.9	5.6–36.4	3.3–11.7
Surface water treatment unit				
Coagulation/flocculation – alum	0	0–3.1	0	8.7–58.6
Lime softening	0	1.0–1.9	0	9.1–19.9
Filtration	0	0	0	0
Chlorination – NaOCl	3.5–24.6	3.4–18.9	5.6–36.4	3.3–11.7

To clarify the performance of the water treatment process, we monitored the performance of each treatment unit and reported the removal percentage of each pesticide in Table 15.6. Several units can remove chlorpyrifos in both groundwater and surface water treatment unit. This is in good agreement with our results shown in Table 15.5 that the detected chlorpyrifos in clean water was lower than detection limit in many study sites. In contrast, atrazine and glyphosate can be removed efficiently by activated carbon and chlorination using sodium hypochlorite.

The highest removal percentages of above 60% were obtained for herbicides (glyphosate, paraquat, and atrazine) as well as insecticide chlorpyrifos with activated carbon adsorption. This process can provide the highest performance for 61.7%, 94.5%, 90.0%, and 86.0% for glyphosate, paraquat, atrazine, and chlorpyrifos, respectively. From our study, activated carbon adsorption unit had been used in only some specific sites (8 of 21 water treatment plants). This method can be an effective treatment technology for the removal of pesticide residues for both groundwater and surface water in upper Nan watershed and neighborhood areas.

The chlorination using sodium hypochlorite was applied in all investigated water treatment plants. Although chlorination can remove the pesticides with some certain degree in water treatment plants such as 24.6%, 18.9%, 36.4%, and 11.7% for glyphosate, paraquat, atrazine, and chlorpyrifos, respectively, this process has a high possibility of providing a disinfection by-product in the treated water. Chlorine is a strong oxidizing agent that reacts with a variety of organic compounds and pesticides and convert the compounds to the by-product that may pose harmful effects in humans. Organophosphate pesticides (including chlorpyrifos) containing P = S bonds can be readily degraded by chlorine and produce oxon (P = O bonds) as a primary by-product (Magara *et al.*, 1994). Chlorpyrifos contains a thione group, -P = S-O- bond, and can give the oxon and other chlorination degradation by-products as well. The oxon can bind permanently to the enzyme acetylcholinesterase, preventing this enzyme from deactivating acetylcholine in the human synapse. Hence, the addition of chlorine to the influent water containing pesticides should be carefully monitored and controlled.

15.3.3 Management practice and mitigation approach

Management practices could be implemented to reduce concentrations of pesticides in the environment and water supply in upper Nan watershed and neighborhood areas. This would require the cooperation of the farmers who grow the maize, rubber, and other perennial vegetables around the river basins and could include introducing organic farming, and knowledge management with good practices in using pesticides for highland agriculture. Organic farming is considered as an environmental innovation, which can reduce environmental damage or improve ecological sustainability by reducing the usage of pesticides (Saint-Ges & Bélis-Bergouignan, 2009). The pesticide concentrations in the runoff to streams and rivers could be reduced using pesticides with lower water solubility. These alternative pesticides are pyrethroids that are more efficient than organophosphorus pesticides due to their low environmental impact at the usage stage (Zhu *et al.*, 2014).

A best practice program in operation and maintenance of water treatment plants is required. Installation of adsorption units with activated carbon is urgently required for the study sites with a hazard quotient higher than 1. This treatment process can help in reducing not only the chlorpyrifos but also several types of pesticides in the drinking and household water. Monitoring and controlling the chlorine addition in water treatment plants are required to reduce the chlorination by-products that are harmful substances to human health. In addition, supportive policies from the government in reducing the pesticide consumption are also needed.

15.4 SUMMARY

The intensive uncontrolled use of pesticides for highland agriculture has resulted in the release of large amounts of pesticides to the environment, and the contamination of the water supply and drinking water in upper Nan watershed. The risk assessment of pesticides in water was incorporated in the life cycle analysis of pesticides. We demonstrated the incorporation of life cycle analysis of pesticides from the starting point of pesticide usage by farmers, the contamination of pesticides in the environment especially water media, the removal of pesticides by water treatment plants, and, finally, the risk analysis of pesticide in water supply and bottled drinking water. The pesticides glyphosate, paraquat, atrazine, and chlorpyrifos were detected in water samples from 8 sub-river basins, 65 associated surface waters, 15 groundwaters, and 21 community water treatment plants. These four chemicals are hardly removed by the existing water treatment plants. The proposed effective technology that can remove more than 60% of initial concentration of pesticides is activated carbon adsorption. Chlorination can remove pesticides to a certain degree, but harmful health by-products can occur. The residual of pesticides in water supply can cause a human health risk. An agriculture best practice program, maintenance of water treatment plants and the installation of adsorption units with activated carbon is urgently required. In addition, the supportive policies from government in reducing the pesticide consumption is needed.

ACKNOWLEDGEMENT

This work is supported by the National Higher Education Science Research and Innovation Policy Council (NXPO) Contract No. C16F630255, the Thailand Research Fund (Grant No. IRN62W0005).

REFERENCES

Alavanja M., Hoppin J. and Kamel F. (2004). Health effects of chronic pesticide exposure – cancer and neurotoxicity. *Annual Review of Public Health*, **25**, 155–197, https://doi.org/10.1146/annurev.publhealth.25.101802.123020

Chalermphol J., Bastakoti G. B. and Bastakoti R. C. (2014). Adoption of improved varieties of vegetable crops with pesticide use in chiang Mai province, northern Thailand. *Procedia Environmental Sciences*, **20**, 418–424, https://doi.org/10.1016/j.proenv.2014.03.053

Dinis-Oliveira R. J., Remiao F., Carmo H., Duarte J. A., Navarro A. S., Bastos M. L. and Carvalho F. (2006). Paraquat exposure as an etiological factor of Parkinson's disease. *Neurotoxicology*, **27**, 1110–1122, https://doi.org/10.1016/j.neuro.2006.05.012

Franz J. E., Mao M. K. and Sikorski J. A. (1997). Glyphosate: A Unique Global Herbicide. American Chemical Society Monograph, USA, **189**, ch. 4, pp. 65–97.

Guyton K. Z., Loomis D., Grosse Y., Ghissassi F. E., Benbrahim-Tallaa L., Guha N., Scoccianti C., Mattock H. and Straif K. (2015). Carcinogenicity of tetrachlorvinphos, parathion, malathion, diazinon, and glyphosate. *Lancet Oncology*, **16**, 490–491, https://doi.org/10.1016/S1470-2045(15)70134-8

Kajitvichyanukul P., Ananpattarachai J., Boonyannupap C. and Pakmart A. (2012). Integrated Research in Enhancing Community Capacity for Prevention and Management of Toxic Substance Contamination in Nan Watershed. Final Report to National Research Council of Thailand, Thailand.

Linders J. B. H. J., Jansma J. W., Mensink B. J. W. G. and Otermann K. (1994). Pesticides: Benefaction or Pandora's Box? A Synopsis of the Environmental Aspect of 243 Pesticides. RIVM Report #67101014. National Institute of Public Health and Environment, Bilthoven, Netherlands.

Magara Y., Aizawa T., Matumoto N. and Souna F. (1994). Degradation of pesticides by chlorination during water purification. *Water Science and Technol*, **30**(7), 119–128, https://doi.org/10.2166/wst.1994.0323

Praneetvatakul S., Schreinemachers P., Pananurak P. and Tipraqsa P. (2011). Economic Impact of Chemical Pesticide Use in Agriculture. Report Prepared for the Thai Health Promotion Foundation. Kasetsart University, Bangkok (in Thai).

Praneetvatakul S., Schreinemachers P., Pananurak P. and Tipraqsa P. (2013). Pesticides, external costs and policy options for Thai agriculture. *Environmental Science & Policy*, **27**, 103–113, https://doi.org/10.1016/j.envsci.2012.10.019

Räsänen K., Mattila T., Porvari P., Kurppa S. and Tiilikkala K. (2015). Estimating the development of ecotoxicological pressure on water systems from pesticides in Finland 2000–2011. *Journal of Cleaner Production*, **89**, 65–77, https://doi.org/10.1016/j.jclepro.2014.11.008

Saint-Ges V. and Bélis-Bergouignan M.-C. (2009). Ways of reducing pesticides use in bordeaux vineyards. *Journal of Cleaner Production*, **17**, 1644–1653, https://doi.org/10.1016/j.jclepro.2009.05.006

Schreinemachers P., Sringarm S. and Sirijinda A. (2011). The role of synthetic pesticides in the intensification of highland agriculture in Thailand. *Crop Protection*, **30**, 1430–1437, https://doi.org/10.1016/j.cropro.2011.07.011

Simpkins J. W., Swenberg J. A., Weiss N., Brusick D., Eldridge J. C., Stevens J. T., Handa R. J., Hovey R. C., Plant T. M., Pastoor T. P. and Breckenridge C. B. (2011). Atrazine and breast cancer: a framework assessment of the toxicological and epidemiological evidence. *Toxicological Sciences*, **123**, 441–459, https://doi.org/10.1093/toxsci/kfr176

Tawatsin A., Thavara U. and Siriyasatien P. (2015). Pesticides used in Thailand and toxic effects to human health. *Medical Research Archives*, **3**, 1–10.

US EPA. (1989). Risk Assessment Guidance for Superfund, vol. I, Human Health Evaluation Manual. Part A (Interim Final). EPA/540/1-89/002. Office of Emergency and Remedial Response, US Environmental Protection Agency, Washington DC, USA.

Watts M. (2011). Paraquat, Pesticide Action Network, Asia & the Pacific.

Zhu Z., Wang K. and Zhang B. (2014). Applying a network data envelopment analysis model to quantify the eco-efficiency of products: a case study of pesticides. *Journal of Cleaner Production*, **69**, 67–73, https://doi.org/10.1016/j.jclepro.2014.01.064

doi: 10.2166/9781789061123_241

Chapter 16

Recognition of diffuse pollution impacts on water utilities

T. Kawata[1,*], S. Fujii[2] and H. Li[3]

[1]Graduate School of Advanced Integrated Studies in Human Survivability, Kyoto University, Kyoto, Japan
[2]Graduate School of Global Environmental Studies, Kyoto University, Kyoto, Japan
[3]International Water Association, Den Haag, The Netherlands
*Corresponding author: kawata.tetsuya.62u@st.kyoto-u.ac.jp

Keywords: Diffuse pollution impacts; water utilities; questionnaire

16.1 INTRODUCTION

Diffuse pollution is when pollution enters the environment from non-point sources. It is generally distributed by weather, including precipitation or wind, and significantly influences aquatic environments. The pollutants released into the aquatic environments would depend on the land use of the basins (D'Arcy *et al.*, 2018). For instance, the Naoli River receives eroded soil from forested areas, as well as fertilizer nutrients through surface runoff from cultivated areas (Haibo *et al.*, 2015). Excessive nutrients have adverse effects on algal growth (Davis & Koop, 2006). Agricultural land was reported as one source of faecal pathogens (Tyrrel & Quinton, 2003). Heavy metals and chemicals enter water bodies from urban areas (Shen *et al.*, 2015) and affect taste and odour (Dima *et al.*, 2006). Considering chemicals, PAHs are discharged from urban traffic area (WHO & ICPS, 1998). Pesticide contamination from agriculture in China has been observed since the 1970s (Li & Zhang, 1999). Pharmaceuticals in particular have drawn international concern because of their persistence after being discharged into the environment (aus der Beek *et al.*, 2016a, 2016b).

Thus, pollutants from diffuse pollution are considered to influence water bodies and water utilities. The extent of the influence depends on the land use of the basin and the water source of utilities. However, there are few reports on diffuse pollution issues related to water utilities. Therefore, as a precursor to countermeasures against diffuse pollution in water utilities, a questionnaire survey was conducted among water utility professionals to determine their level of recognition of pollutants that seriously influence water utilities.

16.2 MATERIAL AND METHODS

A questionnaire survey, as shown in Table 16.1, was conducted by an on-line questionnaire service (SurveyMonkey) during 7–20 February 2018. The questionnaire was sent both to people who were International Water Association (IWA-Connect) members in 2015 (1057 people) and attendees from the water utility sector at an IWA event (165 people). The numbers (Q1–11) in the table show the order

Editor(s): Puangrat Kajitvichyanukul and Brian D'Arcy

Table 16.1 Questions in the questionnaire and response numbers.

	Questionnaire contents [(a)- sub-question items] (Answer choice1/Answer choice2/...)	*TR** (*AA***)
(1)	*Attribution of the respondents*	
Q1	Which water sector do you work in? (Water supply business/Wastewater treatment/ Stormwater management/Others)	97 (97)
Q2	Do you mainly work as? (Engineer, Operator/R&D, consultancy/General staff/Senior, Chief, Manager/Others)	97 (97)
(2)	*Characteristics of the respondent utilities*	
Q3	What is your water source? (Reservoir/Groundwater/River/Seawater/Lake/Rain, Stormwater/Wastewater/Others)	97 (96)
Q5	How frequently does your organisation check the resource condition? (Several times a year/ Several times a month/Several times a day/Several times an hour/Every time, real-time monitoring/Others)	97 (87)
Q8	Do you have a programme of catchment protection work? (Yes/No)	95 (84)
Q11	How frequently does your organisation update/reconsider compliance to water standards? (Once over 10 years/Once 3–10 years/Once 1–3 years/Once a half–one year/More a half year/ Others)	84 (68)
(3)	*Effects of landscapes on diffuse pollution*	
Q6	How protected by surrounding natural vegetation buffer landscapes are your catchment areas? (Not at all protected/Slightly protected/Protected/Fairly protected/Very protected)	95 (86)
Q7	How important are landscape measures to protect raw water quality? (Not at all important/ Slightly important/Important/Fairly important/Very important)	95 (89)
(4)	*Influences of pollutants on raw water quality and water utilities*	
Q4	How important are the following issues of (a)–(f) for your raw water supplies? [(a) taste; (b) odour; (c) algal toxins; (d) sediment; (e) faecal pathogens (*E. coli* and *Cryptosporidium* etc.); (f) chemicals (pesticides etc.)] (Not at all important/Slightly important/Important/Fairly important/Very important)	97 (88–96)
Q9	How important are the following raw/influent water issues of (a)–(g) on the operation in water utility? [(a) sediment load; (b) algal or plant growth; (c) faecal pathogens; (d) pesticides; (e) veterinary medicines; (f) heavy metals; (g) urban pollutants (hydrocarbons and other organic trace pollutants)] (Not at all important/Slightly important/Important/Fairly important/Very important)	84 (69–79)
Q10	How important are intermittent chemical contamination incidents affecting raw water supplies? (Not at all important/Slightly important/Important/Fairly important/Very important)	84 (80)

Numbers of Q1 to Q11 show the sequences of these questions in the original question sheet.
*TR**: No. of Total responses, *AA***: No. of Available answers.

of each question in the questionnaire. A total of 107 survey responses were obtained; however, 10 people answered only attributive questions; these responses were thus not analyzed in this study. As a result, the total number of responses for each question ranged from 84 to 97, as shown in the table. Additionally, all responses, excluding the answers 'I don't know' and 'No opinion', on each question were used in the analysis, and the sample size (=*AA* (the number of available answers)) varied from 69 to 97.

The questionnaire consisted of sets of multiple answer questions (Q1, Q3), and multiple-choice questions (Q2, Q4–11). In Q1–3, Q5, and Q11, 'Others' was set as a choice to enable the respondents to write out their answers in a comment column.

The questions were classified into four types: (1) attribution of the respondents (Q1, Q2); (2) characteristics of the respondent utilities (Q3, Q5, Q8, Q11); (3) the effects of landscapes on diffuse pollution (Q6, Q7); and (4) the influences of pollutants on raw water quality and utilities (Q4, Q9, Q10). Questions Q4–7 and Q9–11 had five evaluation choices (=scale); 'Not at all Important/Protected (I/P)' (= 1), 'Slightly I/P' (= 2), 'I/P' (= 3), 'Fairly I/P' (= 4) and 'Very I/P' (= 5); in addition to these, 'No opinion/I don't know' (= *NA*) was also a choice. A weighted average (*WA*) was calculated from these five response scales (= 1 to 5), excluding *NA*, in calculation.

Personal information of respondents, such as their countries, company names, and contact information, was solicited through voluntary questions. From the 53 answers obtained 12 responses were from 3 African countries, 3 were from 3 American countries, 5 were from 5 Asian countries, 27 were from15 European countries, and 1 was from an Oceanian country.

16.3 RESULTS AND DISCUSSION

16.3.1 Respondents (available answers (*AA*)/total responses (*TR*) = 97/97, Q1, Q2)

Q1 and Q2 examined the characteristics of the respondents' working sectors and positions. Q1 had three choices: water supply, wastewater management, and rainwater management. Only three respondents chose 'Other' and detailed their job descriptions in comments. However, these were found to match the choices provided and were categorized accordingly (e.g., 'drinking water' → water supply sector).

As shown in Figure 16.1, 81% of the respondents were involved in water supply sectors, 59% were in the wastewater management sector, and 18% were in the rainwater management sector; many chose multiple sectors. Considering the possibility of multiple answers in three sectors, seven groups could have been formed (= 2^3–1); however, almost all the respondents (94/97) were classified into four working sector groups, namely, the water supply (WS) group (38%), the wastewater (WW) group (16%), the water supply and wastewater (SW) group (28%), and the triple-sector (TS) group (14%).

Regarding their working position (single answer), 65% of respondents were seniors or managers, 15% were in R&D or consultants, 14% were working as engineers or operators, and 5% were general staff. Four respondents selected 'Other', but these too could be classified under one of the four positions (for instance 'CEO' → senior or manager class). The proportion of managers was the highest of all the working sector groups.

In summary, the questionnaire respondents were divided into four groups (WS, WW, SW, and TS), with more than half of each group belonging to the senior or manager class.

16.3.2 Characteristics of water utilities

Respondents were asked about the types of water sources, the frequency of water source monitoring, the catchment area protection programme, and the water standards update frequency at their respective water utilities.

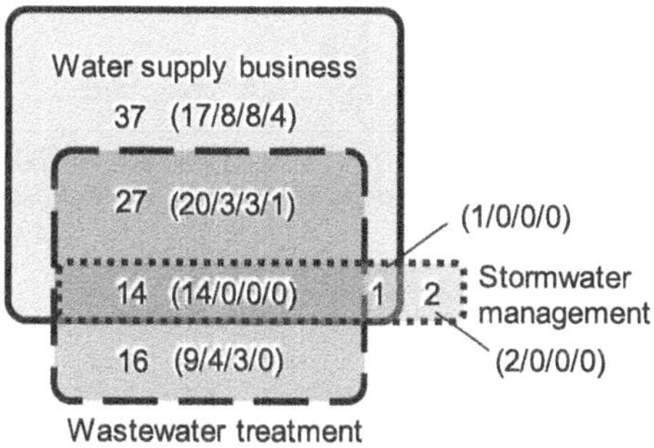

Figure 16.1 Affiliative distribution of respondents in the questionnaire. Values in the figure show numbers of answers (senior, manager/R&D, consultancy/engineer/general staff).

16.3.2.1 Water sources(AA/TR = 96/97, Q3)

Figure 16.2 shows the percentage of each water source (multiple answers) used. Overall, the percentages were 60, 45, 43, 21, 21, 14, and 6% for groundwater, river, reservoir, wastewater, lake, seawater, and rainwater, respectively. The average number of selections (*ANS*) for water sources per respondent was 2.19, which indicates that it was common for multiple water sources to be utilized.

The same analysis was conducted for each of the four previously selected groups. The average number of water sources was the lowest (1.88) in the WW group, while the TS group had the highest number of sources (2.43), which might reflect the increase in the targeted working sectors. The WS and WW groups exhibited different water source patterns. For example, groundwater, rivers, and reservoirs were utilized in the former group, but wastewater was the most utilized source in the latter group.

Water usage percentage differed among working groups for each water source or among water sources in each working group. The differences were then statistically tested in two groups. It was assumed that two sets of A and B had answers of (Y, N) and (y, n), respectively for the sample numbers

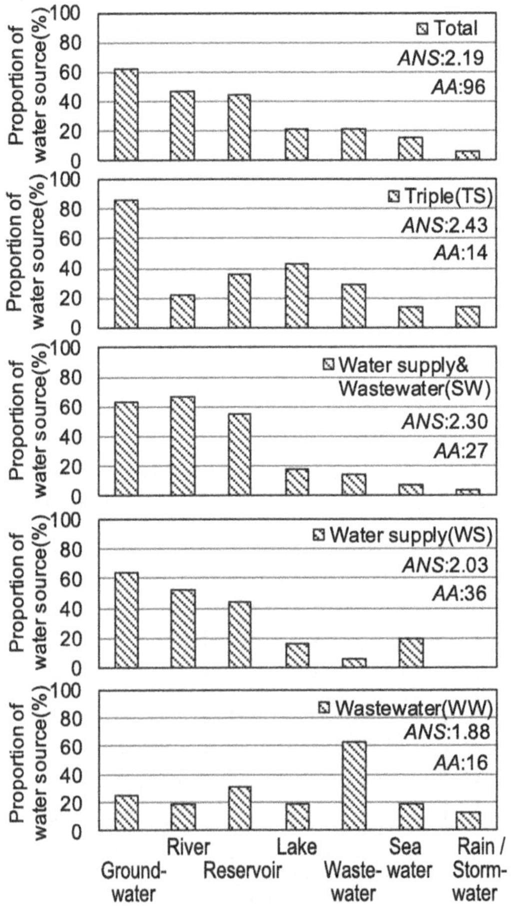

Figure 16.2 Usage percentage of water sources in each water utilitysector. *ANS*: The average number of selections; *AA*: available answers.

of 'Yes' and 'No', and Set A was less positive than B ($Y/N < y/n$). When the sample number of $Y + N$ (= equivalent to Set A) was chosen from a mixed set of $(A + B)$ (= $Y + N + y + n$), the probability P, which was the probability of the number of 'Yes' responses equal to or less than Y, could be deduced from a theory of combination as follows:

$$P = \sum_{k=\max(0,Y-n)}^{Y} \left(\frac{{}_{(Y+y)}C_k \, {}^{*}{}_{(N+n)}C_{(N+n-k)}}{{}_{(Y+N+y+n)}C_{(Y+N)}} \right)$$

If P is particularly small, it is reasonable to consider that sets A and B have different ratios of Yes/No. P indicates the risk level (= significant level) on the first type of error.

Results of the analysis (refer to Table 1s and Table 2s in the supplementary information, available with the online version of this paper) indicate that the probability P was scored under 0.001 in sets that compared groundwater, rivers, and reservoirs *versus* any of the other water sources in the SW and WS groups, indicating their stochastic differences. This was also seen in comparisons of wastewater and other sources in the WW group and in those of groundwater and the other sources in the TS group. On the other hand, a probability of 0.01 was scored in groundwater between the WW and TS groups, in the river between the WW and SW groups and between the SW and TS groups, and in wastewater between the WS and WW groups and between the WW and SW groups. Thus, those answers were indicated to be stochastically different.

16.3.2.2 Frequency of checking water resource condition (AA/TR = 87/97, Q5)

Figure 16.3a shows the monitoring frequency of the water sources. Continuous monitoring was 33%, multiple times per hour was 2%, and several times a day was 26%. Thus, 62% answered that water source monitoring was conducted at least once a day, while 32% reported monitoring several times a month, and 6% reported several times a year. One respondent commented that the monitoring frequency of the water sources depended on the target pollutants and water sources. For example, pH and turbidity were monitored in real time, and pesticides were measured once a month; reservoirs were monitored in real time, and other sources of water were checked several times per month.

16.3.2.3 Programme for catchment protection work (AA/TR = 84/95, Q8)

There were 88% of the participants who reported that they had a programme to protect catchments.

16.3.2.4 Frequency of updating/reconsidering their applied water standards (AA/TR = 68/84, Q11)

Figure 16.3b shows the ratio of the update frequency to the reconsideration of water standards at the utilities: 44% updated or reconsidered their water standards multiple times every six months,

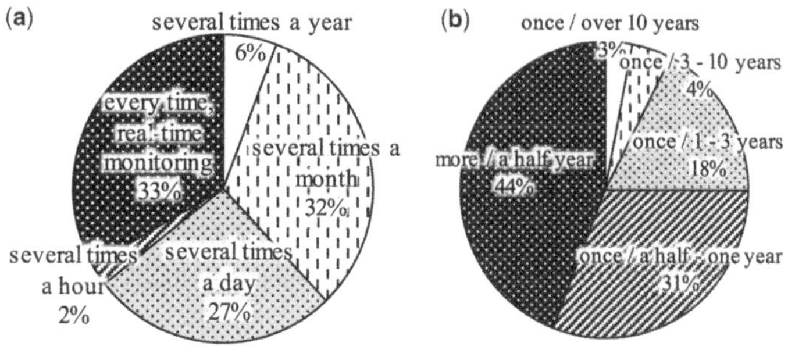

Figure 16.3 (a) The number of answers for frequencies of monitoring water sources, (b) The frequency of updating or reconsidering water standards in utilities.

while 31% only did so once per six months or per year. Thus, 75% updated or reconsidered the water standard at the facilities at least once a year. There were 14 comments on this question and three comments indicated that the frequency of updating depended on the compliance and guidelines of their respective companies.

16.3.3 Effects of land use on diffuse pollution

Landscape (land use) management is expected to alleviate diffuse pollution and reduce the burden of diffuse pollution on the environment (Haibo *et al.*, 2015). In order to investigate the concerns of respondents from the water utility sector regarding land use, the influences of landscape measures on the catchment area and raw water were examined.

16.3.3.1 Land use protection on catchment area (AA/TR = 86/95, Q6)

Figure 16.4 shows the effects of surrounding natural vegetation buffer landscapes on catchment areas. It indicates how land use management, including vegetation control, influences the respondent's catchment areas. Of the respondents' answers, 56% commented that the catchment areas were 'Fairly' or 'Very protected' by natural vegetation (*WA* = 3.44), while 29% answered that it was 'Slightly' or 'Not at all protected'.

The degree to which catchment protection programmes affected the answers to the survey was observed and they were compared by the differences of answers to Q8 (See Figure 1s in the supplementary information, available online). From the comparison, the answers from the two groups were statistically different, and a strong connection between the land use protection concerns in the catchment areas and their programme was proven.

The concerns of land use effects among working groups were also analysed (see Figure 2s in the supplementary information). The different concerns indicated the highest *WA* (= 3.92) in the triple sector group and the lowest *WA* (= 2.85) in the wastewater group. Furthermore, the statistical difference between the WW group and TS group were indicated. The other relationships were not observed at a significance level of 5%. Therefore, the landscape effects on the catchment area were not recognized by the WW group but were recognized by the WS group.

16.3.3.2 Land use effects on raw water quality (AA/TR = 89/95, Q7)

Figure 16.5 illustrates the importance of landscape in the protection of raw water quality. It indicates how land use management is related to raw water quality in water utilities. Landscape measures are considered 'Very important' by 60% of the respondents for water quality protection, and the *WA* (= 4.17) was also higher than that in the previous question about catchment area (= 3.44). The land use programme seems to be closely related to this evaluation. From the comparison about difference of employing the programme or not (see Figure 3s in the supplementary information), the answers

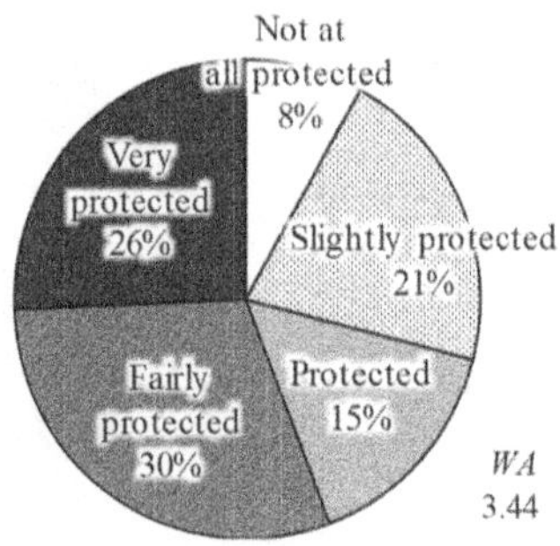

Figure 16.4 The degree of land use protection on the catchment areas.

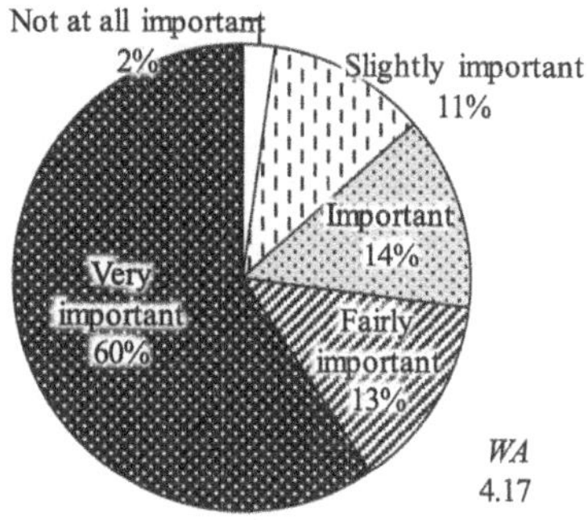

Figure 16.5 The importance of land use on raw water quality.

from the two groups were statistically different. Because land use measures could be an effective way to buffer diffuse pollution in aquatic environments, it is important to periodically monitor the land use and environment connected to the utilities' water sources, and to examine the effects of land use in the future.

Meanwhile, the *WA* on the raw water quality is shown in an opposite order in water supply (= 4.42), water supply & wastewater (= 4.31), and the triple sectors (= 3.71) group (see Figure 2s and Figure 4s in the supplementary information). The WA might have decreased because the respondents related to the rainwater management sector coexist in the triple sector group. From comparisons of the answers using the *t*-test, the answers between the WS and WW groups and between the WS and TS groups were observed to be statistically different. The other relationships were not observed at a significance level of 5%. From the results it is estimated that the landscape effects on raw water quality are extremely important, especially for the WS group.

16.3.4 Influences of each pollutant on water utilities

Many kinds of pollutants derived from diffuse pollution have been reported. The influences of each pollutant on water sources and utilities, such as the occurrence of algal blooms and faecal pathogens, were examined separately. Additionally, the influences of intermittent chemical pollution on water sources were studied.

16.3.4.1 Influences of each pollutant on raw water supplies (AA/TR = 88–96/97, Q4)

Figure 16.6 shows the influences of pollutants on water sources. The highest weight average was obtained for chemicals (pesticides etc.) (= 4.51), and the *WA* of faecal pathogenic organisms was the second highest (= 4.44). Additionally, to analyze significant differences in answers for each pollutant, the relationships among pollutants were evaluated by the *t*-test (refer to Table 3s in the supplementary information). From the comparisons between each answer group, the answers for chemicals and faecal pathogens were statistically different from those of the other pollutants. Because these issues were related to human health, it might be considered significantly important for respondents. On the other hand, the *WA* of sediments was the lowest (= 3.76). The reason for this is assumed to be the frequent occurrence of sediment load throughout a year – it is no longer considered important for water sources because it is already adequately handled and controlled in water utilities.

16.3.4.2 Influences of each pollutant on water utilities (AA/TR = 69–79/84, Q9)

Figure 16.7 shows the influences of each pollutant on the operation of the utilities. The *WA* of faecal pathogenic organisms was the highest (= 4.22), which suggested that faecal pathogens were strongly recognized as an important pollutant for operation at the utilities. On the other hand, the *WA* of veterinary medicine related to pharmaceutical contamination from diffuse pollution was the lowest

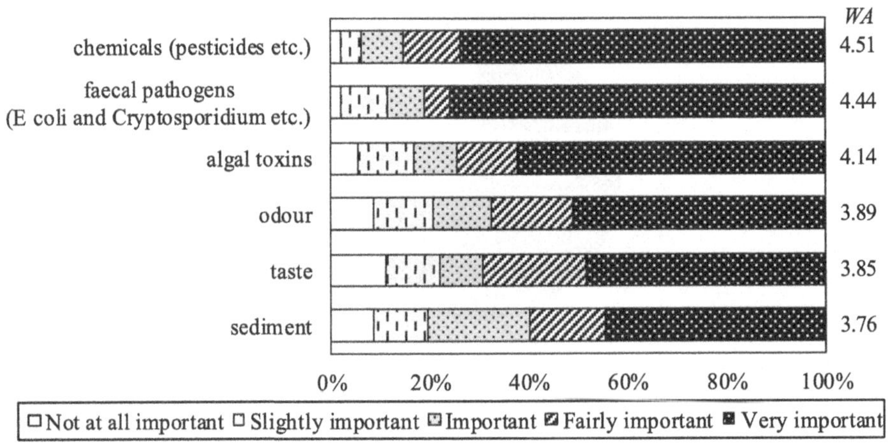

Figure 16.6 The influences of each pollutant on raw water quality.

of all the pollutants (= 3.42). To figure out significant differences of answers for the above pollutants, the relationships among pollutants were evaluated by the *t*-test (refer to Table 4s in the supplementary information). Between veterinary medicines and faecal pathogens, between veterinary medicines and pesticides, and between faecal pathogens and urban pollutants the answers were statistically different by a significant level (1%). In addition, significant differences of 5% were found on faecal pathogens *versus* heavy metal, sediment load and algal or plant growth. Similarly, significant differences of 5% were observed between urban pollutant and pesticides, between veterinary medicines and heavy metals, and between veterinary medicines and algal or plant growth.

Pharmaceutical contamination in the environment was adopted as an emerging policy issue under the context of the Strategic Approach to International Chemicals Management at the International

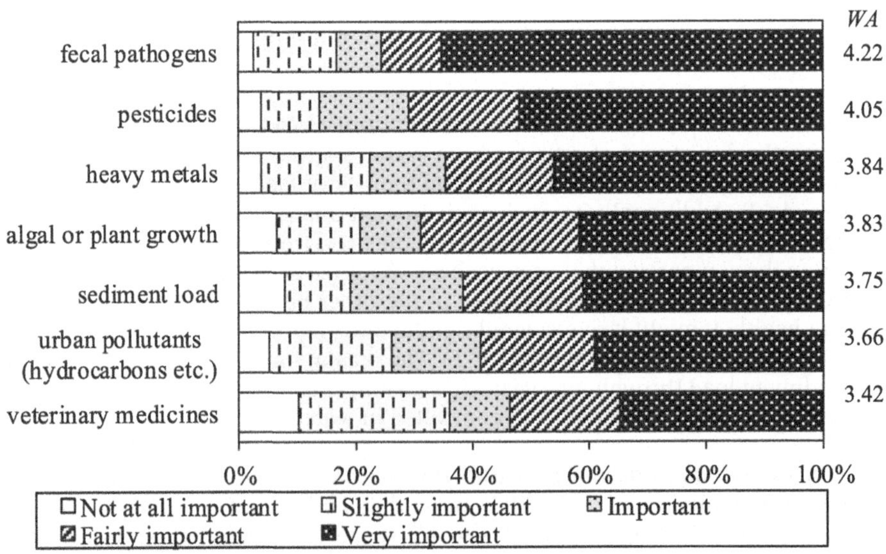

Figure 16.7 The influences of each pollutant on utilities.

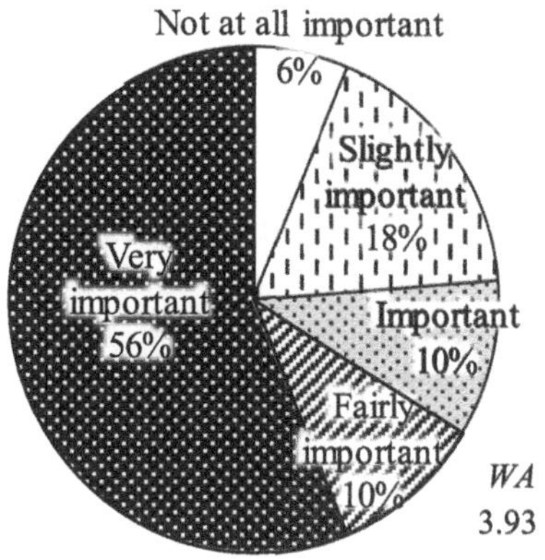

Figure 16.8 The influences of intermittent chemical contamination on raw water quality.

Conference on Chemicals Management in 2015 (aus der Beek *et al.*, 2016a, 2016b). Thus, this issue is expected to receive more attention from stakeholders of water utilities in the future. To clarify the different degrees of recognition for pharmaceuticals per each working group, this was calculated individually from their answers (refer to Figure 5s in the supplementary information). The interest in veterinary medicine was highest in the TS group. This suggests that respondents who belong to comprehensive working position are sensitive to global concerns. To analyse these scores, the answers related to pharmaceuticals were separated into four working sector groups and were evaluated by the *t*-test. However, differences at a significance level of 5% were not observed between any working groups.

16.3.4.3 Influences of intermittent chemical contamination on raw water quality (AA/TR = 80/84, Q10)

Figure 16.8 shows the influence of intermittent chemical contamination on raw water quality. Of the respondents, 56% answered that intermittent chemical contamination was 'Very important' to the water utility sectors. Intermittent chemical contamination is difficult to anticipate in advance. Therefore, frequent monitoring is considered an indispensable means of proper management to counter sudden chemical contamination accidents.

16.4 CONCLUSION

A questionnaire survey on the influences of diffuse pollution on water utilities was sent to more than a thousand professionals in the water utility sector, of which 97 respondents submitted a completed survey. The results of the survey showed the following:

- Respondents could be divided into four groups: water supply, wastewater, a combined water supply and wastewater, and a triple sector group (water supply, wastewater, and rainwater). All groups mainly comprised senior or management level workers.
- 68% of the utilities monitor the water source at least once a day, 88% possessed a program to protect the catchment area, and 75% updated or reconsidered water standards at the utilities at least once a year.
- The land use effects on diffuse pollution affecting raw water quality were recognized by the sectors, especially in 56% of the utilities. In addition, for respondents who had a catchment

protection programme, land use effects were highly recognized in the catchment areas ($WA = 4.31$).

- The interest in chemical substances, such as pesticides ($WA = 4.51$, 4.05) and pathogens ($WA = 4.44$, 4.22), was higher in water sources and utilities. In addition, emerging pollutants, including pharmaceuticals, were of the lowest concern among all the groups ($WA = 3.42$).
- 56% of the respondents considered intermittent chemical contamination of water sources as very important.

The questionnaire did not receive very high response rates, but useful data was obtained as a pilot study regarding the perception of diffuse pollution by water utilities. The proposed countermeasures and diffuse pollution issues should continue to be highlighted and improved through interview and other questionnaire surveys at workshops.

ACKNOWLEDGEMENT

This research was kindly supported by the International Water Association, and we are sincerely grateful for the responses to our survey.

REFERENCES

aus der Beek T., Weber F. A., Bergmann A., Hickmann S., Ebert I., Hein A. and Küster A. (2016a). Pharmaceuticals in the environment – global occurrences and perspectives. *Environmental Toxicology and Chemistry*, **35**(4), 823–835, https://doi.org/10.1002/etc.3339

aus der Beek T., Weber F. A., Bergmann A., Grüttner G. and Carius A. (2016b). Pharmaceuticals in the Environment–Global Occurrence and Potential Cooperative Action Under the Strategic Approach to International Chemicals Management (SAICM), Texte 67/2016. Umweltbundesamt, Dessau-Roßlau, Germany, 95pp.

D'Arcy B., Kim L. H. and Maniquiz-Redillas M. (2018). Wealth Creation Without Pollution – Designing for Industry, Ecobusiness Parks and Industrial Estates. IWA Publishing, London, UK.

Davis J. R. and Koop K. (2006). Eutrophication in Australian rivers, reservoirs and estuaries – a southern hemisphere perspective on the science and its implications. *Hydrobiologia*, **559**, 23–76, https://doi.org/10.1007/s10750-005-4429-2

Dima G., Popescu I. V., Stihi C., Oros C., Dinu S., Manea L. and Vlaicu G. (2006). Fe, Mn and Zn concentrations determination from Ialomita River by atomic absorption spectroscopy. *Romanian Journal of Physics*, **51**(5/6), 667–674.

Haibo Y., Le Z., Wei H., Yi W., Dan W., Shuwen Z. and Kun B. (2015). Cumulative effect of land use and cover changes in Naoli River Basin in Sanjiang Plain on agricultural nonpoint source pollution load. *Sustainable Water Resources Management*, **1**(4), 355–362, https://doi.org/10.1007/s40899-015-0028-1

Li Y. and Zhang J. (1999). Agricultural diffuse pollution from fertilisers and pesticides in China. *Water Science and Technology*, **39**(3), 25–32, https://doi.org/10.1016/S0273-1223(99)00029-3

Shen Z., Liu J., Aini G. and Gong Y. (2015). A comparative study of the grain size distribution of surface dust and stormwater runoff quality on typical urban roads and roofs in Beijing, China. *Environmental Science and Pollution Research International*, **23**(3), 2693–2704, https://doi.org/10.1007/s11356-015-5512-5

Tyrrel S. F. and Quinton J. N. (2003). Overland flow transport of pathogens from agricultural land receiving faecal wastes. *Journal of Applied Microbiology*, **94**, 87–93, https://doi.org/10.1046/j.1365-2672.94.s1.10.x

World Health Organization & International Programme on Chemical Safety (1998). Selected Non-Heterocyclic Polycyclic Aromatic Hydrocarbons, Environmental Health Criteria, **202**. WHO-IPCS, Geneva, Switzerland.

doi: 10.2166/9781789061123_251

Chapter 17

Diffuse pollution and drinking water quality – cost–benefit of reducing the source to improve the environment

J. F. S. Knightbridge[1], I. V. Ribeiro[1] and A. L. Cooke[2]

[1]Mott MacDonald Ltd, 22 Station Road, Cambridge CB1 2JD, UK
[2]Severn Trent Water, Severn Trent Centre, 2 St Johns Street, Coventry CV1 2LZ, UK

Keywords: Agriculture; catchment management; ecosystem services; eutrophication; SWAT; water quality

17.1 INTRODUCTION

The European Drinking Water Directive (Council Directive 98/83/EC) requires that drinking water quality must adhere to a standard of 0.1 μg/l of any individual plant protection product (PPP) in treated water (European Commission, 1998). A plant protection product had been defined as a pesticide product intended for preventing, destroying or controlling any pest causing harm during or otherwise interfering with the production, processing, storage, transport or marketing of food, agricultural commodities, wood and wood products (FAO, 2006). Total global pesticide usage has increased steadily from approximately 2.4 million tonnes in 1990 to 4.1 million tonnes in 2018 (FAO, 2020). The impacts of pesticides on ecosystems and human health are well documented globally (Sharma *et al.*, 2019).

Throughout the UK, water companies have recorded PPP concentrations above the regulatory drinking water standard in surface waters supplying their reservoirs and river intakes. Conventional water treatment can readily remove many PPPs but often not to the target concentrations required. Additional, advanced treatment technologies typically have varying levels of effectiveness (Dillon *et al.*, 2011).

The Water Framework Directive (WFD) (Council Directive 2000/60/EC) aims to protect inland surface waters, transitional waters, coastal waters and groundwaters and sets out objectives for all waterbodies (European Commission, 2000). Article 7 of the WFD obligates member states to move away from end-of-pipe treatment and encourages upstream thinking. Catchment management is widely being considered as the solution to reducing PPP concentrations in drinking water and there have been multiple successful schemes throughout the UK. (Mohamad Ibrahim *et al.*, 2019)

PPP usage throughout the UK varies both annually and geographically driven by demand for certain crops and climate. The potential impacts of PPPs on human health and the environment is widely discussed and documented (Castle *et al.*, 2017) and has led in recent years to restrictions or bans on usage; including in the UK for metaldehyde (from 2020).

The agricultural sector is a key contributor of water pollution in the UK (Defra, 2018). Aligned with these directives and the push for sustainable use of PPPs in the agricultural sector the future will be a challenging one whereby the sector will need to be less reliant on the active ingredients of widely used PPPs and adopt alternative catchment management approaches, such as integrated pest management (Hillocks, 2012) and product substitution.

The objective of this study was to assess the costs and benefits associated with reliably reducing PPP concentrations to below the prescribed concentration value (PCV), which is 0.1 μg/l at the Ogston Reservoir (UK) raw water abstraction. The secondary objective was to reduce the risk of eutrophication of the Ogston Reservoir through in-catchment measures. Benefits generated through in-catchment measures to ecosystem services are considered as secondary or 'additional' benefits to the primary aim of improving the quality of the raw water transferred to the Ogston WTWs from the abstraction.

The study focused on Ogston Reservoir and its natural catchment. Ogston Reservoir receives water from its own natural catchment as well as by transfer from the River Derwent (Ambergate abstraction) and Carsington Reservoir. The study area therefore comprised three catchments; the River Derwent to Ambergate, Carsington Reservoir catchment, and Ogston Reservoir catchment. The catchments are all located in Derbyshire and are shown in Figure 17.1.

17.2 METHODOLOGY

The study was informed by a detailed catchment characterisation to develop a hydrological understanding of the study catchment and eutrophication assessment of the reservoir, including identification of potential sources of pollution and pathways to the drinking water abstraction. For that purpose, surface water catchment investigations were carried out to consider the potential for reducing pollution loading to the drinking water catchment, through the implementation of different agricultural catchment management intervention measures (up-stream thinking), along with the relative costs and benefits of these interventions. This included an ecosystem services assessment to quantify those benefits (Defra, 2013).

The study was divided into three phases: initial assessment (phase 1), detailed catchment modelling (phase 2) and benefit-cost assessment (phase 3).

Phase 1 comprised an initial assessment of each catchment. This initial assessment included a catchment characterisation and conceptualisation, including identification of potential sources of pollution and pathways to the abstraction (receptor). This involved the collection of readily available spatial and temporal datasets, catchment walkovers and a survey of landowners in the catchment. The Phase 1 investigation for the Ogston catchment concluded that the PPPs of concern in the raw water at Ogston WTW are metaldehyde, MCPA and 2;4D. Metaldehyde is a molluscicide used which attracts and kills slugs and snails present in the soil. MCPA and 2;4D are used in the formulation of selective herbicides.

The main outputs from the Phase 1 investigations were risk maps which combined existing datasets. The base for the maps was the existing field risk mapping carried out by Severn Trent Water Ltd. (STWL), which incorporated soil type, slope and proximity to watercourses. In Phase 1, land use was added to further refine the risk maps for different PPPs. The maps were also informed by the site visits completed as part of Phase 1.

Phase 2 of the present study included a detailed modelling of the Ogston Reservoir catchment to estimate the contribution to the metaldehyde, MCPA and 2;4D loadings from the natural reservoir catchment. Modelling of the Ogston Reservoir catchment has been carried out using the Soil and Water Assessment Tool (SWAT), in order to further investigate whether undertaking catchment management in target areas within the Ogston Reservoir catchment could sufficiently reduce the risk of PPP exceedances of the PCV in the raw water.

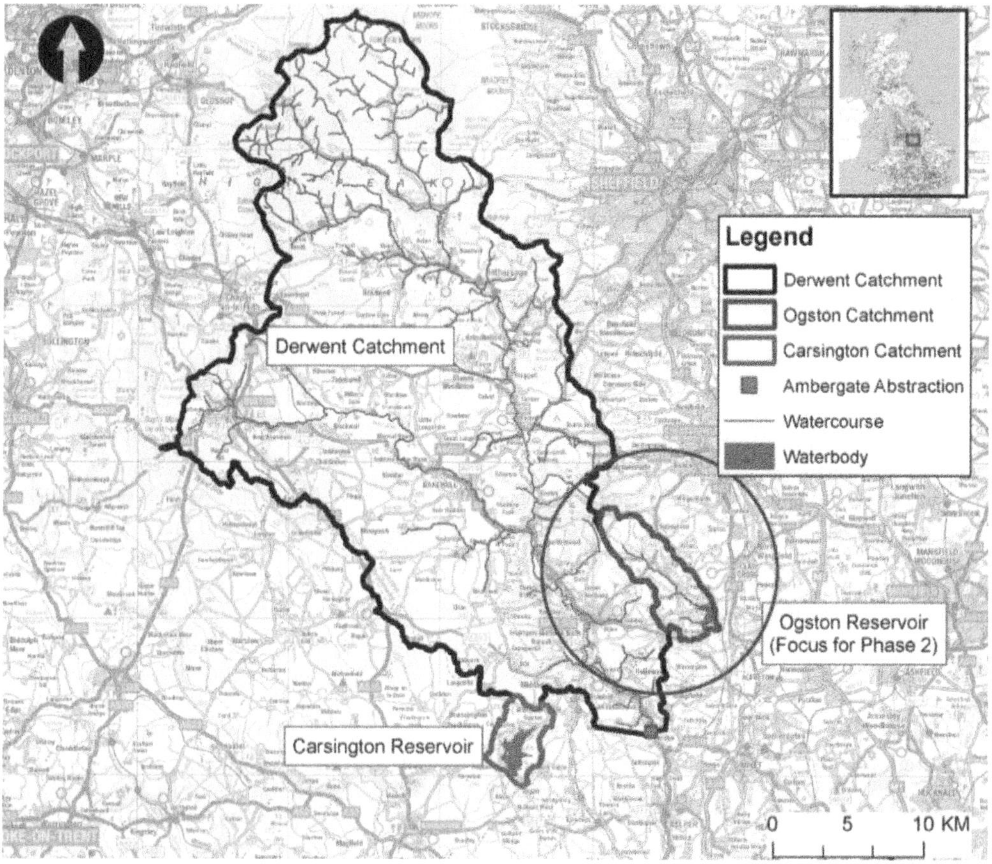

Figure 17.1 Location of Derwent, Ogston and Carsington catchments. Contains Ordnance Survey data Crown copyright and database right © 2016.

To assist in the model build, the graphical user interface for SWAT, ArcSWAT (an ArcGIS extension) was used with ArcGIS 10.3 software to give clear representation of the model throughout its setup. A digital terrain model (DTM), spatial land use data and soil data have been input into the Ogston Reservoir catchment SWAT model. SWAT processes the DTM, land use, soil layers and slope to define a watershed, sub-basins and hydrological response units (HRU) which become the basis of the model input files.

The number of HRUs in the model affects its run-time. To enable model efficiency, the number of HRUs was rationalised by only including the dominant combinations of land use, soil and slope within each sub-basin. This is achieved by specifying threshold percentages or areas to retain the dominant combinations of land use, soil and slope class. Any land use/soil/slope combination that covers a percentage or area of a sub-basin that is less than the specified threshold is eliminated, and the area within these HRUs is reapportioned so that 100% of the land area in the sub-basin is modelled. Thresholds were adjusted to achieve a compromise between the level of model detail and model run-time. A 10% threshold was assumed for land use, soil class and slope class areas. HRU rationalisation represents a simplification of the land use, soil types, and slopes in a catchment. A range of thresholds

Table 17.1 Climate data series model inputs.

Series	Source	Period Included in Model
PET	MOSES weekly data – square 107	2004–2015
Rainfall	Daily rainfall at Ashover station	2004–July 2015
	Daily rainfall at Barbrook station	August–October 2015
Climate data	Loughborough University weather station and CET series	2004–2015

were tested to achieve the smallest practical simplified area; this results in a compromise between eliminating large areas of some sub-basins or generating a high number of HRUs that would adversely impact model development and run-time. The loss of detail in the representation of the catchment properties is not considered to be too significant, given the uncertainties in the parameterisation of values representing PPP properties and catchment processes and recorded data used for model calibration. The HRUs that were eliminated in the rationalisation cover a small area (approximately 12.5%) of the Ogston Reservoir catchment.

Following the initial model set-up using spatial data for the catchment, hydrological inputs are added to the model. Hydrological inputs broadly fall into two categories: spatial parameters describing the catchment; and time-series model inputs. The time-series inputs to the model are data series and are fixed at model setup; spatial hydrological parameters are used to describe the model's routing of rainfall and can be adjusted during calibration.

Climate data used in the model include rainfall, potential evapotranspiration (PET), wind speed, relative humidity, solar radiation and temperature. Table 17.1 summarises the data series which have been input to the model.

Reservoir inflows and outflows (2007–2015) were included in the SWAT model to ensure flows routed through the Ogston Reservoir were consistent with measured values. This data was used as part of the hydrology calibration and was of particular importance given that no gauged data was available for the catchment. Reservoir behaviour is controlled in the SWAT model by parameters describing the reservoir dimensions, properties and operation. The storage volume and surface area of the reservoirs was based on data provided by STWL. The reservoir leakage and evaporation coefficient were initially set to the SWAT default values.

The public water supply abstraction from Ogston Reservoir to the WTW was incorporated into the reservoir in the SWAT model as a daily time series and using monthly average values. There are no other consented abstractions within the Ogston Reservoir catchment. Discharges to surface water are included in the SWAT model as point source inputs. The two discharges included in the model are both from sewage treatment works (STW) and are presented in Table 17.2. The recorded daily data is included in the model and the table includes the consented Dry Weather Flow (DWF) to give an indication of the size of the discharges.

Tile drains have been assumed to be present within certain HRUs, according to their soil types and land use. Most of the soil in the catchment is loamy and is considered to be well drained (NERC, 2012). Drain inputs to the SWAT model were amended to reflect ground conditions, in accordance with literature values.

Table17.2 Discharges included in the model.

Discharge Name	Consented Dry Weather Flow (m^3)	Data Included in Model
Ashover (STW)	510	Daily recorded data (2007–2015)
Kelstedge (STW)	100	Daily recorded data (2007–2015)

Source: STWL.

Land use is defined in the SWAT model at the model build stage. Land cover is then applied by assigning crop types and management operations to each HRU based on the land cover characteristics, and for arable land, the crops grown.

Crop growth influences the model hydrology by affecting catchment runoff, evapotranspiration and soil moisture dynamics. SWAT contains a database describing the properties of each plant, which determines how the plant growth is modelled and its dependence upon climatic conditions, soil nutrients, among others. The default crop parameters have been left largely unaltered during this study, although some effort has been made to ensure parameters are appropriate for the UK. Crop types have been assigned to each HRU, taking into account crop rotation information obtained during Phase 1 of this study.

PPP application is input into the SWAT Model as a load (kg/ha) which can be defined for each HRU. The product is transferred via the hydrological pathways and degrades on route. The degradation is dependent on the rate constants assigned in the model.

The properties of metaldehyde, MCPA and 2;4D were added to the SWAT model pesticide database using the physical and chemical properties identified in Phase 1. These values were used in the initial model parameterisation. Since the actual values of these pesticide parameters have a large uncertainty, and are likely to be site specific, they have been treated as calibration parameters.

The hydrological calibration was undertaken by comparing simulated and recorded reservoir volumes for Ogston Reservoir. In-stream flows in the catchment (for which there is no gauged data) were consistent with the existing HYSIM hydrological model for the catchment, and this provides confidence in the calibration. Water quality calibration of all three PPPs was achieved with a good fit of magnitude and timing of sampled concentrations, which provides a good level of confidence in how the hydrology and water quality are represented within the model. The model performance provides a good level of confidence in the model inputs and outputs. Parameters adjusted to achieve effective calibration were kept within realistic boundaries. Outputs from this section provide a solid basis to complete scenario runs to determine the extent to which PPP concentrations at public water supply abstractions can be reduced by various catchment management interventions.

The simulated catchment management intervention scenarios are defined in Table 17.3. The primary objective of the intervention scenarios is to reliably reduce PPP concentrations to below the

Table 17.3 Summary of catchment management scenarios.

Scenario	Description	Model Area Targeted	
Metaldehyde		km²	% of arable land in model
1	No application across percentage of arable land		
1a	100% of arable land in reservoir catchment	5.55	100%
1b	50% of arable land in reservoir catchment	2.80	50%
2	No application on high-risk areas		
2a	No application on high-sensitivity arable areas	2.83	51%
2b	No application on medium- or high-sensitivity arable areas	4.91	89%
2;4D and MCPA		km²	% of improved pasture in model
1	No application across percentage of improved pasture		
1a	100% of improved pasture in reservoir catchment	15.12	100%
1b	50% of improved pasture in reservoir catchment	7.56	50%
2	No application on high-risk areas		
2a	No application on high-sensitivity improved pasture areas	7.93	52%
2b	No application on medium- or high-sensitivity improved pasture areas	13.58	90%

Table 17.4 Ogston reservoir catchment intervention scenarios.

Scenario	Description
1	Product substitution
1.1	Metaldehyde substitution in high-risk areas of the Ogston Reservoir catchment (51% of arable land or 2.83 km^2)
1.2	Metaldehyde substitution in high- and medium-risk areas of the Ogston Reservoir catchment (89% of arable land or 4.91 km^2)
1.3	MCPA and 2;4D substitution on all improved pasture in Ogston Reservoir catchment (improved pasture covers an area of 15.12 km^2 or 54% of the catchment)
1.4	MCPA and 2;4D substitution on high-risk areas in Ogston Reservoir catchment (7.93 km^2, 52% of improved pasture)
2	Farming advice
2.1	PPP management in Ogston Reservoir catchment: catchment advisor to run information events, hold one-to-one meetings with farmers, advise on good practice
2.2	Nutrient management in Ogston Reservoir catchment: catchment advisor to run information events, hold one-to-one meetings with farmers, advise on good practice
2.3	PPP & nutrient management in Ogston Reservoir catchment: catchment advisor to run information events, hold one-to-one meetings with farmers, advise on good practice
3	Abstraction management
3.1	No abstraction at Ambergate during November and December
4	PPP treatment at WTW
4.1	Metaldehyde removal by advanced oxidation at Ogston WTW
4.2	MCPA and 2;4D removal by existing processes (GAC Filtration) at Ogston WTW with regeneration frequency aligned to PPP peak concentrations

PCV, which is 0.1 μg/l. Article 7 of the WFD requires water companies to address water quality issues through in-catchment measures, as opposed to additional treatment.

The scenario testing also informed the likely effectiveness of different catchment management scenarios. The modelled interventions consisted of three categories of measures; product substitution, farming advice and additional water treatment.

The objective of the Phase 3 investigation was to assess the costs and benefits associated with the selected intervention scenarios. The cost benefit analysis followed the framework set out in UKWIR project WR26A – Benefit Assessment Framework (UKWIR, 2012). This included a counterfactual scenario (which is 'business as usual').

The following intervention scenarios categories were considered in the assessment: product substitution, farming advice and grants, abstraction management and treatment at the WTW. Table 17.4 presents all intervention scenarios considered in phase 3 of the present study.

In order to undertake the cost-benefit analysis, a list of benefits for each of the catchment schemes, including environmental, operational, and financial benefits was produced. This list was informed by the Environment Agency's Water Appraisal Tool (WAT) (EA, 2016), with data provided by STWL, and the project team's knowledge and research of catchment management benefits. This qualitative assessment considered the significance of change in ecosystem service provision for each intervention scenario.

The categories of ecosystem services included in the WAT and considered in this study are: freshwater provision, water for non-consumptive use provision, air quality regulation, climate regulation, water regulation (flow), pest regulation, erosion regulation, water purification, pollination, recreation and tourism, aesthetic value, intellectual, scientific and education value, social relations, nutrient cycling, photosynthesis and habitat provision.

Where it was possible, the benefits of each of the catchment schemes were subsequently quantified (using the Environment Agency's WAT). The assessment considered the value of ecosystem services based on customers' willingness to pay. Where it was not possible to quantify the benefits, anecdotal evidence from the STWL team was used.

The value of the benefits for each intervention was compared to the cost of implementation of the intervention to determine the benefit–cost ratio (BCR) (EA, 2016) for each scenario. The BCR is a measure of present value (PV) benefits divided by PV costs. BCR values lower than 0.5 indicate that interventions are unlikely to be worthwhile in economic terms; between 0.5 and 1.0 indicates that additional benefits could be worthwhile in economic terms; BCR values greater than 1.0 represent value for money, and interventions are worthwhile in economic terms if the BCR is stable following sensitivity analysis. A sensitivity analysis was undertaken using the WAT tools (EA appraisal summary table and EA Stage 1 valuation spreadsheets).

17.2.1 Eutrophication assessment

The secondary objective of the study was to reduce the risk of eutrophication of the Ogston Reservoir through in-catchment measures (in accordance with Article 7 of the WFD). Analysis of existing water quality and ecological data was undertaken to assess the present trophic state of the reservoir in phase 2. This assessment included an analysis of nutrient loadings to the reservoir, as the relationship between the quantity of nutrients applied to the land and the quantity arriving in the watercourse is highly complex, due to the influence of the nitrogen and phosphorus cycles.

In phase 3, the benefits generated through in-catchment measures to ecosystem services were considered as secondary or 'additional' benefits to the primary aim of improving the quality of the raw water transferred to the Ogston WTWs.

17.3 RESULTS AND DISCUSSION

The main outputs from the Phase 1 investigations were risk maps, presented in Figures 17.2 to 17.5.

Phase 1 investigations identified the Ogston Reservoir catchment as a target area for catchment management for metaldehyde and quinmerac. For MCPA and 2;4D, target areas in the upper Derwent catchment and the Ogston Reservoir catchment were identified. The high-level analysis of nutrients in the Derwent and Ogston catchments in Phase 1 concluded that there had been no recent reports of nutrient issues in the reservoir.

Conclusions from the scenario modelling (Phase 2) indicate that catchment interventions to prevent metaldehyde application focused on the Ogston Reservoir catchment would be sufficient to reduce recorded metaldehyde concentrations to below the PCV, and therefore catchment management for metaldehyde at the Ogston WTWs is unlikely to be required in the River Derwent catchment or Carsington Reservoir catchment.

All modelled scenarios reduced metaldehyde concentrations to below PCV for the period 2010 to 2015. The mapped results of the scenario modelling are presented in Figure 17.6. Scenarios 2a and 2b have been taken forward for this cost benefit analysis as the model results suggested that they would be more effective (by area) than scenarios 1a and 1b.

2;4D and MCPA are both herbicides that are typically applied at similar times of the year, often together as one formulated product. Therefore, they are considered together as part of this analysis. Scenarios 1a and 2a from Figure 17.7 were taken forward. Scenario 1a was shown to reduce PPP concentrations to below the PCV, whilst Scenario 2a bought PPP levels to just above the PCV. MCPA and 2;4D can be removed by existing treatment processes at Ogston WTWs, so Scenario 2a is included to allow the costs and benefits of different levels of intervention to be compared.

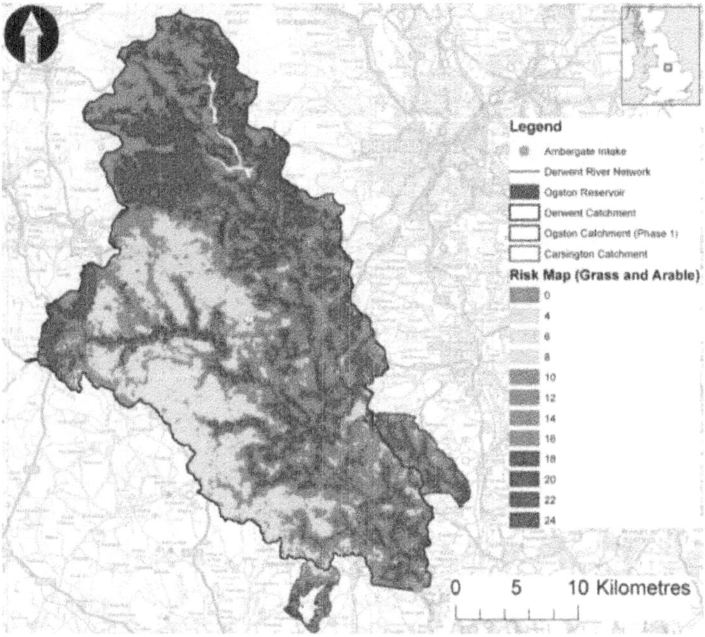

Figure 17.2 Phase 1 herbicide risk map (all catchments). Contains Ordnance Survey data Crown copyright and database right © 2016.

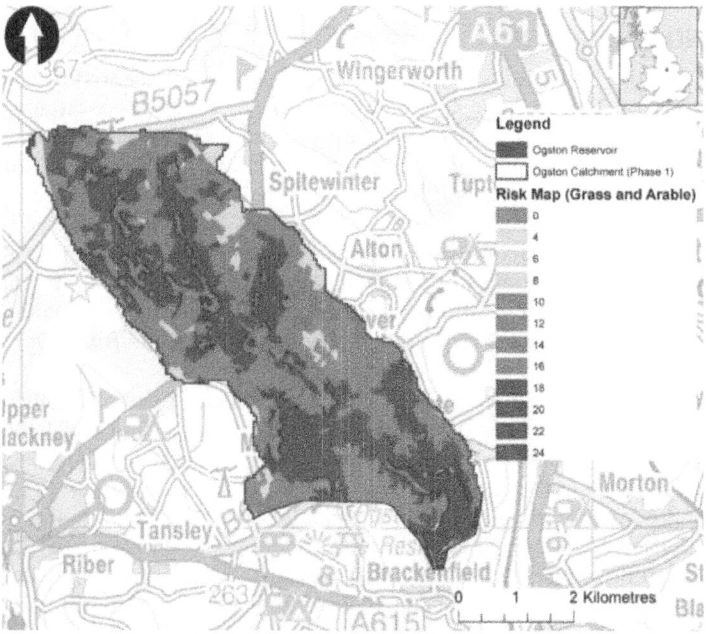

Figure 17.3 Phase 1 herbicide risk map (Ogston catchment). Contains Ordnance Survey data Crown copyright and database right © 2016.

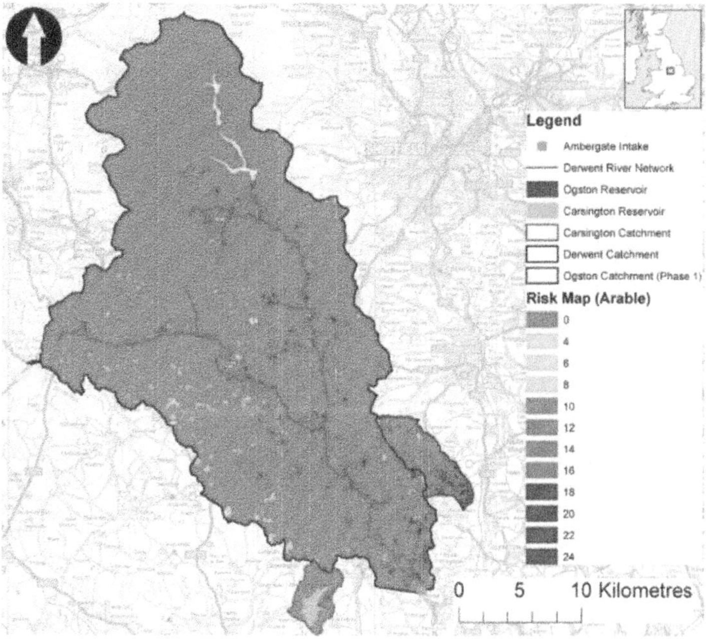

Figure 17.4 Phase 1 metaldehyde risk map (all catchments). Contains Ordnance Survey data Crown copyright and database right © 2016.

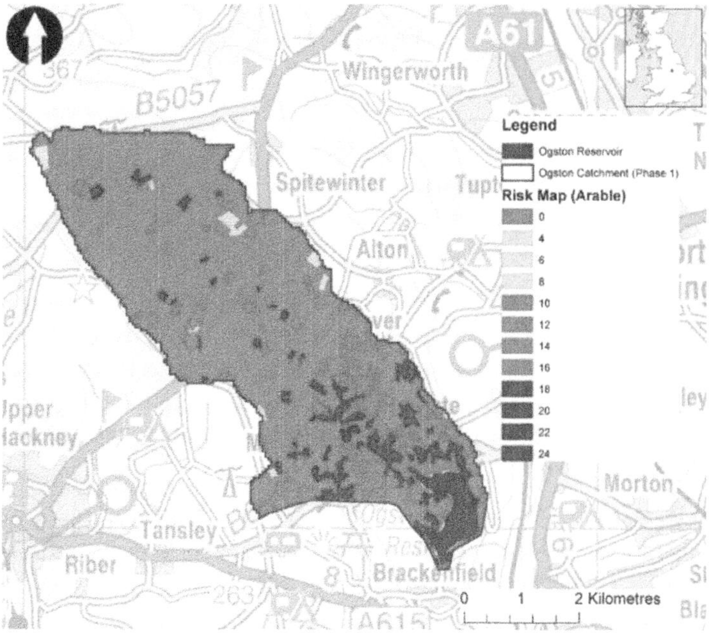

Figure 17.5 Phase 1 metaldehyde risk map (Ogston catchment). Contains Ordnance Survey data Crown copyright and database right © 2016.

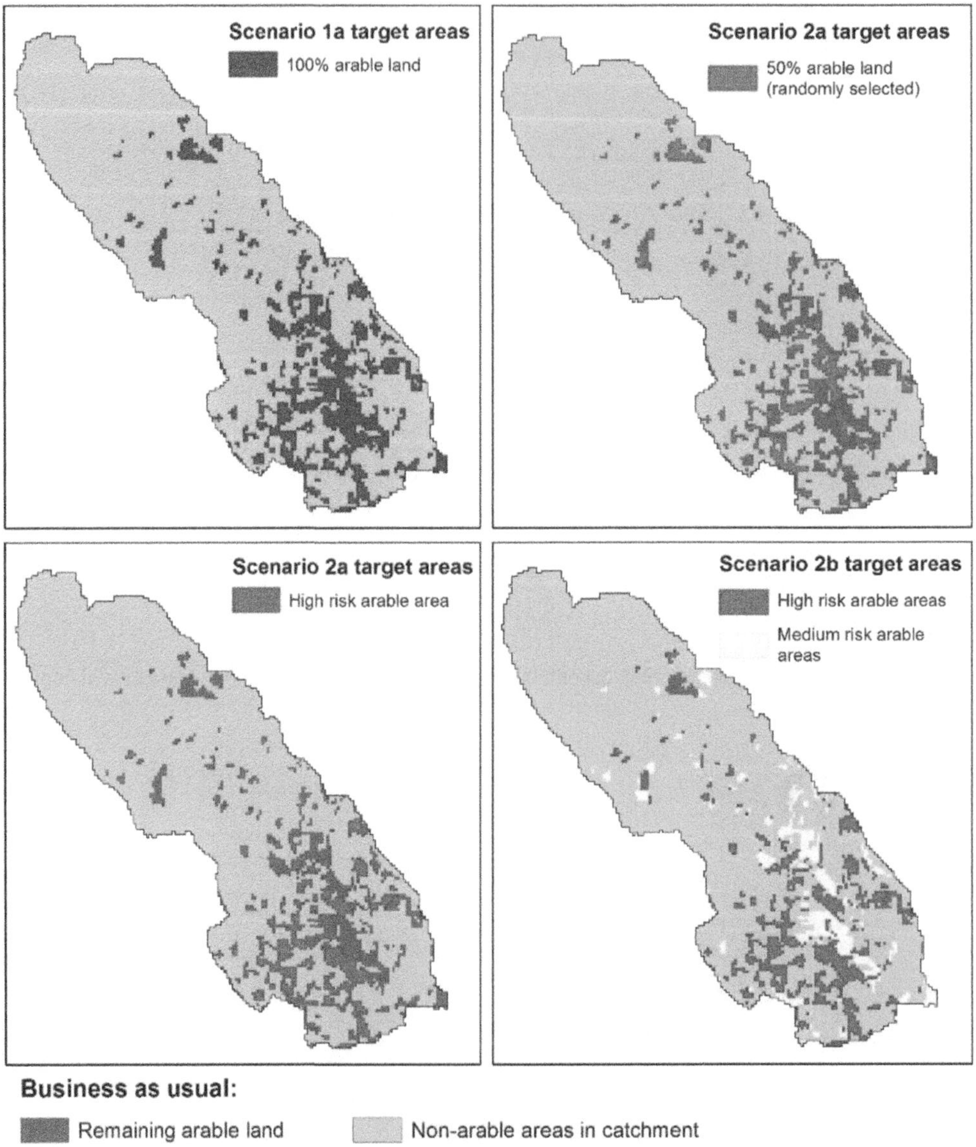

Figure 17.6 Scenarios (Phase 2) and target areas for metaldehyde product substitution. Mott MacDonald, Contains Ordnance Survey data Crown copyright and database right © 2016.

In phase 3, the qualitative assessment informed on the impacts of each scenario on the provision of each ecosystem service detailed within the WAT-AST, which are:

- For 'freshwater provision', long-term positive impact observed of scenarios 1.1 and 1.2 on freshwater due to supply for 232 488 people having metaldehyde levels reliably below the PCV. Improvements are possible for scenarios 1.3 and 1.4 but considered that these will result in

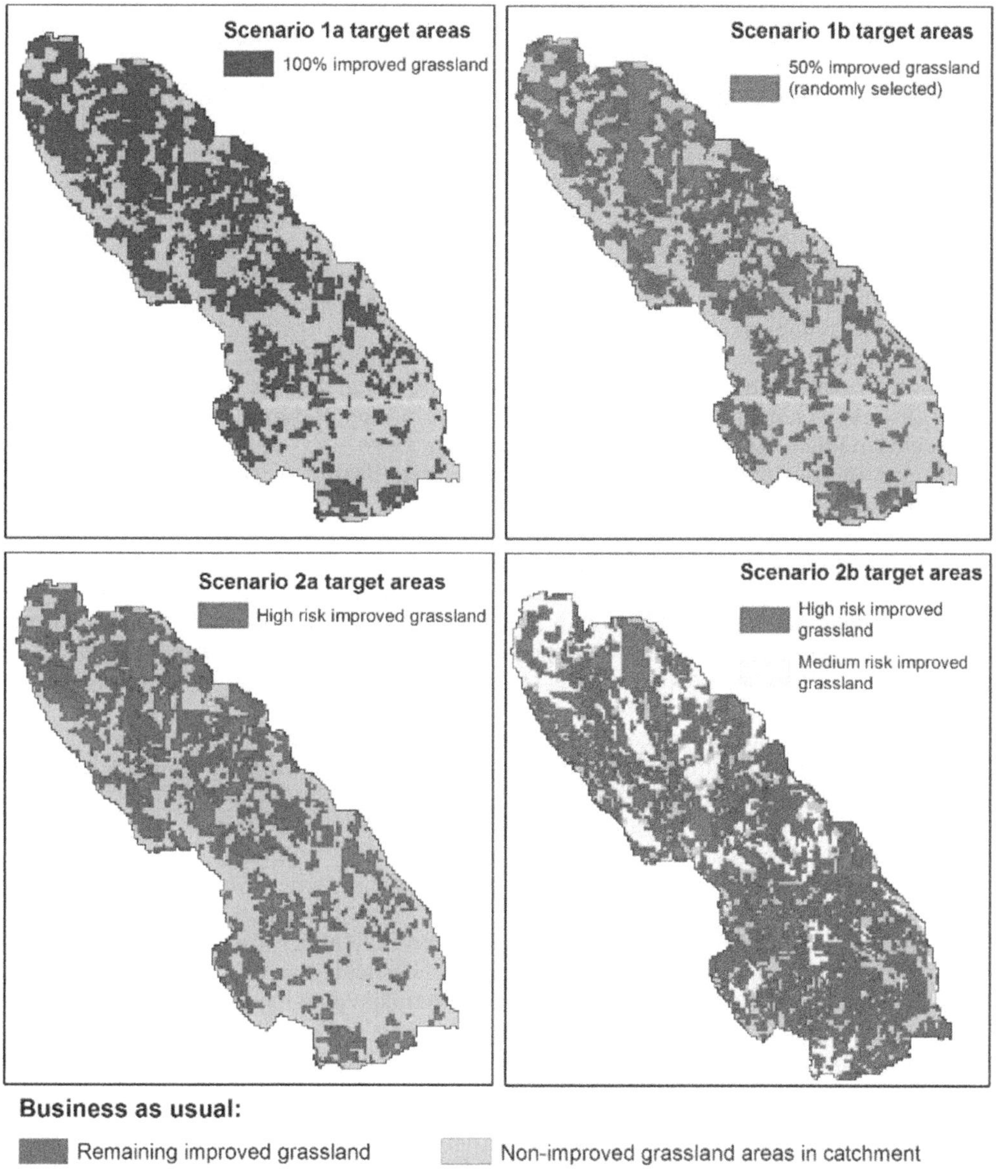

Figure 17.7 Scenarios (Phase 2) and target areas identified for 2;4D and MCPA product substitution. Mott MacDonald, Contains Ordnance Survey data Crown copyright and database right © 2016.

'pollution swapping'. Farm advice scenarios have potential to result in positive impacts due to reduced PPP and nitrate runoff. Treatment options will have a positive impact as freshwater provision will be improved by additional treatment.

- For 'water for non-consumptive use provision', substitution and advice scenarios, if implemented in the catchment, are likely to have a positive impact on catchment watercourses. There is a good probability that there will be some uptake of substitution and/or advice.

- 'Air quality regulation' could be improved by farm advice which results in better timed PPP applications (i.e., reduced spray drift). However, this would require compliance across the catchment and is not the target of catchment interventions.
- Impacts of climate change ('climate regulation') are likely to be influenced by factors outside of the catchment (i.e., global and national scale policy).
- For 'water regulation', there is a possibility that if there are enough farmers receptive to farm advice, there could be improvements to channels/river flow within the catchment. However, due to the number of farmers required for there to be a meaningful impact it is considered to have a medium risk of failure.
- For 'pest regulation', advice will be provided regarding natural pest regulation; however widespread implementation is not expected as farmers are likely to maintain using methods that they consider reliable (i.e., slug pellets).
- Improvements to 'erosion regulation' are likely to require significant amounts of the areas currently in arable land to be converted to woodland/hedgerows. It is considered unlikely that STWL would be able to convince large numbers of farmers to significantly reduce the amount of arable land they use through simple advice scenarios.
- Natural 'water purification' is unlikely to have a great deal of success due to limited areas to expand wetland/reedbeds areas in the catchment.
- 'Recreation and tourism' is likely to see some benefits for existing users in the vicinity of Ogston Reservoir, particularly for farm advice options which reduce the chance of eutrophication of the reservoir. Interventions are not likely to increase the number of recreational users or tourists.
- 'Aesthetic value' services would see a positive impact as there would need to be implementation of buffer strips and hedgerows across the catchment for improvements to be observed in the aesthetics of the catchment. Over a longer time-scale changes could be observed, but noticeable changes are unlikely to occur within a short-term scale.
- 'Intellectual, scientific and education value' service would see some improvement. Whilst farm advice will be provided under some of the scenarios, farmer innovation would require additional funding unlikely to be sourced from STWL.
- Treatment options would have an adverse impact on 'social relations' as increased treatment could result in higher costs.
- 'Nutrient cycling' is contingent upon widespread uptake of farm advice and land management interventions.
- For 'Photosynthesis' services, some herbicides potentially affect a wide range of other plants in the catchment (i.e., aquatic plants). Based on the Pesticide Property Database (University of Herfordshire, 2012) it is considered that substitutes are likely to have a similar impact on plants within the catchment. There may be some improvements as a result of reduced spray drift.
- Use of good agricultural practice will provide additional habitat within the landscape, therefore improving the 'provision of habitat'. However, it is also assumed that agriculture will intensify in the future providing less area for habitat.

The qualitative assessment was followed by the quantitative assessment using the EA Stage 1 valuation tool. The tool monetises the ecosystem services benefits associated with improvements to a set of ecosystem components. The value of these benefits is added to the costs avoided through the implementation and success of the catchment interventions, which results in a BCR value for each intervention scenario. Table 17.5 presents a selection of the interventions that achieved BCR values above 0.5 (BCR values greater than 0.5 suggest that the scenario may be economically beneficial).

The cost-benefit analysis of the intervention scenarios showed that product substitution, combined with farm advice, would achieve the prescribed target, would be the most cost beneficial option, and would achieve more additional ecosystem and social benefits than implementing additional end-of-the-pipe water treatment.

Table 17.5 Benefit-cost ratio (BCR) summary for the ogston reservoir catchment (selection of results).

Scenario	BCR	Reasoning
1.2 – Substitution metaldehyde: medium and high-risk areas	2.06	Second highest BCR due to avoided outage and costs of investigation after an exceedance, and additional valued benefits to mammals, birds and pollinating insects. Highest BCR because lower failure risk as effective over a larger area.
1.4 – Substitution MCPA and 2;4D: high risk areas	2.28	Highest BCR due to GAC filtration regeneration savings but no additional valued benefits to mammals, birds and pollinating insects. AST identified disbenefit to fish, pollinators and aquatic plants but these could not be quantified using the Stage 1 tool.
2.3 – Farming advice: Nutrients and PPPs	0.70	Combined BCR >0.5 assuming improvements to WFD status but does not pass sensitivity testing. Despite high risk of failure (80%), significant and multiple benefits have been identified in the AST.

17.4 CONCLUSIONS

The methodology used to determine the most cost beneficial option to address the water quality problems identified in the Ogston Reservoir, was applied specially for this project. The study was developed in three stages: initial assessment of the catchments (phase 1), detailed catchment modelling (phase 2) and benefit-cost assessment (phase 3). The assessment of the ecosystem services was an innovative and key feature of the study. The outcome provided a hierarchy of a set of catchment interventions, which supported the decision making in order to not only decrease the PPP concentration in the drinking water supply (in compliance with the WFD), but also to generate wider environmental benefits in the catchment.

This study has shown that among all options considered, product substitution with an alternative PPP would provide multiple benefits beyond improvements to water quality in the catchment through knock-on effects of engagement with farmers. This option achieved a greater BCR due to water treatment costs savings and the lower probability of failing to meet the WFD requirements in the reservoir. The greatest identified environmental benefits include reduced soil erosion, reduced nutrient runoff and benefits for pollinating insects.

This case study shows how ecosystems services assessments can be used as part of cost benefit analysis for water company projects in the UK, and potentially internationally. This approach has fed into future business and asset planning, demonstrating how ecosystem services can be captured within decision making frameworks, and provides a template for other catchment management studies, such as for nutrient management and offsetting.

ACKNOWLEDGEMENTS

The authors of this paper wish to acknowledge Severn Trent Water Limited; the SWAT model (http://swat.tamu.edu/); Information providers: Environment Agency, MetOffice, NSRI, UKWIR.

REFERENCES

Castle G., Mils G., Gravell A., Jones L., Townsend I., Cameron D. and Fones G. (2017). Review of the molluscicide metaldehyde in the environment. *Environmental Science: Water Research and Technology*, **3**, 415–428, https://doi.org/10.1039/C7EW00039A

Defra. (2013). Developing the Potential for Payments for Ecosystem Services: an Action Plan. Report of the Department for Environment, Food & Rural Affairs, London, UK.

Defra. (2018). A Green Future: Our 25 Year Plan to Improve the Environment. Report of the Department for Environment, Food & Rural Affairs, London, UK.

Dillon G., Hall T., Jönsson J., Murrell K., Shepherd D., Song J. and Stanger M. (2011). Treatment for New and Emerging Pesticides. Report Ref. No. 11/DW/14/4. UK Water Industry Research Limited.

Environment Agency. (2016). Water Appraisal Guidance: Assessing Costs and Benefits for River Basin Management Planning. Environment Agency, London, UK.

European Commission. (1998). Council Directive 98/83/EC of 3 November 1998 on the Quality of Water Intended for Human Consumption.

European Commission. (2000). Directive 2000/60/EC of the European Parliament and of the Council of 23 October 2000 Establishing a Framework for Community Action in the Field of Water Policy.

FAO. (2006). International Code of Conduct on the Distribution and Use of Pesticides. Guidelines on Efficacy Evaluation for the Registration of Plant Protection Products. FAO, Rome.

FAO. (2020). Pesticides Use Statistics. Global, Regional and Country Trends 1990–2018. FAOSTAT Analytical Brief Series No 9. FAO, Rome.

Hillocks R. (2012). Farming with fewer pesticides: EU pesticide review and resulting challenges for UK agriculture. *Crop Protection*, **31**, 85–93, https://doi.org/10.1016/j.cropro.2011.08.008

Mohamad Ibrahim I., Gilfoyle L., Reynolds R. and Voulvoulis N. (2019). Integrated catchment management for reducing pesticide levels in water: engaging with stakeholders in east anglia to tackle metaldehyde. *Science of the Total Environment*, **656**, 1436–1447, https://doi.org/10.1016/j.scitotenv.2018.11.260

NERC. (2012). The Advanced Soil Geochemical Atlas of England and Wales. NERC, London, UK. ISBN 9780852727096.

Sharma A., Kumar V., Shahzad B., Tanveer M., Sidhu G. P., Handa N., Kohli S. K., Yadav P., Bali A. S., Parihar R. D., Dar O. I., Singh K., Jasrotia S., Bakshi P., Ramakrishnan M., Kumar S., Bhardwaj R. and Thukral A. K. (2019). Worldwide pesticide usage and its impacts on ecosystem. *SN Applied Sciences*, **1**, 1446, https://doi.org/10.1007/s42452-019-1485-1

UKWIR. (2012). Quantifying the benefits of water quality catchment management initiatives. Volume 2 – a benefit assessment framework – overview report. UKWIR, London, UK.

University of Hertfordshire. (2012). Pesticide Properties Database (PPDB). Developed by the Agriculture & Environment Research Unit (AERU), 2006–2013. University of Hertfordshire, Hatfield, UK.

doi: 10.2166/9781789061123_265

Chapter 18

Landscapes for pollution risk management

B. J. D'Arcy[1], P. Morison[2], B. C. Braskerud[3] and P. Kajitvichyanukul[4]

[1]Independent Environmental Consultant, Kinross, UK
[2]VIC Water, Melbourne, Australia
[3]Oslo City, Norway
[4]Department of Environmental Engineering, Faculty of Engineering, Chiang Mai University, Chiang Mai, Thailand

Keywords: Diffuse pollution; land use; landscape; multiple benefits; rural best management practices (BMPs); low impact development (LID); sustainable urban drainage systems (SUDS); water sensitive urban design (WSUD)

18.1 INTRODUCTION

Pollution is not simply associated with chemicals which could, at least in theory, be banned and replaced with harmless substitutes. The control of industrial and other process effluent discharges has improved dramatically since the inefficient use of resources during the Industrial Revolution 200 years ago, driven by a combination of greater business efficiency to minimise waste and environmental legislation (D'Arcy *et al.*, 2018). Where that is not the case suggests short-term business planning and weak government, perhaps in both instances driven by poor appreciation of the modern possibilities for long-term sustainable manufacturing. In parallel with progress revolutionising resource recovery and managing industrial effluents (D'Arcy *et al.*, 1999; Kim *et al.*, 2018), the technological advances for sewage treatment have been immense over the past 100 years, including *u-v* treatment (e.g., at recreational waters) and nutrient removal. The challenge now is diffuse pollution: unintended consequences of everyday activities and practices from farming and forestry to building houses, roads and commercial and industrial developments.

The previous chapters in this book described a broad spectrum of land use and water quality problems and issues. Pollution of the water environment by weather-driven processes is a common factor. In Chapter 1 that was recognised in a new definition of diffuse pollution, building on the earlier work by Novotny and Olem (1994). The defining characteristic here is that diffuse pollution involves:

> 'The movement of contaminants from the land or air into the aquatic environment, driven by weather conditions'

For this chapter, we wish to consider the case for a radical reappraisal of landscape design, to strive for a multi-benefits new paradigm for considering landscapes and water, in the interests of managing water pollution risks, achieving natural flood risk management, habitat corridors and habitat protection for biodiversity and for amenity, human health and well-being. For each of those interests there are many other actions and behaviours which can further the work to enhance and protect human interests. But if landscape is left to be advocated piecemeal as one action in separate case-by-case circumstances, the potential for catchment-scale landscape measures, sufficient to address even one of those issues effectively, is unlikely to be achieved. The multiplicity of benefits, as well as the costs of environmental impacts, are otherwise externalities to the development or status quo situation seeking to comply with a narrow regulatory requirement (Novotny, 2003).

18.2 LAND USE AND POLLUTION CONTROL

Different land uses are associated with different pollutants and problems. Some problem chemicals (e.g., PCBs, some fire retardants, many pesticides, plastic which cannot be recycled, etc.) usually cause pollution for decades until evidence of their adverse impacts mounts and they are eventually restricted in use or finally banned. As shown below, the timescale required involves prolonged pollution and associated damage to health and ecosystems. Better guidance and processes for design and application of industrial chemicals and of course pesticides are an important need for the future of ecosystems and human health, a requirement for sustainable manufacturing, and agriculture and so on. The process of achieving reductions in the scale and severity of pollution from just one pesticide for example is summarised by the Case Study in Text Box 18.1 below.

Text Box 18.1. Restricting the use of a problem pesticide

Case study from Thailand: steps leading to ban on use of the pesticides

Over the past decades, the use of pesticides in Thailand has increased significantly over time. From the report of Office of Agricultural Economics – Ministry of Agriculture and Cooperatives, a total of 170 932 tons of pesticides were imported into Thailand in 2018. About one third of them are either WHO class I or WHO class II pesticides.

The most imported herbicides were paraquat, glyphosate, 2,4-D, ametryn, and atrazine. The most commonly imported fungicides were dithiocarbamate, mancozeb, carbendazim, and captan (Laohaudomchok *et al.*, 2020). Water, soil, and biota contamination by pesticides was detected in many regions of Thailand. Inevitably, the residual pesticides were found in the community water supply, in edible plants, and in the human food chain. In the environment, surface runoff is the major factor in pesticide mobility from the topsoil to the surface water bodies including rivers, lakes, and streams. Moreover, the bioaccumulation of herbicides also posed critical issues in the human food chain, as they were found in edible plants and aquatic liveforms. However, little work in Thailand has been done to look at the chronic health effects of pesticides or the impact of being exposed to a mixture of pesticides, as is the case for virtually all Thai farmers. Laboratory studies have shown that common pesticides like glyphosate and chlorpyrifos may be endocrine disrupters that could increase risk of cancer in humans (Laohaudomchok *et al.*, 2020).

In the late 1970s, organochlorine pesticides in Thailand were phased out and eventually banned (Thirakhupt *et al.*, 2006). These pesticides were replaced by other insecticides like the organophosphates and carbamates. In addition, the use of herbicides has also risen. Groundwater was contaminated with several types of pesticides including atrazine, 2,4-D and carbofuran (Thapinta & Hudak, 2003). In the rainy season glyphosate, atrazine, and organophosphate pesticides such as fenitrothion, fenthion and diazinon were detected in drinking water samples (Klaimala, 2019).

In 2016, the National Bureau of Agricultural Commodity and Food Standards (ACFS) in the Ministry of Agriculture and Cooperatives issued a revised notification entitled the Thai Agricultural Standards on Pesticide Residues: currently, maximum Residue Limits (TAS 9002-2016) cover

53 pesticides commonly used in Thailand (Government of Thailand, 2016). The standard states that the method used for the sampling and analysis of pesticide residues for compliance with MRLs shall comply with the latest Recommended Methods of Sampling established by Codex Alimentarius Commission, Joint FAO/WHO Food Standard Programme. It is recommended that intervention studies to evaluate the impact of stricter regulations on the sale of pesticides as well as training and licensure of sellers and users are needed to ensure efficacy of the policy. Provision of safe pesticide disposal sites in all agricultural districts is needed.

An example for a different chemical and product, is copper pollution associated with brake pad wear (Brake Pad Partnership, 2021). In the 1990s this was identified as a significant contributing source to the exceedance of a California Toxics Rule criterion of 3.1 ug/l soluble copper in the Lower South Bay of San Francisco Estuary, California (San Francisco Estuary Institute, 2007). A joint industry and environmental group was set up with the Brake Pad Partnership, who entered into a collaborative investigation. Costing over $1M, the studies concluded that between 10% and 35% of the soluble copper measured in the San Francisco Estuary could be accounted for by brake pad erosion around the estuary (midpoint scenario 23%). In a fifteen-year programme of investigations, including industry research and testing for alternatives, agreement of all parties was reached on legislation to set limits on the copper content of brake pads. In 2010 California Governor Schwarzenegger signed into law Senate Bill (SB) 346 which required brake pad manufacturers to reduce the amount of copper in brake pads sold in California to no more than 5% by 2021 and no more than 0.5% by 2025 (Brake Pad Partnership, 2021). In parallel, similar measures were enacted in 2010 by Washington State. In 2015 on a voluntary basis a national agreement to phase out copper in brake pads was agreed with the manufacturers (USEPA, 2015).

18.3 POLLUTION CONTROL AND LANDSCAPES

The remaining environmental pollution problems are, however, far more extensive than a list of pesticides or problem industrial chemicals, daunting though they are (e.g., currently a focus on nicotinoid pesticides, microplastic and nanoparticles of many kinds). The big challenges involve the movement of those and many other substances and materials from the terrestrial environment and the atmosphere in rainfall wash-out, driven by weather conditions – diffuse pollution. Some pollution issues involve substances measured in micrograms (phosphate, toxic metals), other in tonnes (sediment in rivers in storm conditions). For plastic the problem extends from litter/trash to micro or even nanoparticles. Not least amongst the facets of the challenge is the load delivered to the water environment in very large storm events; a driver for a complex of controls from sources to immediate, local and regional features, with capture and removal maintenance required for landscapes.

The diagram in Figure 18.1 represents the shifting focus of pollution control as technological and risk management progress has allowed great improvements in effluent management, from resource recovery, spillage and waste minimisation to effluent control and sensible discharge standards to protect the environment. [As the focus shifts to diffuse pollution, obviously achieved progress must not be neglected].

In many instances, despite improvements in major point sources (sewage and trade effluents), the quality of runoff from the catchments of rivers is still too contaminated to reach good river quality status. Table 18.1, modified from Campbell *et al.* (2004), summarises the approaches to managing agricultural pollutants, showing landscape measures alongside planning and management actions. Landscape measures are key features of diffuse pollution control efforts (Braskerud, 2001a; Cestti

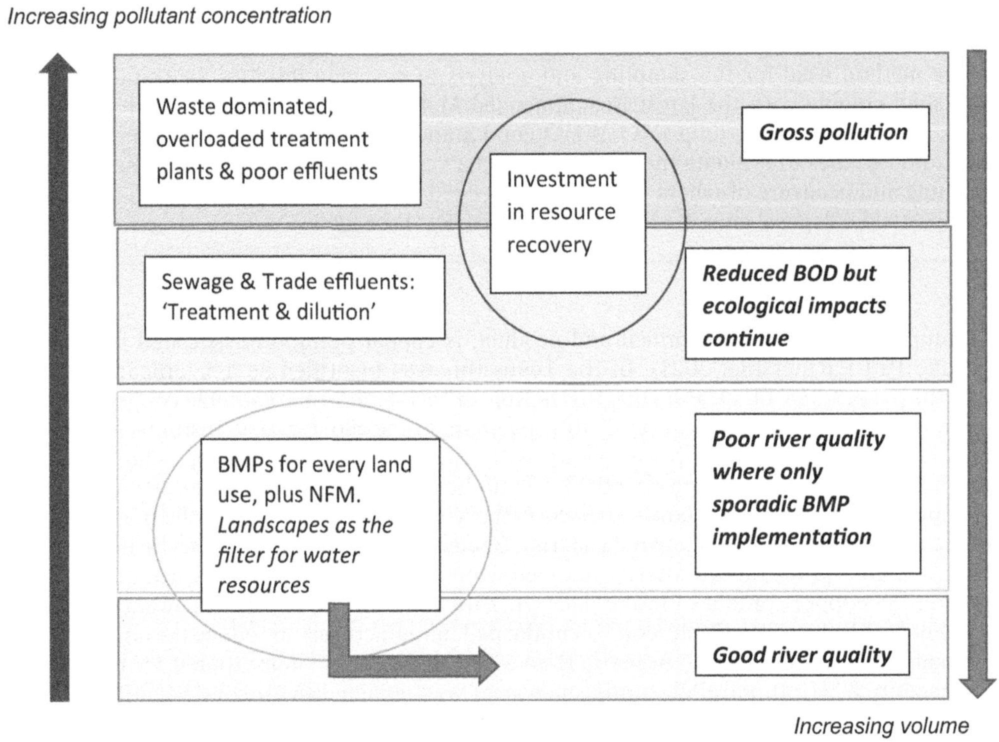

Figure 18.1 Historical progressive shifts (simplified) in focus for water pollution control, from untreated effluents and diffuse pollution to landscapes for water resources (with resource recovery for industrial and sewage effluents allied to treatment technology). *Key terms*. BMPs: best management practices (D'Arcy & Frost, 2001; Novotny, 2003); NFM: natural flood management (Environment Agency, 2017); BOD: biochemical oxygen demand.

et al., 2003; Committee on Long-Range Soil and Water Conservation, 1993 and underpin equally important practices such as nutrient budgeting and animal waste management (Audsley, 2003).

Table 18.2 offers an equivalent focus for diffuse pollution in villages, towns and cities. The tables illustrate the scope for infrastructure and especially landscape interventions as key actions in striving for a clean water environment. For both categories (Tables 18.1 and 18.2), for resolution of land use and water quality problems the challenge is not just to stop at source to promote best practice ways to minimise losses, but in addition, *to accept the need for land-use measures which can reduce mobilisation and loss of potential pollutants to the water environment.* In terms of physical measures there are three categories:

- minimise mobilisation
- capture mobilised contaminants
- enhance self-purification (biodegradation of degradable contaminants, development of vegetation)

As well as capture persistent pollutants plus contaminants such as eroded soil and sediment from the built environment, the landscape features should support biodegradation of contaminants *in situ* (e.g., Napier *et al.*, 2009). In Table 18.2 river restoration is identified as part of the set of measures which can address diffuse pollution. That is because exposure of culverted watercourses to light and

Table 18.1 A four-point focus for managing diffuse pollution risks from agriculture and forestry (after campbell *et al.*, 2004).

Planning Tools	In-field Measures	River Margins	Built Environment, Including Roads and Tracks
Nutrient budgets	Conservation tillage	Buffer zones	Pesticide stores and pre- & post-spraying activities indoors
Manure application plans	Ploughing across field slopes, rather than downslope	Exclusion fencing with alternative water supply systems	Oil tanks bunded
Pesticide targeting	Leaving stubble *in situ* over winter	Wetland margins for denitrification	Farm steading hotspots under cover to exclude rainfall (livestock standing areas, tractor washing)
Contingency plans for pesticides accidents	Crop residue mulches	Woodland shelter belts	Individually minor seepages & contamination managed with adjacent grass-soil features
Contingency plans for livestock slurry or oil accidents	Grassed areas for runoff erosion hotspots	Leave a native vegetation margin either side of runoff channels in plantations	Grass swales
Harvesting timber by blocks, not clear-felling	Tree shelter belts to encourage livestock away from cooling in watercourses.	Site-specific features for soil stabilisation at hotspots	Grass filter strips
Compliance with planning conditions	Stocking density and grazing management	Soil stabilisation by vegetation on riparian slopes	Runoff from tracks diverted at multiple locations into vegetation
Compliance with codes of good agricultural	Location of field gates, bale-feeder units & other hot-spot features	River restoration to slow runoff, create habitats & allow self- purification	Grass swale drainage for general forestry yard areas
Compliance with codes of good forestry practice	Frequent sediment sumps in plantations		Pond or wetland for final polishing (not for heavy contamination)
Irrigation scheduling	No downslope ploughing in plantations		

sediments to oxygen, enhances the degradation of pollutants and the restoration of self-purification processes. This is part of the strategy of Oslo in Norway.

To achieve such landscapes requires a radical change in mindset for planners, businesses, public sectors and the general public (Figure 18.2). It is not how new development is built currently. It is also not the routine mindset for farming and plantations of many kinds from coffee to conifers, including tea, fruit, rubber, oil palm and hardwoods. Radical yes, but minor compared to the challenges of sanitation and water supply in cities, yet such actions would also work towards safer more secure water supplies too (Kovacs *et al.*, 2012; National Audit Office, 2010; Wild, 2020). Benefits need to be stacked against the passive motivations to not take the necessary actions (Figure 18.2, and D'Arcy and Frost, 2001). It is important to recognise that significant changes in land use patterns (e.g., livestock

Table 18.2 Scope for managing diffuse pollution in towns and cities.

Planning	Infrastructure & Landscape			
	Roads and Rail	**Housing**	**Commercial/ Industrial**	**Parks & River Margins**
Reducing pesticide & nutrient application by switching to more natural parkland landscapes	Biofiltration features	Biofiltration features	Biofiltration features	Natural vegetation
Public transport	Grass swales	Grass swales	Grass swales	Buffer zones
Sewer misconnection checks & policies for prevention/ remediation	Gravel filter drains	Gravel filter drains	Gravel filter drains	Woodland margin/s
Vehicle checks for emissions	Permeable surfaces	Raised bed raingardens	Raised bed raingardens	River restoration
Regulations, inspections & enforcement	Sanitation systems for trains	Street trees	Landscape trees	Woodland areas
Clean air programmes	Road sweeping	Green roofs	Green roofs	Wet meadows
Contingency planning for fires & fire-water disposal	2nd stage treatment features (biofiltration, pond, basin)	Grass filter strips	Basins or ponds which can function as holding areas for firewater or extreme flood events	Ponds and basins for exceedance flows from developments
Dog fouling, fly-tipping, refuse collection & recycling drop-off points	Raingardens (e.g., in Copenhagen & planned for Oslo)	Permeable surfaces above soil level of gardens	Permeable surfaces for cars, not trucks	Wetlands with margins for flood events
Maintenance programmes for infrastructure and BMPs	Grassed tram ways (e.g., Lyon, France)	Driveways to drain into gardens (building design)	Inspection chamber with suspended oil absorbent cushion on last point on the private drainage system	–
Traffic reduction	–	Soakaways	–	–
Public engagement & information provision	Embankments as habitats	Raingarden soakaways	Grass filter strips	–
Policy for pollution preventions on construction sites	Retain grass margins to filter runoff	Retain grass margins to filter runoff	Retain grass margins to filter runoff	–

density) occur for many reasons; the landscape focus for flood-pollution-climate resilience has to fit a changing world (e.g., Sansom, 1999; Walcott *et al.*, 2014).

For a landscape approach to be taken up sufficiently widely and routinely to enable diffuse pollution management and control, it needs to be fully integrated with public demands for actions on other issues (a common cause approach):

- biodiversity protection and enhancement
- natural flood risk management
- sustainable urban drainage, water sensitive urban design, low impact development and so on
- biodiversity

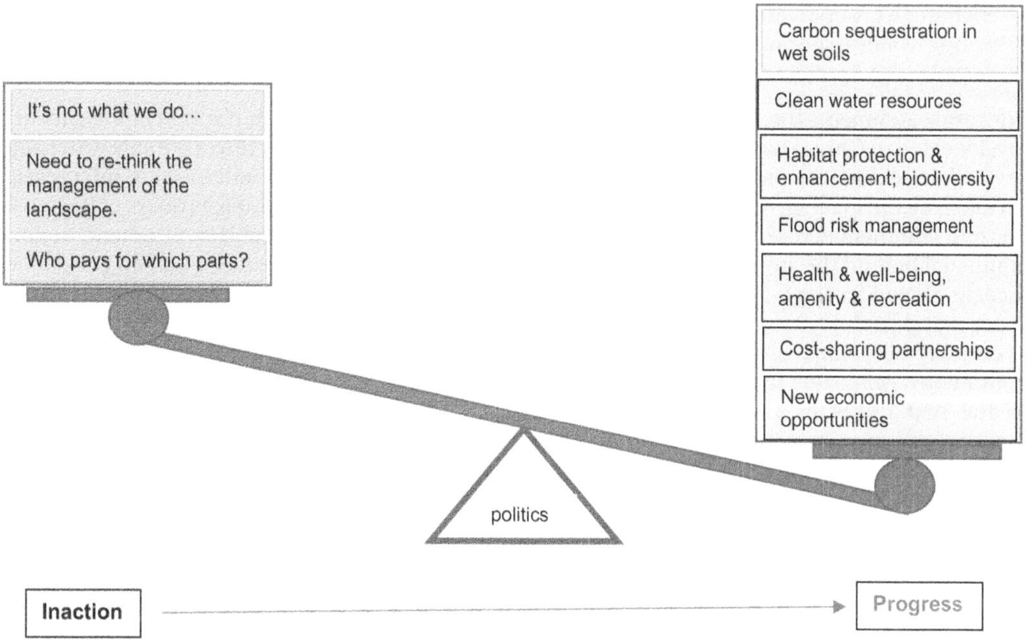

Figure 18.2 The arguments for routine, multipurpose landscape measures to manage diffuse pollution – balancing costs and benefits (modified from D'Arcy & Frost, 2001).

- human health and well being
- local and national economies (not vested interests in status quo).

In the UK, the economics and business case for landscape scale diffuse pollution and flood risk management in a city has been explored for London (Ainley, 2021), and the benefits of sustainable drainage systems were found to be greatest when socio-economic benefits (physical and mental health, land values, etc) were significant. The potential benefits of a landscape approach were equally clear in studies in Africa (Alemu *et al.*, 2017; Iwugo *et al.*, 2003).

An integrated approach is becoming a recognised necessity in pilot projects in various countries. One of the countries facing diffuse pollution and flood risk challenges on a very large scale is China, where there are more than 100 cities with a population greater than 1 million people (Migiro Atlas, 2018) and at least 5 megacities (more than 10 million people). To manage the multiple environmental problems that generates, including pluvial flooding and diffuse pollution, China has developed a concept which puts multi-purpose landscapes at the heart of developing designing cities for the future and the present: the Sponge City concept (Yu, 2015, 2016). The Sponge City approach has been progressed from concept and policy to action, by initiating sixteen pilot Sponge City projects (e.g., She, 2018; Zhao *et al.*, 2018).

Professor Kongjian Yu (2014) summarised the actions and benefits:

> 'The Sponge City will slow down and tame the water and enable downward movement that will help infiltrate and supply groundwater, benefit wildlife habitats, and purify water to better serve urban environments.'

The key elements of the Sponge City concept (Yu, 2015), include the following:

- slow down, rather than speed-up rainfall runoff
- store rather than discharge stormwater

- encourage groundwater recharge
- resolve water issues onsite (flood risk and water conservation)
- decentralised projects are more sustainable, especially if managed by local people.

For rural catchments too, the mindset for local landscape design for rainfall runoff management has been advocated in China. For example, reducing diffuse sources of nutrients into a potable water supply reservoir for Beijing was pioneered in early 2000's, which involved many micro-scale interventions, undertaken by the local farmers, with added benefits for them too. There, arable farmers collected field stones and laid them across seasonal channels where soil erosion was a problem. Gradually sediment accumulated behind the stones and the gradient driving the erosion was reduced and an easier, flatter terrace developed for the farming. Lower down the catchment, erodible drainage pathways were sown with grass and useful trees planted there too. For continuous runoff events, removal rates for TS and TP were over 60%, and for soluble P exceeded 40%. Much higher removal rates were measured for discontinuous events due to reduced runoff (Wang *et al.*, 2004; Yin *et al.*, 2006).

In the Text Boxes 18.2 and 18.3 below, two examples of other 'joined-up' international policy and action programmes, rural and urban, demonstrate what can be done in seeking long-term measures for more climate resilient landscapes to deliver diffuse pollution control together with other key benefits. The Australian examples in Text Box 18.2 illustrate the diffuse pollution–climate change-flood risk management nexus referred to in Chapter 1 of this book (D'Arcy *et al.*, 2022). The philosophy in Victoria was originally set out in their *Integrated Water Management Framework for Victoria* (DELWP, 2017), which includes at its core the aim to 'make our communities a better place to live' (Lisa Neville in the Minister's Foreword). The report indicates that one of the key ways to achieve that is to not just address the risks which drive the actions, but to look creatively for the opportunities that actions can create. In New South Wales, the Department of Environment and Climate Change (2009) in their strategy for managing diffuse pollution state that no single management action can be sufficient to prevent pollution: combinations are required. The strategy is taken forward for landscape planning to protect waterways for example from urban runoff (Dela-Cruz *et al.*, 2017).

In Norway, innovators have also sought to develop landscape solutions to diffuse pollution in rural and urban landscapes (Braskerud & Blankenberg, 2005; Blankenberg *et al.*, 2015; Braskerud, 2001b, 2016). As an urban example of integrated actions, progress in Oslo is outlined in Text Box 18.3.

The good practice examples are well documented, but they still represent the exceptions, not the rule. Whether in England or Thailand, Netherlands or New Zealand, it is an everyday observation that missed opportunities for routine resilience design for catchments are commonplace.

Text box 2. Greater Melbourne and Victoria, Australia

Case study: Landscape measures for water multiple benefits in Victoria and New South Wales, Australia

Water is a vital water resource, and Australia is a water scarce country. Landscape measures have long been recognised for their importance in protecting water resources in New South Wales and Victoria. A review by Roser and Ashbolt (2005) compared the quality of raw water supplies in a series of reservoirs with different degrees of protection by undeveloped forests in their immediate catchment areas. Greater Melbourne had a greater degree of catchment protection by undisturbed forest land than the equivalent water resource reservoirs in Sydney, which in turn was better protected than Adelaide (also reported in that review).

Now those forested lands are recognised for biodiversity value and the wooded landscapes are valued as assets for informal recreation and enjoyment, as well as for their effectiveness at protecting raw water quality and stabilising soils and preventing diffuse pollution which would increase with alternative land uses.

Supplies have been critically reduced during prolonged periods of drought. A 10 year drought in Greater Melbourne led to a reappraisal of rainfall runoff as a potential asset, rather than simply an inconvenience to be drained from the urban settlements. Melbourne Water established an innovative *10 000 Raingardens Campaign*, to encourage the urban population to harvest rainfall directly in raingardens of various sizes on their own property, and to accept the provision of raingardens of various kinds and sizes across the municipal landscapes too.

The 10 000 Raingardens Campaign was successful in all its aims (e.g., using rainwater and groundwater recharge). The final numbers of various types of landscape features were:

Downpipe diversion	969
In-ground raingarden	263
Infiltration raingarden	2426
Planter box raingarden	396
Std. rain garden	3189
Ponds & basins	303
swale	2237
Porous paving	46
wetland	360
Total	**10 565**

Source: Data from Melbourne Water, 13.4.2017.

Climate change is a significant issue for Australia, not just for water resources: its relationship to extreme events including bushfires is a source of water supply contamination. Pickrel (2020) reported that Warragamba Dam, to the West of Sydney, provides drinking water for 3.7 million people, but vast bushfires in and around its catchment area have burned more than 148 000 acres of forest. The traditional water resource measures taken for Melbourne and Sydney were far-sighted actions: now an equally far-sighted strategy of climate change action is required. Hybrids of alternative supply options according to usage and risk are being considered. Thus protection of ever more scarce water resources, in a seasonally very dry landscape with rising temperatures, is a top priority.

Text Box 18.3. Oslo, Norway

Case study: Integrating flood risk management with amenity in Oslo City

Surface water runoff is managed from source control to wetlands and townscapes in Oslo, as an aspiration to being a sustainable blue–green city. A key part of that aim is that blue and green areas should be useful for other purposes when not needed for stormwater management.

Three levels of intervention for city sustainability make up the *Stormwater 3 step approach* (S3SA):

(1) keep smaller precipitation impacts localised to the site
(2) delay runoff from precipitation with higher intensities as long as possible
(3) transport overflow and excess surface runoff after heavy rainfall safely on the surface.

Typical step 1 measures are permeable surfaces, green roofs, downpipe disconnection, raingardens, and so on.

Step 2 measures in addition include grass swales, wet or dry ponds, inundation of parks, and so on. Challenges finding sufficient space can be difficult but where successfully implemented this level of intervention can hold up the flow of excess rainwater and so help alleviate the worst effects of heavy rainstorms by minimising the risks of water immediately flashing off impervious open surfaces.

Step 3 typical measures are green streams (unpolluted with riparian habitat zones giving space for water); secure flood paths; reopening culverted watercourses. Defined flood paths are planned where damage to infrastructure is least likely, and key roads can still be used by emergency vehicles.

The building code in Oslo includes the use of the S3SA two first steps for all new and upgraded buildings. Stormwater measures need to be 'open', that is part of the urban hydrological cycle. The municipality is responsible for the floodway system and has started mapping and modelling the possible water flow for a 200-year return period rainfall event, including a climate factor of 1.5.

Key steps towards achieving sustainability aspirations are:

- Recognising that heavy cloudbursts are rare; consequently the landscape measures should be positive attributes for the city, creating an inviting landscape which citizens can experience and enjoy every day.
- Consequently the city recognising that open SUDS features are desirable, rather than underground storage systems.
- Above-ground features, by virtue of using vegetation and soil systems, also facilitate capture and breakdown of many diffuse pollutants such as hydrocarbons, and atmospheric pollutants in precipitation and runoff.
- The use of vegetation systems in above-ground systems also provides amenity value as well as biodiversity benefits. As a result, a strategy on green walls and roofs is being prepared.
- The City of Oslo has started a 'plant 100 000 trees before 2030' project.
- Urban agriculture/horticulture are highlighted. The city sponsors several 'city farmers' helping inhabitants who like farming small plots and on roofs to be successful.
- Actions for restricted use of cars are ongoing, with an increased construction of bicycle lanes; hence fewer contaminants are being released to atmosphere and road surfaces.

Schemes have been implemented or under construction for level 2. To achieve level 3, Oslo has mapped the city surface drainage pattern using high-resolution terrain data and GIS modelling using ArcGIS. That identified practicable flood paths. As important as the high-level strategic flood routes, has been the effort to encourage stormwater management on a plot-by-plot basis, using water in private raingardens. Distributed measures offer maximum collective benefit but require a catchment landscape approach.

18.4 CONCLUSIONS

18.4.1 Pollution prevention

(a) Persistent or other seriously problematic pollutants will continue to need restrictions on their use.
(b) Creation of new problem chemicals needs to be avoided at the design stage by manufacturers. It is not just pesticides; serious pollution has been caused by chemicals designed for widespread use but with insufficient consideration of associated environmental contamination (e.g., PCBs, plastic).
(c) Globally, diffuse pollution is often the unintended environmental consequences of everyday activities; there are well researched behaviours and techniques for minimised pollution in the course of many activities and information needs to be more effectively disseminated once awareness of the extent of the problems is raised sufficiently. Important pollutants include suspended solids, nutrients, faecal pathogens, as well as toxic chemicals.
(d) There will always be a pollution risk where behavioural change is the only action sought. Therefore, it needs to be managed in conjunction with integrated landscape design, appropriate to each particular land use.

18.4.2 Landscapes and inter-related problems

(a) The inclusive definition supported here, which is simply based on key physical environmental characteristics (*contamination of air or landscape, mobilised and transported into the water environment by weather conditions*) has direct implications for landscapes if diffuse pollution is to be effectively managed.
(b) Climate change will exacerbate diffuse pollution problems, as more severe rainfall patterns mobilise contaminants in greater quantities. Designing landscapes for flood and fire resilience has emerged as a major challenge.
(c) The association of diffuse pollution with runoff and high-flow conditions, means that the application of surface-water runoff management techniques needs to be applied as close to source as possible, and as widespread as possible, if there is to be an effective control of diffuse pollution. Landscape interventions for runoff management need to become routine in rural as well as urban contexts.
(d) For that to be possible, landscape interventions need to be designed for multiple benefits: climate change resilience, flood risk management, aesthetics and place-making, biodiversity and habitat connectivity.
(e) For diffuse pollution, the relation with landscape and people's perceptions of how landscapes should look, as well as practical sector considerations regarding design of urban/rural landscapes, requires resource for sector and public engagement by governments and environmental protection agencies.
(f) Clear presentation to governments of the importance of diffuse pollution (as reported in papers in this publication, and elsewhere), citing adverse impacts across the spectrum of problems, is a pre-requisite for recognition of the need for a radical approach to landscapes as a fundamental step for environmental resilience.
(g) Restricting the use of individual problem pollutants is only a partial remedy for diffuse pollution and allows decades of pollution whilst engaging with sectors and regulators; collecting and evaluating evidence. There are many pollutants for which such a remedy is inappropriate too (e.g. nutrients, sediment) – some of the most serious causes of environmental pollution. Management at source, by integrating good practices with multi-purpose landscapes, is necessary for meeting Millennium Goals and sustainability aspirations.

ACKNOWLEDGEMENT

This work is partially supported by National Higher Education Science Research and Innovation Policy Council (Contract No.C16F630255) and Thailand Research Fund (Grant No. IRN62W0005).

REFERENCES

Ainley S. (2021). The benefits of distributed SuDS in London. Susdrain Briefing, May 2021, pp. 1–4. Arcadis. www.susdrain.org (accessed 27 March 2022).

Alemu T., Bahrndorff S., Alemayehu E. and Ambelu A. (2017). Agricultural sediment reduction using natural herbaceous buffer strips: a case study of the east African highland. *Water and Environment Journal*, **31**, 522–527, https://doi.org/10.1111/wej.12274

Audsley R. (2003). The 4 Point Plan (4PP) - Farming and Water Scotland (online version), https://www.farmingandwaterscotland.org/managing-water-on-your-farm/the-4-point-plan-4pp/ (updated 1 April 2020, accessed 26 March 2022).

Blankenberg A. G. B., Haarstad K. and Paruch A. M. (2015). Agricultural runoff in Norway: the problem, the regulations, and the role of wetlands. In: The Role of Natural and Constructed Wetlands in Nutrient Cycling and Retention on the Landscape, J. Vymazal (ed.), Springer, Switzerland, pp. 137–147, https://doi.org/10.1007/978-3-319-08177-9_10 (accessed 27 March 2022).

Brake Pad Partnership. (2021). [Online] https://suscon.org/project/brake-pad-partnership/

Braskerud B. C. (2001a) The influence of vegetation on sedimentation and resuspension of soil particles in small constructed wetlands. *Journal of Environmental Quality*, **30**(4), 1447–1457, https://doi.org/10.2134/jeq2001.3041447x

Braskerud B. C. (2001b). Sedimentation in small constructed wetlands. Retention of particles, phosphorus and nitrogen in streams from arable watersheds. Doctor Scientiarum Theses 2001:10, Agricultural University of Norway, As, Norway. ISSN: 0802–3220.

Braskerud B. C. (2016). Oslo: protecting the blue-green city. *The Geographer*, Autumn 2016, p. 22.

Braskerud B. C. and Blankenberg A.-G. B. (2005). Phosphorus retention in the Lier wetland. Is living water possible in agricultural areas? *Jordforsk book* nr. 48/05. 145, 126–128. ISSN/ISBN:82-7467-537-1.

Campbell N., D'Arcy B., Frost A., Novotny V. and Sansom A. (2004). Diffuse Pollution – An Introduction to the Problems and Solutions. IWA Publishing, London. ISBN: 1 900222 53 1

Cestti R., Srivastava J. and Jung S. (2003). Agriculture Non-Point Source Pollution Control. Good Management Practices – The Chesapeake Bay Experience. World Bank Working Paper No. 7. The World Bank, Washington DC, USA. ISBN: 0-8213-5523-6.

Committee on Long-Range Soil and Water Conservation. (1993). Soil and Water Quality: an Agenda for Agriculture. Board on Agriculture, National Research Council, National Academy of Sciences, National Academic Press, USA.

D'Arcy B. J. and Frost A. (2001). The role of best management practices in alleviating water quality problems associated with diffuse pollution. *Science of the Total Environment*, **265**, 359–367, https://doi.org/10.1016/S0048-9697(00)00676-8

D'Arcy B. J., Kim L.-H. and Morison P. (2018). Industrial pollution and the water environment: a historical perspective. In: Wealth Creation without Pollution. B. D'Arcy, L.-H. Kim and M. Maniquiz-Redillas (eds), IWA Publishing, London, pp.1–24. ISBN: 9781780408330

D'Arcy B. J., Kay D., Napier F., Haygarth P., Sun Y. Y. and Wada K. (2022). Land use and diffuse pollution: are perceptions part of the problem? In: Land Use and Water Quality, P. Kajitvichyanukul and B. J. D'Arcy (eds), IWAP, London, pp.1–23.

D'Arcy B. J., Todd R. B. and Wither A. W. (1999). Industrial Effluent Control and Waste Minimisation: case studies by UK regulators. *Water Science and Technology*, **39**(10), 281–287.

Dela-Cruz J., Pik A. and Wearne P. (2017). Risk-Based Framework for Considering Waterway Health Outcomes in Strategic Land-use Planning Decisions. Office of Environment and Heritage and Environment Protection Authority, Sydney.

DELWP (2017). The State of Victoria Department of Environment, Land, Water and Planning. Melbourne, Australia. ISBN 978-1-76047-645-8

Department of Environment and Climate Change. (2009). NSW Diffuse Source Water Pollution Strategy. State of NSW and Department of Environment and Climate Change, Sydney. ISBN: 978 1 74122 961 5.

Environment Agency. (2017). Natural flood management – part of the nation's flood resilience – Environment Agency, Bristol, UK, www.gov.uk (Accessed 30th July 2021)

Government of Thailand. (2016). Thai Agricultural Standard (TAS 9002-2016) Pesticide Residues: Maximum Residue Limits [in Thai]. *The Royal Gazette* **133**(288).

Iwugo K., D'Arcy B. J., Heath R. G. M. and Andoh R. (2003). Breaking the pollution cycle in poor African settlements. Paper presented at 29th WEDC conference, 22–26 September, Abuja, Nigeria.

Kim L.-H. and Maniquiz-Redillas M. (2018). Green industry concept and practices. In B. J. D'Arcy, L.-H. Kim and M. Maniquiz-Redillas (eds), Wealth Creation without Pollution – Designing for Industry, Ecobusiness parks and Industrial Estates, IWAP, London.

Klaimala P. (2019). Pesticide Residues in Soil and Water. Pesticide Research Group – Dept of Agriculture. A presentation at the GEO Health Hub Networking Meeting, 2019.

Kovacs A. S., Fulop B. and Honti M. (2012). Detection of hot spots of soil erosion and reservoir siltation in ungauged Mediterranean catchments. *Energy Procedia*, **18**, 934–943, https://doi.org/10.1016/j.egypro.2012.05.108

Laohaudomchok W.,Nankongnab N., Siriruttanapruk S. and Woskie S. R. (2020). Pesticide use in Thailand: current situation, health risks, and gaps in research and policy. *Human and Ecological Risk Assessment: An International Journal*, **27**(1), 1–23, https://doi.org/10.1080/10807039.2020.1808777

Migiro G. (2018). Countries by number of cities over one million. World Atlas, https://www.worldatlas.com/articles/countries-by-number-of-cities-over-one-million.html (accessed 27 March 2022).

Napier F., Jefferies C., Heal K. V., Fogg P., D'Arcy B. J. and Clarke R. (2009). Evidence of traffic-related pollutant control in soil-based sustainable urban drainage systems (SUDS). *Water Science and Technology*, **60**(1), 221–230, https://doi.org/10.2166/wst.2009.326

National Audit Office. (2010). Tackling Diffuse Water Pollution in England. Report by the Comptroller and Auditor General. National Audit Office, London, UK. HC 188 Session 2010–2011.

Novotny V. (2003). Water Quality – Diffuse Pollution and Watershed Management. John Wiley, New Jersey, USA. ISBN 0-471-39633-8.

Novotny V. and Olem H. (1994). Water Quality: Prevention, Identification and Management of Diffuse Pollution. Van Nostrand Reinhold, New York.

Pickrel J. (2020). Australia's raging fires will create big problems for fresh drinking water. [Online]. Australia's wildfires will create big problems for fresh water (nationalgeographic.com) (Accessed 7 April 2020).

Roser D. J. and Ashbolt N. J. (2005). Source Water Quality Assessment and the Management of Pathogens in Surface Catchments and Aquifers. Cooperative Research Centre for Water Quality and Treatment, Salisbury, SA, Australia.

Sansom A. (1999). Upland vegetation management: the impacts of overstocking. *Water Science and Technology*, **39**(12), 85–92, https://doi.org/10.2166/wst.1999.0533

SFEI. (2007). The Pulse of the Estuary: Monitoring and Managing Water Quality in the San Francisco Estuary. SFEI Contribution 532. San Francisco Estuary Institute, Oakland, CA. see www.sfei.org

She N. (2018). Green industrial park practice: a case study of green infrastructure in wenling, China. In: Wealth Creation Without Pollution – Designing for Industry, Ecobusiness Parks and Industrial Estates, B. J. D'Arcy, L.-H. Kim and M. Maniquiz-Redillas (eds.), IWA Publishing, London, UK, pp. 99–107.

Thapinta A. and Hudak P. F. (2003). Use of GIS for assessing groundwater pollution potential by pesticides in central Thailand. *Environment International*, **29**, 87–93, https://doi.org/10.1016/S0160-4120(02)00149-6

Thirakhupt K., Sitthicharoenchai D., Keithmaleesatti S. and Siriwong W. (2006). Organochlorine pesticides and their usages in Thailand: a review. *Journal of Scientific Research, Chulalongkorn University*, **31**, 1–15.

USEPA. (2015). Memorandum of Understanding on Copper Mitigation in Watersheds and Waterways, https://www.epa.gov/sites/default/files/2015-11/documents/copper_brakepads_mou.pdf (accessed 25 March 2022).

Walcott J. J., Zuo H., Loch A. D. and Smart R. V. (2014). Patterns and trends in Australian agriculture: a consistent set of agricultural statistics at small areas for analysing regional changes. *Journal of Land Use Science*, **9**(4), 453–473, https://doi.org/10.1080/1747423X.2013.807312

Wang X., Yin C. and Shan B. (2004). Control of diffuse P-pollutants by multiple buffer/detention structures by yuquiao reservoir, north China. *Journal of Environmental Sciences*, **16**(4), 616–620.

Wild T. (2020). Nature-Based Solutions – Impacts on Water Quality & Waterbody Conditions. An analysis of EU Funded Projects. Directorate-General for Research and Innovation, European Commission, Luxembourg. ISBN 978-92-76-18198-9.

Yin C., Wang X. and Shan B. (2006). Retention of pollutants by the sink structures in catchments – studies to reduce pollution in China's rural areas. In: Managing Rural Diffuse Pollution, Agriculture and Environment VI, L. Gairns, K. Crighton and W. Jeffrey (eds.), Proceedings of the SAC-SEPA Biennial Conference, 5–6 April 2006, Edinburgh, pp. 111–120. ISBN 1-901322-63-7.

Yu K. (2015). Sponge philosophy. *Landscape Architecture Frontiers*, **3**(2), 8–9. ISSN 2095-5404 (print), 2095-5413 (online).

Yu K. (2016). Sponge City: theory and practice. Proceedings of the International Conference on Green Infrastructure and Resilience, 16–19 March 2016. Water Environment Federation (WEF) and Ecological Urban Planning Academic Committee of Chinese Urban Planning Society. Shenzhen University, Shenzhen, pp. 4–6.

Zhao H., Changliang Z., Jiang Z. and Xuyong L. (2018). Role of low-impact development in generation and control of urban diffuse pollution in a pilot sponge city: a paired-catchment study. *Water*, **10**(7), 51–63, https://doi.org/10.3390/w10070852

Index

A

Agricultural area, 86–87, 89, 115, 119, 129, 156, 158–159, 165–167
Agriculture, 1–3, 15–16, 25–26, 28, 35, 45–46, 76–77, 112–113, 115, 119, 129, 140, 142, 145, 148–149, 158, 167, 169, 172–173, 175, 177–179, 191, 213–214, 220, 222, 227–228, 231, 235, 237–238, 241, 262, 266, 269, 274
Algal blooms, 15–16, 46, 115, 119–120, 126–127, 152, 214, 217–218, 220–222, 224, 247
Aquatic ecosystems, 51, 75, 170, 187–188

C

Catchment, 1–2, 5–6, 8–9, 11–13, 17, 35, 73, 76–79, 107–108, 117, 121, 139, 142–145, 147–149, 154–155, 159, 161–162, 164, 170–182, 188, 197–200, 202–204, 206, 211, 213–215, 217, 219–222, 224, 228, 242–243, 245–246, 249–259, 261–263, 266–267, 272–274
Catchment management, 6, 17, 73, 181, 251–252, 255–257, 263
Coastal areas, 55–56, 59, 62, 65–66
Contaminants, 1, 8, 11, 13–14, 16–17, 73, 83–84, 88, 115, 124, 129–130, 145, 155, 162, 164, 174, 176–180, 265, 268, 274–275

D

Definitions, 4, 7
Diffuse pollution, 1–17, 25–35, 46, 49, 52, 73, 75, 77, 111, 139–142, 144–145, 147–149, 153, 169–170, 176–181, 213, 216–217, 219, 223, 225, 227, 241–242, 246–247, 249–250, 265, 267–273, 275

E

Ecosystem services, 91, 155, 170, 181–182, 213, 252, 256–257, 262–263
Emerging pollutants, 5, 73, 76, 83, 85, 87–89, 91, 93, 95, 97, 167, 250
Environmental regulation, 17, 213, 215, 217, 221, 225
Eutrophication, 4, 45–47, 49, 51, 53, 119, 124, 130, 145, 153, 170, 218, 252, 257, 262

F

Faecal indicators, 73, 75–76, 197
Farming, 9–11, 75, 142, 145, 170, 172, 177, 215, 220, 227, 235, 237, 256, 263, 265, 269, 272, 274
Forestry, 2, 8, 35, 46, 112–113, 115–116, 130, 141–142, 145, 147–149, 158, 169–170, 172–174, 176, 214, 227, 265, 269

G

GC–MS, 57–58

H

High-frequency sampling, 187–188, 194

I

impacts, 1–2, 4–6, 11–12, 14–17, 25–26, 28, 35, 48, 62, 64–65, 73, 75, 77–78, 83, 89, 92, 99, 104, 107, 112, 116, 120, 122–124, 126–127, 130, 139, 153, 156, 158–159, 169–170, 172–176, 178–179, 181–182, 190, 213–217, 221–222, 224, 229, 231, 238, 241, 251, 254, 260–262, 266–268, 270, 274–275
Inkomati, 172

K

Kruger National Park, 169–171, 173–175, 177, 179, 181, 183, 185

L

Lake Pátzcuaro, 45, 47, 49, 51–53
Land management, 77, 147, 262
Land use, 1–4, 6, 8–10, 14–16, 65, 76–78, 83, 86, 88–89, 111–112, 114, 119–120, 140–141, 145, 149, 151–152, 154, 159, 160–161, 163–166, 170, 173–175, 180, 187–191, 214, 221, 227, 241, 246–247, 249, 250–255, 265–266, 268–269, 273, 275
Land use and land use changes (LULUC), 151, 159
Landscape, 1–2, 5–7, 10–11, 13–17, 76, 78, 86, 119, 140, 149, 151, 169–170, 173–174, 182, 214, 222, 227, 242, 246–247, 262, 265–275, 277
Limpopo, 170, 174–175, 177–179
Litter, 6, 13–14, 99–100, 105, 107–108, 218, 221, 267
Livestock, 2, 6, 8, 10–11, 46, 74, 76–78, 85–87, 89, 119, 140, 154, 156–159, 166, 170, 174, 177, 220–221, 269
Loch Leven, 213–222, 224
Low impact development (LID), 153–155, 164–166, 270

M

Marine waste, 99
Mitigation, 35, 39, 75, 77, 111, 142, 181–182, 190, 230, 237
Mozambique, 170–176
Multiple benefits, 142, 263, 272, 275

N

Nature-based solutions, 154, 156
Neonicotinoid pesticides, 187–188, 190–191, 194–195
Nonpoint source, 3, 111–113, 115–117, 119–120, 122–123, 126, 128–130
Nutrients, 1, 5–6, 9, 11–17, 46, 50, 52, 76–77, 86, 112, 115–120, 123–124, 126, 129–130, 141, 144, 147, 149, 156, 170, 175, 177, 213, 241, 255, 257, 263, 272, 275

P

Pathogens, 1–2, 4–6, 9, 13–16, 73–77, 112, 115–120, 127, 130, 141, 149, 170, 197–198, 206, 241–242, 247–248, 250, 275
Pesticides, 5–6, 8, 14, 16–17, 25–28, 30–39, 76–77, 85–86, 112, 115–119, 121–122, 125, 141, 145–146, 149, 170, 173, 176, 178, 180, 187–195, 221, 227–239, 242, 245, 247–248, 250–251, 266–267, 269, 275
Phage tracer, 198–200, 206–209, 211
Plastic waste, 101
Point source, 2–4, 6, 8, 13, 76–77, 86–89, 92, 107, 111–113, 115–117, 119–120, 122–123, 126, 128–130, 151–153, 161–163, 166, 172, 214, 241, 254, 267
Policy development, 151
Polycyclic aromatic compounds, 55, 112
Protected areas, 141, 158, 169–171, 173–177, 179, 181–183, 185

Q

Questionnaire, 216, 218, 241–243, 249–250

R

Restoration, 144, 148–149, 166, 213, 215, 218, 268–270
Risk assessment, 56, 64–66, 188, 193–195, 229, 238
Rivers, 2, 9, 11, 15, 33, 77–78, 89, 101, 111–113, 115–122, 124, 130, 140–141, 147, 149, 152–153, 155, 161, 163, 165–166, 169–183, 185, 187–188, 192, 195, 202,

213, 216, 221, 228, 236, 238, 244–245, 266–267
Rural best management practices (BMPs), 5, 216–217, 217–219, 268, 270

S

Socio-economic issues, 213
Soil and water assessment tool (SWAT), 77, 252, 263
South Africa, 16, 169–175, 177–179, 181, 183, 185
Species sensitivity distribution (SSD), 188–190, 193–195
Storm events, 15, 112, 117, 120–121, 123–124, 127, 267
Sustainable urban drainage systems (SUDS), 152, 216, 216–217, 217, 219, 219, 222, 222, 274

T

Terrestrial soils, 56–58, 60, 62, 64
Tourism, 14, 16, 46, 140, 149, 169, 175, 177, 188, 215, 220, 222, 224, 257, 262
Trophic state, 45–47, 49–50, 52, 257
Tropical lake, 47

U

Urban land use, 6, 86, 112, 114, 119, 154, 161

W

Waste management, 91, 167, 268
Water contamination, 75, 91, 120, 127, 158
Water monitoring, 197
Water pollution, 2–3, 6, 17, 76–77, 111, 113, 115, 117, 119, 121, 123, 125, 127–131, 133, 135, 137, 252, 265, 268
Water quality, 1–2, 4–6, 8–12, 14, 17, 39, 45–48, 52, 73, 76–79, 86–87, 89, 111–112, 115–116, 119–120, 123, 128–130, 139–142, 144, 146, 149, 151, 152–154, 156, 158–159, 163, 166–167, 170–182, 194, 197–199, 204, 209, 211, 217, 219–221, 224, 242–243, 246–249, 265, 268, 273
Water sensitive urban design (WSUD), 270
Water supply, 14, 74, 77, 89, 91, 141, 158, 227, 229, 231, 233, 235–239, 242–243, 247, 249, 254–255, 263, 266, 269, 272–273
Water utilities, 17, 76, 241–243, 245–247, 249–250

IWA Publishing's authorised EU representative for General Product Safety Regulations is Diane D'Arras, 15 rue Duret, 75116 Paris, France, e-mail: safety@iwap.co.uk.

Printed and bound by CPI Group (UK) Ltd, Croydon, CR0 4YY
16/06/2026
02140367-0004